PROPERTIES OF MATTER
UNDER UNUSUAL CONDITIONS

Edward Teller

PROPERTIES OF MATTER UNDER UNUSUAL CONDITIONS

(In Honor of Edward Teller's 60th Birthday)

EDITED BY

HANS MARK

Department of Nuclear Engineering
University of California at Berkeley

AND

SIDNEY FERNBACH

Lawrence Radiation Laboratory
University of California, Livermore

INTERSCIENCE PUBLISHERS

A DIVISION OF JOHN WILEY & SONS

NEW YORK/LONDON/SYDNEY/TORONTO

Copyright © 1969 by John Wiley & Sons, Inc.
All Rights Reserved. No part of this book may
be reproduced by any means, nor transmitted,
nor translated into a machine language without the written permission of the publisher.

Library of Congress Catalog Card Number 68-21493
SBN 470 569905

Printed in the United States of America

Preface

For those who know Edward Teller, a commonly used practical unit of energy and enthusiasm is the "microteller." Since this unit has not changed noticeably in the time we have known him, it was only by a fortunate accident that we discovered that he would celebrate his 60th birthday in 1968. This volume represents a tribute to him on the anniversary from a few of his many friends, teachers, students and collaborators. To have had contributions from all of his colleagues who would wish to honor him in this way would have resulted in an *Encyclopedia of Applied Science*—a project we intend to leave for his 70th birthday!

The editors have taken the liberty of collecting papers in a particular area of applied science, namely that spoken of in the title: "The Properties of Matter Under Unusual Conditions." This title seemed appropriate to us in that it deals with areas of great interest to Edward Teller, fields in which he himself worked and provided stimulating leadership. Furthermore the title strikes a parallel in representing the properties of a very unusual man.

In soliciting contributions to this volume, we attempted to encompass Edward Teller's scientific endeavors by starting with his university professors, continuing with his early collaborators and finally his students and colleagues at the University of California and the Lawrence Radiation Laboratory. Although we intended to adhere as closely as possible to the title of the book in selecting topics to be treated, the fact of Teller's broad scientific range could not be ignored. We have thus included a number of papers on other topics, all of which reflect his interest in mathematical physics. We have also included papers by Professors Wigner and Heisenberg giving personal appraisals of Teller's career and his scientific style. The ordering of the chapters in the book is somewhat arbitrary; we have tried to place those chapters dealing with practical matters in the beginning and those treating more abstract mathematical topics toward the end of the book.

The topics of the papers themselves only partially cover Teller's scientific interests. Molecular physics was the field in which he made his first important contributions. This was followed rapidly by significant work in cosmology, nuclear physics and solid state physics. Following the war years, Teller developed his interests in high energy and elementary particle physics. He also played a strong role and continues to do so in the application of nuclear technology to defense problems. We have not

v

been able to do full justice to all of these interests, but we have tried to extract the major trends.

Both of the editors have known and worked with Edward Teller for over 15 years at the Lawrence Radiation Laboratory. He was instrumental in founding the Livermore branch of this institution and in guiding its growth and development. The contributions of the laboratory to the national security are largely due to his initiative and imagination. Teller has also been an enthusiastic initiator and supporter of several interesting and important research and development programs at the Laboratory such as the Plowshare effort which concerns itself with the industrial uses of nuclear energy. He recognized before most others the importance of large scale, high speed computing facilities in the development of applied science and helped to build at the Livermore Laboratory one of the finest computer centers in the world. Finally, he has not neglected his academic activities. In 1963, he organized a Department of Applied Science on the Davis Campus of the University of California, and was the first chairman of the department. The department holds graduate studies at Livermore. Working with the Livermore Laboratory's personnel and facilities, graduate students are being given a broad scientific and technical background with the hope that they will develop such broad capabilities as Teller himself exhibits.

It has been a pleasure for us to work on this book. We dedicate it with affection to a very great man who has made monumental contributions to science, his adopted country and to the world.

HANS MARK
SIDNEY FERNBACH

Livermore and Berkeley, California, 1968

Authors

D. W. BAKER, *Institute of Geophysics and Planetary Physics, University of California, Los Angeles, California*

CHARLES L. CRITCHFIELD, *University of California, Los Alamos Scientific Laboratory, Los Alamos, New Mexico*

RUSSELL E. DUFF, *Lawrence Radiation Laboratory, University of California, Livermore, California*

H. P. DÜRR, *Max-Planck-Institut für Physik und Astrophysik, Munich, Germany*

G. GAMOW, *University of Colorado, Boulder, Colorado*

D. T. GRIGGS, *Institute of Geophysics and Planetary Physics, University of California, Los Angeles, California*

W. HEISENBERG, *Munich, Germany*

GARY H. HIGGINS, *Lawrence Radiation Laboratory, University of California, Livermore, California*

FREDERIC DE HOFFMANN, *Gulf General Atomic, Inc., San Diego, California*

C. E. LEITH, *Lawrence Radiation Laboratory, University of California, Livermore, California*

HERMAN F. MARK, *Department of Chemistry, Polytechnic Institute of Brooklyn, Brooklyn, New York*

JOSEPH E. MAYER, *Department of Chemistry, University of California, San Diego, California*

RICHARD F. POST, *Lawrence Radiation Laboratory, University of California, Livermore, California*

KENNETH M. WATSON, *Institute of Geophysics and Planetary Physics, University of California, La Jolla, California*

JOHN ARCHIBALD WHEELER, *Palmer Physical Laboratory, Princeton, New Jersey*

WILLIAM L. WHITTEMORE, *Gulf General Atomic, Inc., San Diego, California*

EUGENE P. WIGNER, *Princeton University, Princeton, New Jersey*

T. T. WU, *Harvard University, Cambridge, Massachusetts*

C. N. YANG, *Institute for Theoretical Physics, State University of New York, Stony Brook, New York*

Contents

AN APPRECIATION ON THE 60TH BIRTHDAY OF EDWARD TELLER.
By Eugene P. Wigner 1
THE CONCEPT OF "UNDERSTANDING" IN THEORETICAL PHYSICS.
By W. Heisenberg 7
ON THE ORIGIN OF GALAXIES. By G. Gamow 11
THE ORIGIN OF DEEP-FOCUS EARTHQUAKES.
By D. T. Griggs and D. W. Baker 23
RESPONSE OF ROCKS TO STRESS. By Gary H. Higgins . . . 43
MATERIAL PROPERTIES AT HIGH PRESSURE. By Russell E. Duff . 73
POLYMERIC MATERIALS FOR EXTREME CONDITIONS.
By Herman F. Mark 105
ULTRA-HIGH NEUTRON FLUXES: THEIR PRODUCTION AND USE.
By Frederic de Hoffmann and William L. Whittemore . . 119
THE PHYSICS OF HIGH TEMPERATURE PLASMA. By Richard F. Post 141
NUMERICAL SIMULATION OF TURBULENT FLOW. By C. E. Leith . 267
A STATISTICAL MECHANICAL TREATMENT OF MACROSCOPIC CHANGE
WITH TIME. By Joseph E. Mayer 273
APPROXIMATE SYMMETRIES IN ATOMIC AND ELEMENTARY PARTICLE
PHYSICS. By H. P. Dürr 301
ON THE THEORY OF NEAR-ADIABATIC TRANSITIONS.
By Kenneth M. Watson 327
SOME SOLUTIONS OF THE CLASSICAL ISOTOPIC GAUGE FIELD
EQUATIONS. By T. T. Wu and C. N. Yang 349
EIGENVALUES OF CASIMIR OPERATORS. By Charles L. Critchfield 355
STRANGE MATTER. By John Archibald Wheeler 365

AUTHOR INDEX 381
SUBJECT INDEX 387

An Appreciation on the 60th Birthday of Edward Teller

EUGENE P. WIGNER

Princeton University, Princeton, New Jersey

It has been said that great men have many admirers, but each also has some passionate opponents. Edward Teller surely qualifies as a great man under this criterion. The present writer is one of a large number of his close and affectionate friends, friends who admire his imagination, his human kindness, his unfailing wit, and the breadth of his interests. On the other hand, a clipping from a paper in Hungary, sent to me for this occasion, accuses him of ill will, greediness, and of a violent temper. All this in a critical review of one of his books which is, actually, "not available" there.

Teller was born in Hungary in 1908, the son of an attorney. His was a closely knit family, and the warmth and affection of its members toward one another was apparent even to the casual visitor. At the time of the young Teller, the high schools in Hungary were as good as any in the world and, to no one's surprise, he completed his studies with distinction. The solid foundation of knowledge, acquired in the high school, served him well in later years, as it did many another young Hungarian.

Teller started his university studies at the Technische Hochschule (Institute of Technology) of Karlsruhe in Germany, one of the best schools of the period. His interest in chemistry and chemical physics goes back to the Karlsruhe years. However, the challenge of the new quantum mechanics and the miracle of its success proved to be too much of an attraction, and he left the Technische Hochschule after two years for München and Sommerfeld and Heisenberg. He followed Heisenberg to Leipzig a year later.

There is nothing that I could say about the rest of Edward's life that would not be common knowledge. It is the fate of a strong personality to act out his life on an open stage, exposed to the not always kind public. He went from Leipzig to Göttingen, from Göttingen to Bohr in Copenhagen, then as Lecturer to the University of London. He came to this country in 1935 as Professor of Physics at the George Washington University, spending a year at Columbia before joining the uranium project. His contributions to that undertaking were virtually unique, and I shall return to them when discussing his work. He joined Fermi in Chicago after the war, then went to the University of California where he could participate more effectively in the development of atomic weapons—he and many of his colleagues felt that he had a mission to fulfill there. He is now Professor-at-Large there and is also Associate Director of the Lawrence Laboratory at Livermore, a weapons research laboratory.

During a long life among scientists, I have not met anyone with a more fertile imagination than Edward's. He is one of the theoretical physicists, nowadays rare, who are not captives of the beauty of the mathematical formalism, but whose attention is captured by the phenomena themselves. The physics in which he is now interested shows, in fact, some traits of natural history; this is concerned not with the species that could exist, but with those that do. In the same way, Teller is interested not only in the basic laws of physics, that is the limitations on the phenomena and objects that *could* exist, but also in finding the structure of those objects and the description of those phenomena that *do* exist.

The trait of Teller's physics which was just described manifested itself most clearly in later years. His early papers were entirely in the spirit of the times: the expanding world of the applications of quantum mechanics. His very first paper, "Hydrogen Molecular Ion," determining the molecular orbitals in the field of two equal charges, can be considered, along with the paper of Burrau and those of Hund and Mulliken, to form the foundation of one of the two theories of molecules—of the one which is, at present, the more successful. Nevertheless, the lengthy calculations which this paper demanded did, even at that time, not suit Teller's taste, and there is a story that, when a friend of his asked why he never came to work before six in the afternoon, he replied that he was not tired enough before. However, there is every evidence that he enjoyed the company of the other young physicists and participated in their jokes, discussions, and ping pong whole-heartedly and with gusto.

The subsequent papers of Teller's "molecular period," extending from 1930 to 1936, show his familiarity with wide areas of chemical physics, his interest in all of them, and his ability to inject new ideas into many of them. They also show his interest in collaborating with his colleagues; in the thirty-odd papers which he wrote in this period, he had almost as many collaborators. Some of these, notably James Franck and K. F. Herzfeld, he considered to be his teachers, and he preserved an unswerving loyalty toward them throughout his life. Others, H. Sponer, Tisza, Herzberg, Heitler, Placzek, Landau, Breit, were more nearly his contemporaries and he had, even at that early age, a number of gifted students.

Around thirty papers originated from the "chemical physics" period, and each has something interesting to say; only one of them did not stand the test of time. In order to review them adequately, as many pages would be needed. Let the Atomic Energy Commission's evaluation stand here instead, prepared for the occasion of the Fermi award: "Perhaps his theory of molecular vibrations and sound distribution is the most substantial, his observation of the magnetic cooling process, published with W. Heitler, the most illuminating, his contribution to the Jahn-Teller theorem the most erudite, and his theory of the ortho–parahydrogen conversion the most ingenious of his accomplishments in the physical chemistry period. Nor should one overlook his work on the adsorption of gases on solids in collaboration with S. Brunauer and P. H. Emmett." One of the papers referred to, that on the magnetic cooling process, deals with behavior of substances under unusual conditions—a precursor of this book—and brought new understanding of that condition. The paper mentioned last gave a new solution to a long outstanding problem.

Whereas in Europe, Heisenberg and James Franck were Teller's most honored friends and teachers, Gamow became his closest collaborator in the early years in America. Nuclear physics moved to the center of his interest, and his intimate familiarity with this subject became of crucial importance later when he became involved in the uranium project. The establishment of the G-T (standing for Gamow-Teller) selection rules for β disintegration, i.e., the recognition that the angular momentum of the nucleus may change by one unit in an allowed β decay, is the best known result of the period. However, other papers, including those on the scattering of neutrons by ortho- and parahydrogen and by molecules in general, with Schwinger and R. G. Sachs, and those on the α-particle model (with Hafstad and with Wheeler) were probably more to his taste. In addition, Teller's later interest in the properties of matter under unusual conditions began to develop and he published, in collaboration with Gamow, several papers on the internal structure of stars. In addition, throughout this period, and indeed throughout his life, he has continued to cultivate his first love, chemical physics.

Teller's activities in pure science came to an abrupt halt with the outbreak of World War II. He felt a deep obligation to the country which provided a new home for him and wanted to serve it to the best of his ability. His first assignment was to the Metallurgical Laboratory in Chicago, where he made significant contributions to the theory of nuclear chain reactions. True to his interest in a detailed description of events, he was not satisfied with the usual global description of neutron populations but tried to trace the history of individual neutrons from their birth to their absorption. His theory provides even now, after 25 years, insights which the standard theories do not furnish.

When the ability of the Metallurgical Laboratory to establish a large-scale nuclear chain reactor appeared to be assured, Teller's attention turned to the problem of using the nuclear explosive which the nuclear reactors were to furnish. He moved to Los Alamos where this problem was to be tackled under J. R. Oppenheimer's leadership.

The years at Los Alamos were not happy. To live in a secluded spot, work within a rigid organization, not to be able to discuss one's work even with one's closest friends, wore on the nerves of most physicists in Los Alamos and very heavily on those of Edward. In addition, the objective of the work, though considered by all to be necessary, was not to the liking of any.

It surely was not to the liking of a person with Edward's inclinations. Nevertheless, he stuck it out and contributed greatly to the success of Los Alamos. Some of the work he carried out there had sufficient scientific importance to be published years later. This applies particularly to the work on shock waves, in collaboration with Bethe, and the properties of matter under very high pressure, on which Feynman, Metropolis, and the Rosenbluths collaborated with him.

Teller left Los Alamos as soon as he could gracefully do so and did not return to military problems until he felt that their pursuit was again necessary in the interest of his adopted country. When he felt this, he did return.

The years after Los Alamos, and until the renewal of his preoccupation with problems of national security, were perhaps Teller's most fruitful years scientifically. Gone were, nevertheless, the carefree days of easy friendship among all. It was difficult also to adjust to the situation when people started to take science and scientists seriously and, instead of smiling when talking about them, attributed a deep responsibility to them. Teller was greatly affected by the changes, and his relations to his colleagues were not the same as before the war. The events at Los Alamos, the disagreements both technical and political, left their scars. (Teller opposed exploding the bomb over Hiroshima but also the total abandonment of Los Alamos.) Scientifically, however, the days were bright and he followed his inclinations more freely than ever before. He wrote on cosmology, judging, as it now appears correctly, that the dimensionless constants (such as the ratio of gravitational and electric forces) had remained the same for billions of years. He speculated with M. G. Mayer on the origin of the elements. One cannot help marveling at the imagination of the authors when reading their report to the Solvay Congress, and at the care with which they followed the consequences of their assumptions. The insights gained in the course of this work contributed significantly to M. G. Mayer's formulation of the j–j coupling shell model of nuclei which brought her the Nobel Prize. Teller also conceived ideas on the origin of cosmic rays and contributed to the theory of nuclear forces.* The articles of this period shared with most of his other articles the characteristic of excelling, not by mathematical elegance, but by the imaginative use of the available empirical information.

Around 1950, Teller's concern with the defense of the country caught up with him. He felt that it was dangerous that all the research on nuclear weapons was concentrated in a single laboratory. He recalled that, even in its heyday, Los Alamos relied heavily on the advice of outsiders and that every monolithic institution is in danger of accepting some ideas uncritically and of disregarding, or not even conceiving, some others. As a result of the exclusive interest of most of our scientists in pure science, Los Alamos became, in Teller's opinion, dangerously isolated from the scientific community. He pressed for the establishment of another weapons laboratory. His counsel played a large role in the formation of the Lawrence Radiation Laboratory at Livermore and he contributed to the success of that laboratory greatly, not only on the technical level, but also by the guidance which he provided.

Teller's activities in the interest of national defense were, and are, not universally popular. Many seem to think that the Livermore Laboratory is unnecessary and, even further, that the hydrogen bomb—in the conception of which Teller had a decisive role—would never have been invented, not even in the USSR, had it not been for Teller's efforts. These views compliment Teller's genius, probably unintentionally, but in this writer's opinion surely in an exaggerated fashion. As to the intent of inventing the hydrogen bomb, Teller reminds us that it is not in every country possible for a scientist to work on the problem of his choice and that, in the words of one of the most eminent physicists of the USSR, the Russian physicists remained "mobilized" well into the 1950's.

However, Teller's interest in the applications of nuclear physics was not confined to the military ones. He was, and is, equally interested in the potential for power production, and much of the initiative for the exploration of controlled thermonuclear processes comes from him. He was also,

---

* That he, nevertheless, did not take those theories too seriously at that time is attested to by the last verse of the poem he wrote for the records of the Solvay Congress:

> From mesons all manner of forces you get,
> The infinite part you simply forget,
> The divergence is large, the divergence is small,
> In the meson field quanta there is no sense at all.
>
>> What, no sense at all?
>> No, no sense at all!
>> Or, if there is some sense
>> It's exceedingly small.

as far as this writer knows, the first to express concern for the safety of power-producing reactors and was the first chairman of the U.S. Atomic Energy Commission's Reactor Safeguards Committee. His concern for the safety of the installations of the burgeoning nuclear industry persists unabated. Similarly, he was the initiator, and is one of the prime advocates, of the use of nuclear explosions for the excavation of harbors and canals.

Thus, Teller's concern for the safety of the nation which adopted him led gradually to his development from one of the most imaginative scientists to one of the most thoughtful statesmen of science with ever-broadening areas of interest. These led him from a preoccupation with, and almost exclusive interest in, specific technical problems to an interest in the objectives of the scientific endeavor and its uses not only for defense but also for all other practical purposes. This interest was responsible for his having assumed the chairmanship of the Department of Applied Science of the University of California at Davis. It would be too much to expect, on the part of the scientific community, a unanimous approval of all his ideas, hopes, and objectives. All are convinced, however, of his absolute sincerity and most have confidence in the soundness of his judgment. This writer surely looks forward to the further development of one of his closest friends under the unusual conditions of our times.

The Concept of "Understanding" in Theoretical Physics

W. HEISENBERG

Munich, Germany

During the time almost 40 years ago, when Edward Teller took part in our Leipzig seminar, the question frequently arose, whether we had understood the behavior of nature in a given group of phenomena; and since Teller and other members of our group liked discussions on philosophical problems, the search was often continued by asking what the word "understanding" means in theoretical physics. Therefore the present volume may be a proper place to formulate some remarks on this problem.

A few historical examples may illustrate the difficulties. In ancient astronomy Ptolemaeus, ignoring the ideas of Hipparchus and Aristarchus, had formed his theory of the motion of planets and stars, by describing their orbits as superpositions of cycles and epicycles. The "predicting value" of this theory was very high—the astronomers were able to predict eclipses of the sun or the moon with considerable precision. Had the astronomers thereby "understood" the motion of the planets? Fifteen centuries later, Newton developed his theory of the planetary motion on the basis of his law of gravitation. Most physicists would agree that only then had one "really understood" these phenomena.

When the first important experiments on electricity had been performed in the 18th century, the physicists gave their attention to the forces between charged bodies and analyzed many problems of this kind in considerable mathematical detail. It was, however, Faraday who gave the problem a slight turning by asking for the "field" of force, by attaching some kind of reality to force as a function not of bodies but of space and time. Only after this change of the concepts was Maxwell able to formulate his famous equations. Should the essential step toward understanding be credited to Faraday or to Maxwell?

The final mathematical formulation of the natural law is not always identical with understanding. The Navier-Stokes equations of hydrodynamics were formulated in the middle of the 19th century. Still the turbulent motion of fluids was not understood until one hundred years later, when statistical concepts were developed which through the work of

Kolmogoroff, Onsager, and von Weizsäcker lead to an interpretation of the spectrum of isotropic turbulent motion.

The phenomenon of superconductivity was not understood, in spite of the mathematical scheme of quantum mechanics, until such concepts as "collective motion" and "degeneracy of the ground state" were formulated and analyzed.

Before going into an investigation about the meaning of the word "understanding" in these various examples, it may be useful to describe the general way which, in a new field of physics, seems to lead toward "understanding" of the phenomena. The first step will always be the collection of data by new experiments. Even at this stage the physicists will try to relate different experiments, to interpolate or extrapolate their results in order to predict the results of further experiments; and they will attempt to get some order into the experimental material by applying the traditional concepts (e.g., by using the laws of Newton's mechanics or nowadays by attaching quantum numbers to the various energy levels in a spectrum).

But this will as a rule not be sufficient for an understanding of the phenomena, even in a case where the fundamental natural laws are already known. Because it will frequently turn out that the traditional concepts or at least some of them don't really work—that one cannot get hold of the phenomena by using this kind of language. Difficulties will appear which indicate that the experimental material somehow requires new concepts. Sometimes considerable progress can be achieved at this stage by carrying out a "crucial experiment," i.e., an experiment which allows one to test just the controversial concepts. A famous example is the experiment of Bothe and Geiger on the coincidences between γ-ray and electron in the Compton effect, which demonstrated the conservation of energy and momentum in the single event. But even such a result is as a rule not sufficient for understanding. What then actually happens may perhaps be described as follows: The knowledge of the totality of the experiments in the new field—not only of a few special experiments—produces in the mind of the physicists a gradual change of the language and of the thinking applied to the new field. This process may take a number of years or even decades, and finally, out of this new way of thinking, somebody will develop and formulate new concepts. The greatest effort is required for the severing from the old concepts; it is usually much simpler to find new concepts than to get rid of the old ones. The final step will be the application or mathematical elaboration of the new concepts and will, if the correct concepts have been found, lead to the conviction that "now one has understood" the phenomena. This may be so even if a quantitative description cannot be reached on account of mathematical complications (as, for instance in quantum chemistry).

What, then, is the criterion for "understanding"? The well-known statement: "When we can foresee the result of any not too complicated experiment in the field concerned we have understood this field" is certainly not correct. It is wrong in both ways. There are fields which have been essentially understood but where we cannot foresee the results of simple experiments (e.g., quantum chemistry). There have been other fields in which precise predictions were possible, but where, however, the essential points had been missed (e.g., Ptolemean astronomy, or as a minor example, Voigt's theory of the anomalous Zeeman effect). A much better definition of "understanding" seems to be: "We have understood a group of phenomena when we have found the right concepts for describing these phenomena." This definition, however, requires a criterion for the correctness of concepts. How can we know that we have found the right concepts? In some cases the mathematical formulation of the concepts may lead to a mathematical scheme of such clarity and fundamental simplicity that it convinces already by its esthetic values. This may be said, for example, for the theory of special relativity. In such cases the results of the mathematical scheme may easily be compared at many points with the experiments and the agreement then demonstrates the correctness of the concepts. In other cases the mathematical formulation may not be so simple. Still, by constructing simplified models and by demonstrating that these models do show the characteristic features of the phenomena, we convince ourselves that the concepts are correct (e.g., in the theory of superconductivity), that we have "understood" the phenomena. But ultimately, I would believe, it is always the simplicity of the concepts in comparison with the great wealth of complicated experimental material, which convinces of their correctness. Usually in a new field many very different experiments can be carried out; and if all these experiments allow a description by the same simple new concepts, these concepts will finally be accepted as the correct ones. The old Latin sentence: *Simplex sigillum veri* (simplicity is the sign of correctness), which was written in big letters in Pohl's lecture room in Göttingen,* may still be the best criterion for the correctness of concepts. Again it should be emphasized that it is usually much easier to accept new concepts than to abandon old ones—which, however, is necessary. The difficulties in understanding, e.g., special relativity, came almost exclusively from the necessity to abandon the old concept of simultaneity. The difficulty in understanding elementary particle physics may be due to the necessity of abandoning the old concept of a "really fundamental" particle.

* Some of Pohl's students in experimental physics of the old times are said to have translated this sentence into German by: "Siegellack ist einfach das Wahre"; but this was not meant.

If in this way simplicity is taken as one of the most decisive criteria for the correctness of concepts, one refers to the fact that understanding usually occurs when we can say: "Yes, this is the same thing as . . . ." We connect the problem concerned with other related problems, and if a concept has the power of combining very many different phenomena under some aspect, under which they appear to be the "same" or closely related, then the concept will be accepted just on account of this power. "Understanding" then means: adaptation of our conceptual thinking to the totality of the new phenomena; or: discovering in the wealth of phenomena some underlying structures, which correspond to fundamental innate structures in our conceptual equipment and which therefore enable us to form concepts.

It is evident from this discussion that narrow specialization is a hindrance for understanding. It is only by looking at the whole field of new phenomena that the correct concepts can be found. Even in a very special problem, "understanding" can frequently be obtained by referring to a similar problem and its solution in a different field of physics.

Mathematical analysis can be an important help after the correct concepts have been found, since it may then enable the physicists to draw precise conclusions and to compare them with the facts. Before the correct concepts have been found it is only of little use. Because then it can only establish a precise connection between assumptions, expressed in the old concepts, and their consequences. But the assumptions are probably incorrect and therefore their consequences need not represent the phenomena. Hence mathematical analysis is usually not the direct way toward understanding; mathematical physics and theoretical physics are very different sciences.

Since the days of our first discussions in Leipzig, Teller has taken an essential part in both of these sciences by a number of brilliant papers. Looking back to the decades when atomic physics was developed, the collaboration with Edward Teller belonged for me to the best parts of this period.

On the Origin of Galaxies

G. GAMOW

University of Colorado
Boulder, Colorado

When Edward Teller was exactly one-half of his present age, he and I published a paper entitled "On the Origin of Great Nebulae,"[1] in which we formulated the conditions under which condensations could have been formed in the originally homogeneous material filling the expanding space. Since that time considerable progress has been made in the understanding of the physical processes in the expanding universe, and some of our early conclusions have to be modified although the main idea remains essentially the same. In the present article I want to reiterate the basic equations which we used thirty years ago, and draw the conclusions obtained by applying these equations to the more recent information concerning the expansion process. We wrote:

Let us consider space uniformly populated by particles (stars or molecules) moving with certain random velocities (v). The condition for the formation of condensation due to gravitational instability (Jeans) can be written in the form:

$$\frac{G\rho(4\pi R^3/3)}{R} \geq \frac{v^2}{2} \tag{1}$$

where G is Newton's constant of gravitation, ρ is the density at the time in question, and R the radius inside of which the condensation will take place. It is clear that this condition can always be satisfied if we choose R sufficiently large. Thus, if space was originally filled with a gas of any given temperature, the condensation of sufficiently large masses was always taking place. The situation, however, will be rather different if we accept that the distribution of matter in space is in a state of uniform expansion given by the formula: $V = \alpha R$. In this case the condensation could be formed only if:

$$v > V = \alpha R \tag{2}$$

since otherwise the resultant velocities of particles on the surface of the prospective condensation will be always directed away from it. Substituting this velocity into Eq. (1), we obtain:

$$\frac{G\rho(4\pi R^3/3)}{R} > \frac{\alpha^2 R^2}{2} \tag{3}$$

or:

$$\rho > \frac{3\alpha^2}{8\pi G} = \rho_0 \qquad (4)$$

This formula no longer contains the radius of the condensing region. It shows that in the expanding space the formation of gravitational condensations can take place only when the average density is above a certain critical density ρ_0.

Using for the Hubble's constant (denoted by α in our article) the then accepted value 500 km/sec/Mpc (which was leading to the wrong eons* value 1.8 for the age of the universe), we obtained for the critical density: 6×10^{-28} g/cm³, which was much larger (by a factor of 600) than the then accepted mean density 10^{-30} g/cm³ in space. Using today's accepted value: $\alpha(=H) = 100$ km/sec/Kpc (which leads to the age of universe of about 10 eons), one finds $\rho_0 = 2.4 \times 10^{-29}$ g/cm³, which is still larger (by a factor 33) than today's accepted mean density 0.7×10^{-30}. From this fact we concluded that:

> under present conditions the formation of condensations in the universe on a great scale is impossible whatever the masses or velocities of particles may be ...

a conclusion which still holds if one uses today's more accurate information.

Furthermore, we have assumed:

> that the expansion is uniform in time, i.e., the relative velocity of two nebulae (galaxies) always remains constant.

from which follows, according to $v = \alpha R$, that

$$\alpha' = \alpha \frac{l}{l'} \qquad (5)$$

where l is the length the change of which characterizes the expansion, and the primed symbols denote the quantities corresponding to a certain past time. Similarly, one writes

$$\rho' = \rho \left(\frac{l}{l'}\right)^3 \qquad (6)$$

Now one can write the condition for the formation of the galaxies in the form:

$$\left(\frac{l}{l'}\right)^3 \rho = \frac{3\alpha^2}{8\pi G}\left(\frac{l}{l'}\right)^2 \qquad (7)$$

* One *eon* is defined as one (American) billion years (10^9 yr).

or

$$\frac{l}{l'} = \frac{3\alpha^2}{8\pi G\rho} = \frac{\rho_0}{\rho} \tag{7'}$$

With the old (wrong) value of α, this meant that, at the era of galactic formation, the linear scale of the universe was 600 times smaller than it is today. With the correct (present) value of α this figure must be reduced to 33.

To calculate the velocities of particles at the moment of formation of galaxies, we used the relation

$$V = \alpha'R = \alpha\left(\frac{l}{l'}\right)R \tag{8}$$

where the radius of the galaxies at the moment of formation (assumed to be unchanged since that time) was calculated from

$$R = \frac{1}{2}\left(\frac{l'}{l}\right)D \tag{8'}$$

D being the present distance between two neighboring galaxies. Using $l/l' = 600$, and Hubble's old value 1.7×10^{-3} hubbles* for the intergalactic distances, we have found from (8) and (8'), $V = 100$ km/sec, and concluded that:

> this velocity is common in the world of stars (but too high for thermal velocities of molecules) so that most of the stars must have been formed before the separation of nebulae.

Using today's information, one finds the same value for V as before since (l/l') and (l'/l) cancel out and the decrease of α is compensated by the corresponding increase of D. However, it is no longer surprising today that, at the era of the formation of galaxies, the molecules of gas could have had such large velocities. In fact, as we will see later, the temperature of the universe at that time is known to have been very high (see Fig. 1).

Moreover, the assumption of gaseous "protogalaxies" is quite necessary in order to explain their regular shapes. Indeed, if the original condensations were made of stars, whose free path exceeds galactic diameters by a factor of millions, it would be impossible to interpret their regular shapes on the basis of the dynamics of the rotating coherent masses. According to the present views, the formation of stars took place *after* the original gaseous protogalaxies attained their regular shapes.

* One *hubble* is defined as one (American) billion light years (10^9 light years). One hubble per eon is the speed of light.

These gaseous shapes were preserved as "fossilized remainders" in their present stellar state.[2]

In my paper with Teller, we had no means to know what the temperature of the universe was at the time of the formation of the protogalaxies.

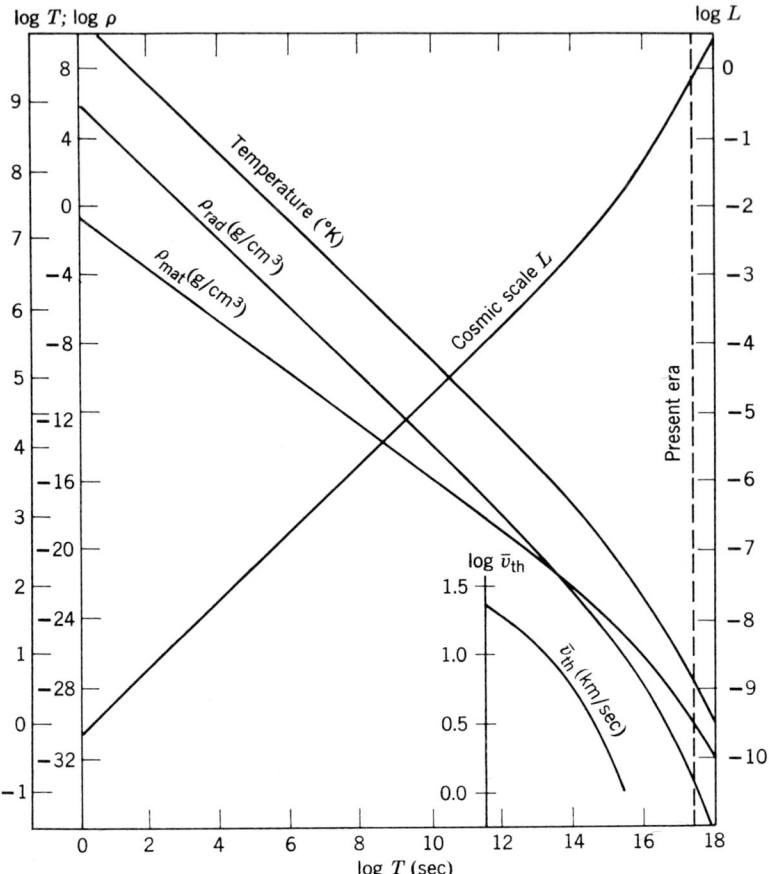

Fig. 1. Time dependence of the cosmic scale (L), temperature (T), matter density (ρ_m), radiation (mass) density (ρ_{rad}), and thermal velocity ($\bar{v}_{th}$). After Alpher, Gamow, and Herman.

However, a decade later, I found that the relativistic equations of the expanding universe led to an inevitable conclusion that, *during the early stages of expansion, the universe was dominated entirely by radiation, the mass-density of which was much larger than the density of matter.*[3] The calculations led to a result that during this radiation stage of expansion, or the "fireball" * as it is often called today, the temperature of matter was given by the expression:

$$T = \sqrt[4]{\frac{3c^2}{32\pi Ga}} \cdot \frac{1}{t^{1/2}} = \frac{1.52 \times 10^{10}}{t^{1/2}} \,°\text{K} \qquad (9)$$

and the radiation (mass) density by the expression:

$$\rho_{rad} = \frac{3c^2}{32\pi G} \cdot \frac{1}{t^2} = \frac{4.4 \times 10^5}{t^2} \frac{\text{g}}{\text{cm}^3} \qquad (10)$$

where t is counted in seconds from the singularity at the beginning of the expansion. For the density of matter the theory led to the expression:

$$\rho_{mat} = \frac{b}{t^{3/2}} \frac{\text{g}}{\text{cm}} \qquad (11)$$

where b is a coefficient to be determined from the considerations pertaining to the formation of elements during the early stages of expansion. To do this, I have considered the formation of deuterium in the reaction: $_1\text{H}^1 + {_1}n^0 \rightarrow {_1}\text{D}^2 + \gamma$ subject to the condition that by the end of the reaction, about one-half of the originally present hydrogen was consumed in the formation of deuterium, and consequently, formed heavier elements. These calculations led to the conclusion that the originally prevailing radiation (mass-) density became equal to the density of matter at $t = 4 \times 10^{15}$ sec, when $T = 340°\text{K}$ and $\rho_{mat} = \rho_{rad} = 3 \times 10^{-26}$ g/cm³. Substituting these values into Jeans' expression of gravitational instability:

$$R \simeq \sqrt{\frac{5\pi kT}{3mG\rho}} \qquad (12)$$

* The use of the term "fireball" for the radiation stage of the expanding universe, or for its present stage, is actually a misnomer. Indeed, the fireball produced by a nuclear explosion is a region of highly ionized air, resulting from the radiation escaping from the region originally occupied by the nuclear component of the bomb. The early state of the universe can be better compared with the very early stage of man-made nuclear explosions (before the radiation leaked out) when the mass-density of radiation within the original core is about 1 mg/cc while the density of the gas formed by fission products is still about 20 g/cc. However, *the most characteristic similarity between the nuclear bomb and the expanding universe is that, in both cases, the heat capacity of the radiation is much larger than that of matter.* Thus, the radiation sucks out (by Teller's "inverse Compton effect") most of the energy from the matter without much change in its (radiation's) temperature. In fact in the case of the expanding universe, the ratio of heat capacities of radiation and of matter (per unit volume) is of the order of 10^{10}. This ratio remains practically constant from the very first minutes of the expansion process all the way to the present era. The pure number 10^{10} represents one of the important dimensionless constants characterizing our universe.

where $m = 1.66 \times 10^{-24}$ g, and k is the Boltzmann's constant, I obtained for the radii of the condensations:

$$R_{gal} \cong 6.5 \times 10^{21} \text{ cm} = 6 \times 10^{-6} \text{ hubbles}$$

and for their mass

$$M_{gal} = 2.7 \times 10^{40} \text{ g} = 1.8 \times 10^{7} \text{ sun-masses}$$

Similar calculations, carried out soon thereafter by Alpher and Herman[4] adjusting the value of b by consideration of the overall element production, led to the values:

$$R_{gal} = 1.0 \times 10^{-5} \text{ hubbles}$$

and

$$M_{gal} = 3.8 \times 10^{7} \text{ sun-masses}$$

In both cases the calculated sizes and masses were closer to those of the galaxies than, for example, of the moon but fell short of the observed values by one order of magnitude in size and three orders of magnitude in the mass. In fact it looked as if the density at the time of the formation of protogalaxies was estimated correctly, whereas the diameters were underestimated by a factor of 10.

The recent discovery of the thermal cosmic radiation[5] corresponding to the temperature of 3°K (which was predicted rather closely by the calculations quoted before) strengthened the views concerning the radiation stage of the expanding universe, and permitted more exact calculation of the temperature regime in its earlier history. In a recent paper, Alpher, Herman, and I[6] used the analytical solution of the space expansion equation to obtain the exact values of the temperature and the matter densities all the way back from the present state of the universe to the very early stages of expansion. The constants of integration were obtained by using the observed present values of Hubble's constant, and the present matter-density and temperature values. One set of these curves, corresponding to $H_p = 100$ km/sec/Mpc; $\rho_{m,p} = 0.7 \times 10^{-30}$ g/cm³, and $T_p = 3°K$ is given in Fig. 1. This figure also gives cosmic scale L defined as $l(t)/l_p$. The equations for the upper limit of R, given by the rate of expansion, and the lower limit of R, given by Jeans' condition, can be rewritten as:

$$R_{max}(t) = \frac{\bar{v}_{th}(t)}{H(t)} \quad : \quad H = \frac{1}{L}\frac{dL}{dt} \tag{13}$$

and

$$R_{min}(t) = \sqrt{\frac{3}{8\pi G}} \frac{\bar{v}_{th}(t)}{\sqrt{\rho_m(t)}} \tag{14}$$

where the velocity of thermal motion $\bar{v}_{\text{th}}$ is calculated from $T(t)$ by the formula

$$\tfrac{1}{2} m_{pr} \bar{v}_{\text{th}} = \tfrac{3}{2} kT \qquad (15)$$

The results of these calculations were somewhat disappointing. While, for L close to 1, we obtained $R_{\max} \ll R_{\min}$, which precludes the possibility of condensations for L close to the point where $\rho_{\text{mat}} = \rho_{\text{rad}}$, the curves $R_{\max}(t)$ and $R_{\min}(t)$ run very close together, the first one being a little bit lower than the second. If, disregarding the fact that, even in this region, $R_{\max}$ is slightly smaller than $R_{\min}$, one used the common value of the radii for calculating the mass, it comes out to be about 2×10^5 sun-masses, not so different from the earliest calculations, and just as much in disagreement with the actually observed galactic masses. The most probable reason for this disagreement is that, in formulas (13) and (14), one should not use the mean thermal velocity of molecules, thus substituting Dirac's δ function for Maxwell's distribution. Indeed, the presence of particles with $v_{\text{th}} > \bar{v}_{\text{th}}$ will increase the values of both $R_{\min}$ and $R_{\max}$ in about the same proportion, and may well lead to the correct values of the galactic masses. However, the analytical treatment of the condensation problem, taking into account Maxwell's distribution of velocities, represents a difficult mathematical problem. An attempt is being planned to solve that problem by using the Monte Carlo method employed, for example, by S. Ulam[7] in his studies of the dynamics of stellar clusters. To do this, one should follow the motion of a large number of gravitating particles scattered uniformly throughout a certain volume of space with the initial velocity distribution given by:

$$H(\mathbf{r} + \mathbf{v}) \qquad (16)$$

where $\mathbf{r}$ is a radius vector from a central point O and $\mathbf{v}$ is the thermal velocities distributed at random in their directions, and according to Maxwell's curve in their numerical values (Fig. 2). It is hoped that such an approach will help to bring into order the long-standing problem of the formation of protogalaxies in the expanding space of the universe.

Another important point in the problem of formation of galaxies is the comparison of the cosmological model leading to the curves shown in Fig. 1 with the actually observed distribution of very distant galaxies known at present as radiogalaxies and quasi-stellar radio sources in the space of the universe. There exists at present a disagreement concerning the nature of these sources. Longsighted cosmologists place them at very large distances, whereas the shortsighted ones prefer to think that they are very close to us. As was indicated by Teller and myself in the old paper discussed here, the curvature radius $\mathscr{R}$ of the universe can be found from the equation:

$$\frac{1}{l}\frac{dl}{dt} = \sqrt{\frac{8\pi G}{3}\rho - \frac{c^2}{\mathscr{R}}} \tag{17}$$

by substituting the observed value of Hubble's constant

$$H = [(1/l)(dl/dt)]_{\text{now}}$$

and the mean density ρ of the universe.

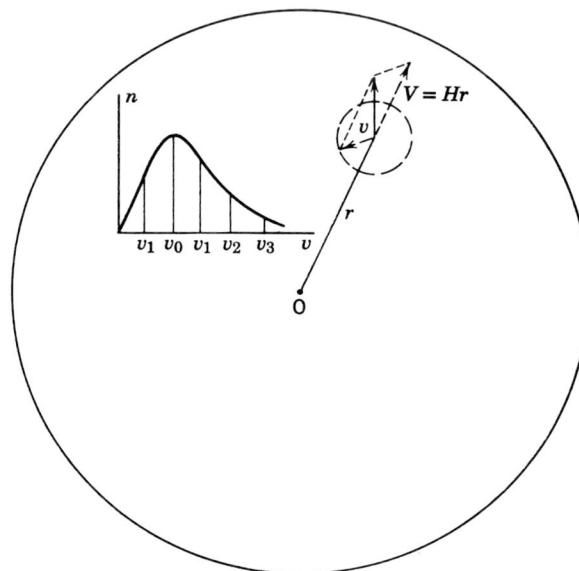

Fig. 2. Selection of the initial velocity distribution of N "particles" for the Monte Carlo problem of the formation of protogalaxies.

Using Hubble's old value $H = 500$ km/sec/Mpc and putting $\rho = 10^{-30}$ g/cm³, we came to the conclusion that $\mathscr{R}$ is an imaginary quantity, so that the space of the universe is infinite, possessing hyperbolic geometry. With the present value of Hubble's constant, one finds for the curvature radius of space:

$$\mathscr{R} = 1.0 \times 10^{10} \, i \text{ cm} = 10i \text{ hubbles} \quad (i = \sqrt{-1})$$

The line element corresponding to this geometry is

$$ds^2 = -\frac{L(t)^2}{[1 - (r/(2|\mathscr{R}|))^2]^2}(dr^2 + r^2 d\theta^2 + r^2 \sin^2\theta \, d\phi^2) - (dct)^2 \tag{18}$$

where $L(t)$ is given in Fig. 1.

ORIGIN OF GALAXIES

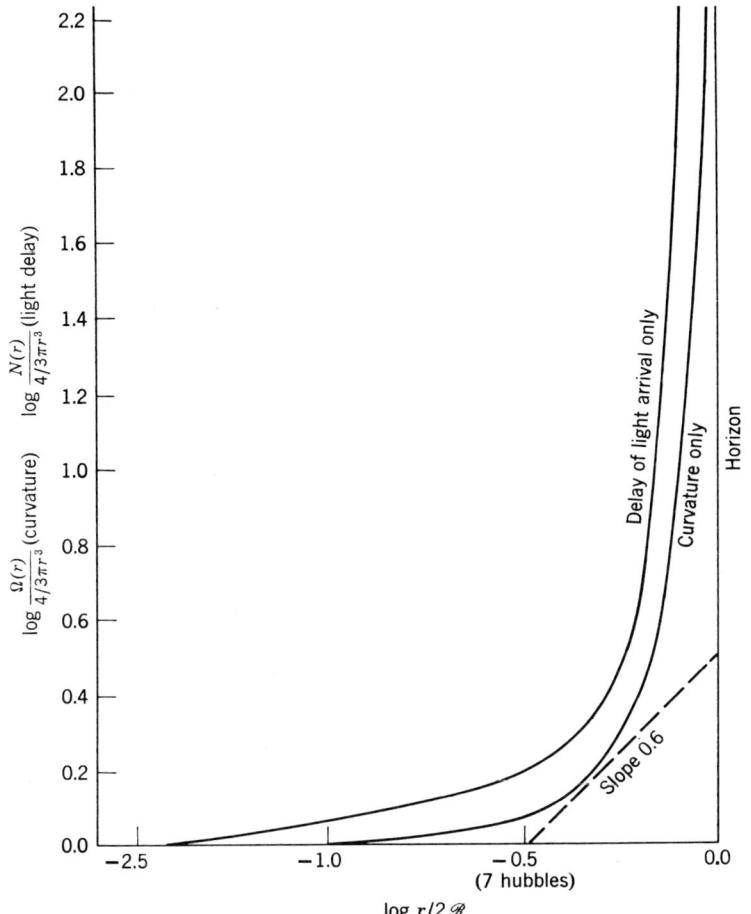

Fig. 3. Calculated dependence of the number of galaxies counted within a certain distance r on that distance.

The volume element is apparently given by the expression:

$$d\Omega = \frac{L^3(t) r^2\, dr}{[1 - (r/(2\,|\mathscr{R}|))^2]^2} \sin\theta\, d\theta\, d\phi \qquad (19)$$

and the volume of the space within a given distance r from the (arbitrary) origin O is:

$$\Omega(r) = L^3(t) \int_{r=0}^{r} \int_{\theta=0}^{\pi} \int_{\varphi=0}^{2\pi} \frac{r^2\, dr}{[1 - (r/(2\,|\mathscr{R}|))^2]^2} \sin\theta\, d\theta\, d\phi \qquad (20)$$

It is easy to see that this integral diverges when $r \to 2\,|\mathscr{R}|$. Thus the space possesses a "horizon," near which, at any given moment t of universal time, the space density of galaxies as observed from O tends to infinity.

Another factor which affects the observed space density of galaxies is that, because of the finite velocity of light, the value of $L(t)$ should be taken for t corresponding to the moment at which the light left the galaxy in question. This leads to the relation:

$$z = \frac{\lambda'}{\lambda} = \frac{l'}{l} = L^{-1} \tag{21}$$

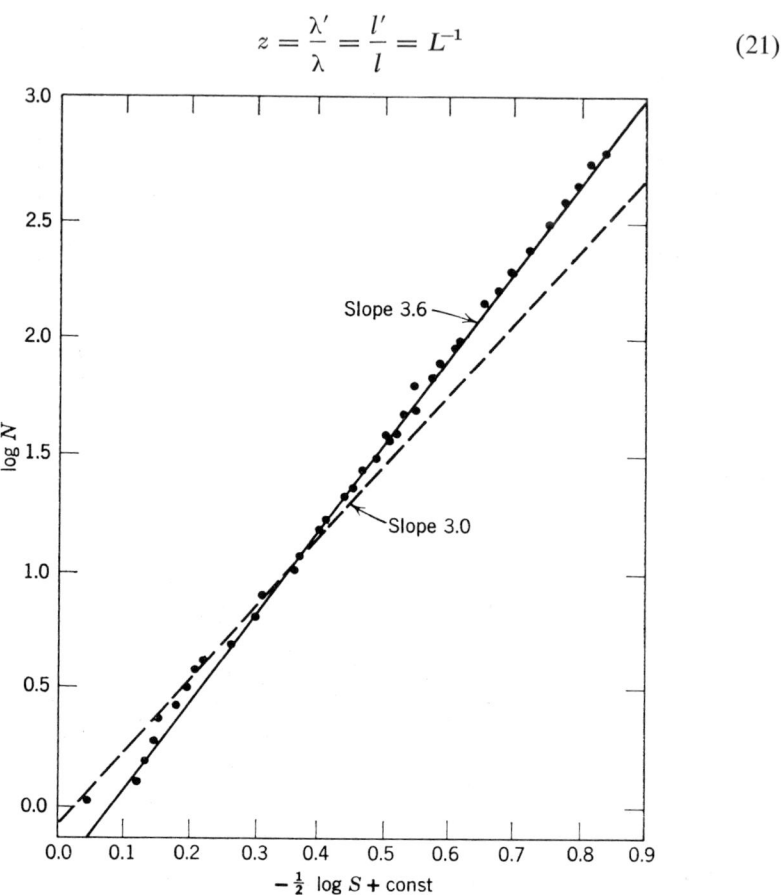

Fig. 4. The results of Ryle's galactic counts plotted against the distance. Slope 3.0 would correspond to the Euclidean or the Steady State cosmology.

Both effects, the negative curvature of space and the finite velocity of light, tend to make the observed distribution-density of the faraway galaxies *increase faster than r^3* as it would be for the Euclidean nonexpanding space.

The curve shown in Fig. 3 gives the results of the calculations based on the cosmological data accepted in the present paper.

This theoretical result can be compared with the observational material concerning "galactic counts" available today. In Fig. 4 are shown the results of galactic counts carried out by Martin Ryle and his co-workers at Cambridge University.[8] The original curve, which was showing the number of galaxies with radiointensities larger than different values of S, is replotted, using $-\frac{1}{2}\log S$ instead of $\log S$ in the original plot.

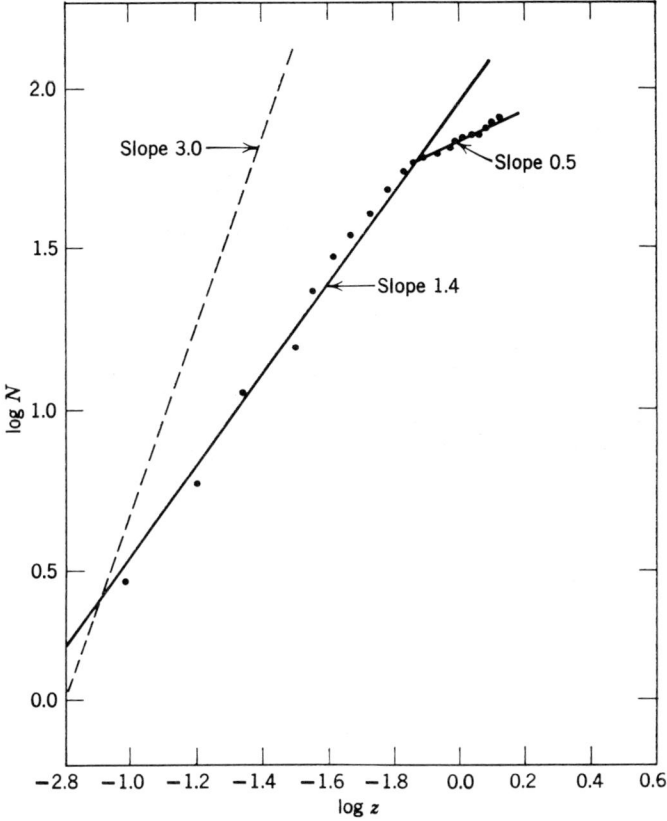

Fig. 5. The number of radiooptical distant sources plotted against the red shift z. Probably affected by strong observational selection.

If one assumes that the intrinsic luminosity of radiosources is independent of time, this curve is comparable with that given in Fig. 3, indicating a faster increase in the number of galaxies than the cube of their distance from us.

After the discovery by Maarten Schmidt[9] of the red shift in the optical spectra associated with the radiosources, it became possible to obtain the distances involved here directly from the optical data. In Fig. 5 is given the plot of $\log N$ vs. $\log z$ for 87 radio objects compiled by Geoffery and Margaret Burbidge in early 1967.[10] Surprisingly, this plot shows that Ω

increases much more slowly rather than faster than r^3. The effect is probably due to the strong observational selection of the brighter galaxies for the study of the red shifts.

References

1. G. Gamow and E. Teller, *Nature*, **143,** 116 (1939) (written in 1938); *Phys. Rev.*, **55,** 654 (1939).
2. J. Belzer, G. Gamow, and G. Keller, *Proc. Sci. Computer Forum*, I.B.M. Corp. 1948; *Astrophys. J.*, **113,** 166 (1951).
3. G. Gamow, *Phys. Rev.*, **74,** 505 (1948); *Nature*, **162,** 680 (1948). See also: *Kgl. Danske Videnskab. Selskab, Mat.-Fys. Medd.*, **27,** 10 (1953).
4. R. A. Alpher and R. C. Herman, *Phys. Rev.*, **75,** 1086 (1949).
5. A. A. Penzias and R. W. Wilson, *Astrophys. J.*, **142,** 419 (1965). P. G. Roll and D. T. Wilkinson, *Phys. Rev. Letters*, **16,** 405 (1966).
6. R. A. Alpher, G. Gamow, and R. C. Herman, *Proc. Natl. Acad. Sci. U.S.*, **58,** 2179 (1967).
7. S. Ulam, *Los Alamos Sci. Lab. Rept.* No. 12345 (1956) (unclassified).
8. M. Ryle, *Monthly Notices Roy. Astron. Soc.*, **122,** 389 (1961).
9. M. Schmidt, *Nature*, **197,** 1040 (1963).
10. G. Burbidge and M. Burbidge, *Quasi-stellar Objects*, Freeman, San Francisco, 1967.

The Origin of Deep-Focus Earthquakes*

D. T. GRIGGS and D. W. BAKER

Institute of Geophysics and Planetary Physics, University of California, Los Angeles, California

Introduction

The occurrence of deep-focus earthquakes presents one of the most intriguing problems of earth science. The sudden release of elastic energy at depths down to 700 km seems superficially similar to fracture, yet fracture as we commonly experience it cannot occur at very high pressure and temperature. Nevertheless, there is some form of instability which results in these seismic events. When this is understood, it will help us to understand the dynamic processes in the mantle which apparently cause continental drift and mountain building.

Earthquakes with focal depths greater than 300 km are almost entirely restricted to four regions on the landward side of the circum-Pacific earthquake belt: Western Argentina, west of the Kamchatka-Japan-Marianas trench, west of the Mindanao-Celebes trench, and west and south of the Tonga-Kermadec trenches. The most detailed work by Sykes[1] shows the distribution of earthquake foci in the Tonga-Kermadec region (Fig. 1).

The radiation pattern of deep-focus earthquakes is similar to that of shallow and intermediate-depth earthquakes which strongly suggests that the mechanism is similar to a shear fracture. The difficulty in explaining deep-focus earthquakes is that a material at such high pressure (250 kb at 700 km) and at temperatures which must approach the melting point would not be expected to fracture, but rather to yield by plastic flow.

In order that fracture may occur, the resistance to displacement after loss of cohesion must be less than the flow strength. In ordinary shear fracture at modest pressures, this resistance is friction. At normal stresses of only a few kilobars, however, the dry friction in rocks becomes equal to the flow strength and fracture is inhibited.

The presence of an interstitial fluid in sufficient quantity will allow fracture to occur at any value of normal stress, provided the viscosity of

* Publication No. 630 of the Institute of Geophysics and Planetary Physics.

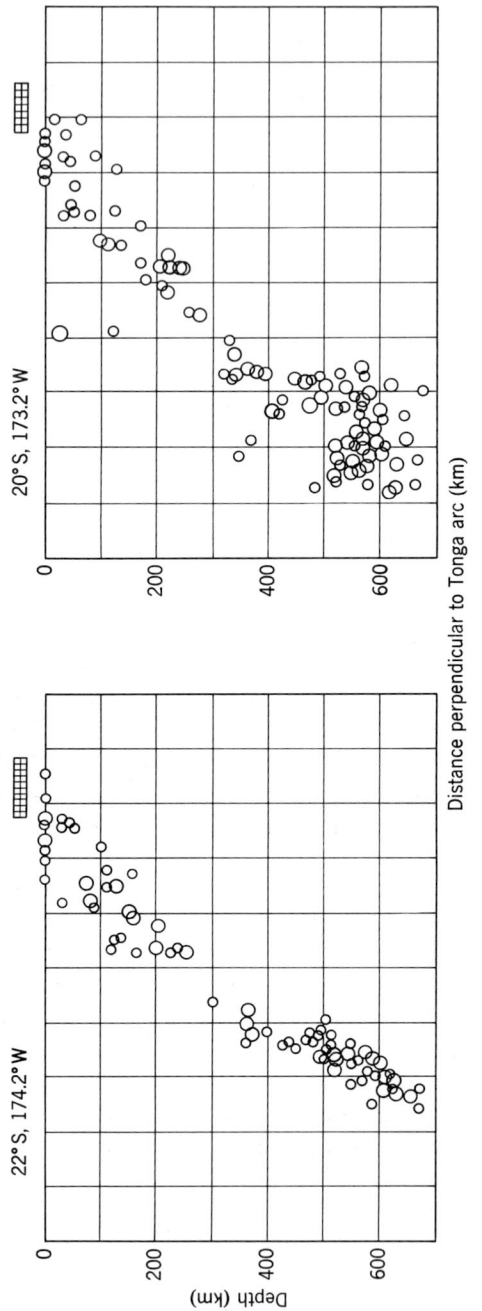

(a)

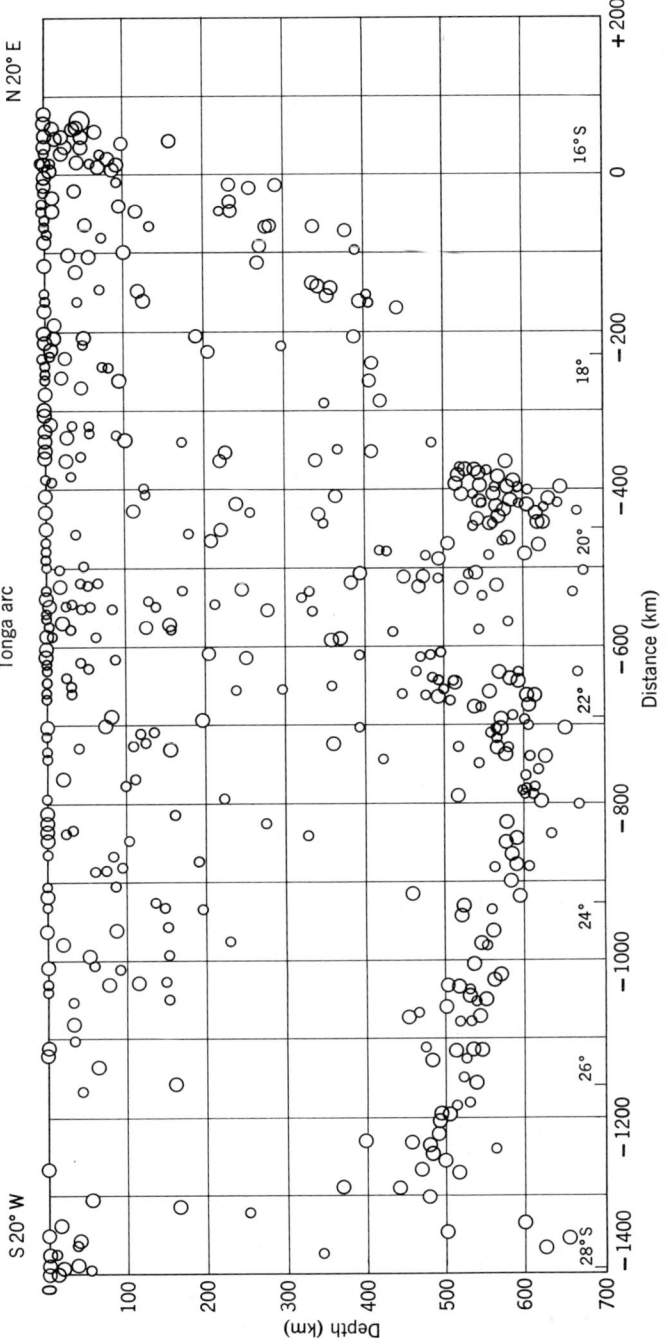

Fig. 1. (a) Vertical sections oriented *perpendicular* to the Tonga arc. Hypocenters projected parallel to the strike of the arc from distances within 125 km of each section. Larger symbols represent more accurate hypocenter locations. (b) Vertical section *parallel* to the Tonga arc. All hypocenters projected onto the vertical section. After Sykes.[1]

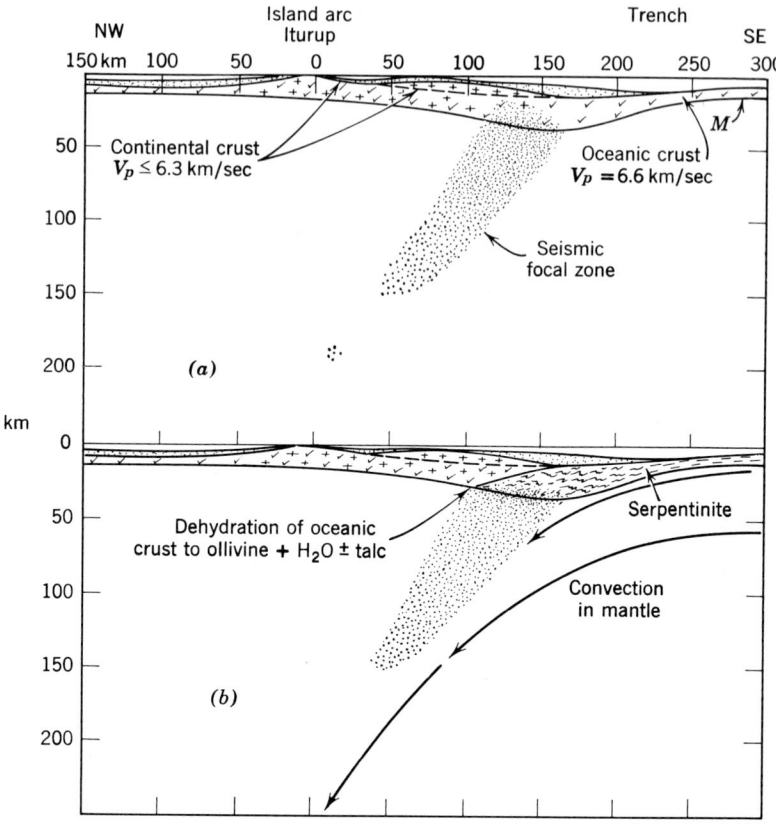

Fig. 2. (a) Cross section through southern Kurile Islands showing crustal structure and distribution of seismic foci. Ocean and mantle are blank; sediments are stippled. Density of stippling in seismic focal zone corresponds approximately to density of foci. (b) Interpretation of structure above showing oceanic crust of serpentinite dragged beneath island arc by convection in oceanic mantle. Serpentinite dehydrates to water + olivine at base of crust and dehydration products are carried beneath island arc. Faulting takes place in weak and brittle dehydration products. After C. B. Raleigh.[3]

the fluid remains sufficiently low at the high pressure. Interstitial water is believed by the authors to be present in sufficient quantity to permit fracture and/or stick-slip friction down to perhaps 30 km. Below that Raleigh[2,3] suggests that the dehydration of hydrous minerals such as chlorite, serpentine, talc, and the amphiboles can release enough water to permit shear fracture. Below about 100 km, however, one expects that hydrated minerals will no longer exist, so a new mechanism is needed.

The tectonic setting of deep-focus earthquakes is of interest and provides the major clue as to the conditions under which they occur and the regional motions and forces which must be responsible for their occurrence. It has recently been demonstrated beyond much doubt that the ocean floor is spreading laterally from the world-wide system of mid-ocean ridges at a rate of 1–5 cm/year.[4-8] This implies large-scale mantle currents which rise under the ridges, spread laterally under the thin crust, and must go down someplace. The obvious place for these downward currents is the circum-Pacific belt.[9] A typical construction of this relationship is shown in Fig. 2.

It is important to recognize that the oceanic sediments must be carried down by these descending currents. If this were not so, tens of kilometers of sediment would have accumulated in the troughs which mark the site of the descending currents. Trenches which are currently active as indicated by seismicity have little or no sediment in them. Inactive trenches—relics of past currents—on the contrary, have deep accumulations of sediments deposited since the currents ceased. It is not clear to the authors how these exceedingly weak and water-saturated sediments of low density are carried down between the moving suboceanic mantle current and the more or less stable subcontinental mantle, but it seems clear that this must be happening. There must be something akin to surface roughness or local eddies at the interface to prevent the light sediments from being skimmed off the descending current.

Whatever the cause of this phenomenon, these sediments must serve to lubricate the interface in the upper regions where the mantle is cool, and to provide the requisite fluid pressure for shallow earthquakes in these zones as suggested above. At depths below about 100 km, however, the predominant mechanism of motion must be flow. It is the purpose of this paper to consider one class of instability which might exist in these deep regions, resulting in earthquakes. The same type of instability is believed to be responsible for the "snapping" which occurred in Bridgman's shearing experiments at high pressure. Discussion of this phenomenon will illustrate the nature of the instability we suggest, which will later be considered as a mechanism for deep-focus earthquakes.

An Explanation of Bridgman's "Snapping" Phenomenon

In a revealing set of experiments, Bridgman subjected a wide variety of materials to intense shearing strain.[10-12] Material was placed between two pistons and an anvil (A, Fig. 3), subjected to pressures up to 50 kb and then the anvil was rotated to induce very high shear strain in the very thin sample wafer. Most materials deformed smoothly, but materials of low thermal conductivity and high melting point exhibited unsteady behavior. This varied from a quiet hissing with imperceptible drops in the applied torque to violent snapping in which the applied force would suddenly drop to zero and then build up at a nearly steady rate until another snap occurred.

Bridgman showed that this was a property of the material and not a characteristic of the piston–sample interface by showing that friction inhibited piston–sample displacement above a pressure of about 10 kb.

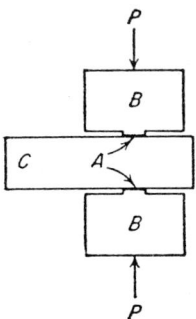

Fig. 3. Bridgman's shearing apparatus. Sample placed at A between pistons B and anvil C. Pressure P subjects thin sample to quasi-hydrostatic pressure. The anvil is then rotated, inducing large shear strains in sample.

He interpreted this snapping as rupture, and suggested that this form of rupture might occur in deep-focus earthquakes.[11] Bridgman calculated the temperature in the sample on the assumption of steady-state heat flow and satisfied himself that the temperature rise could not exceed 30°C under the most extreme conditions.[13]

Griggs suggested local shear melting as the cause of this snapping phenomenon.[14] Griggs and Handin derived the conditions for shear melting on the assumption of propagation of a flaw in a manner similar to the progagation of Griffith cracks.[15] The aspect ratio of the flaws required for this type of shear melting is very high—of the order of 100 for the Bridgman experiment and a million for deep-focus earthquakes, which makes this mechanism somewhat dubious.

We now consider another type of intrinsic instability which would also lead to shear melting in the central zone of Bridgman's wafers. In a material being deformed plastically the work done goes to increasing its internal energy or to heat. In Bridgman's experiments practically all the work done appears as heat since the strain rates are large, the strains are large enough to produce an equilibrium density of dislocations, and the

ORIGIN OF DEEP-FOCUS EARTHQUAKES

stress varies comparatively slowly prior to a "snap." This heat is conducted to the anvil face, and the direction of heat flow is essentially normal to the anvil face, since the diameter is 20 to 50 times the thickness. Hence, to a good approximation, the equation for heat flow in the sample is:

$$\tau\dot{\gamma} = \rho C_p \frac{\partial T}{\partial t} - k \frac{\partial^2 T}{\partial x^2}$$

where τ = shear stress, $\dot{\gamma}$ = strain rate, ρ = density, C_p = specific heat, k = conductivity, T = temperature, t = time, and x is the direction normal to the piston face.

It has been shown that most materials and all rocks so far examined obey flow laws which contain an activation energy and probably belong to one of the two classes:

$$\dot{\gamma} = \alpha \sinh (\tau/\tau_0) \exp (-E/RT)$$

or

$$\dot{\gamma} = \alpha' \tau^n \exp (-E/RT)$$

where E = activation energy, R = gas constant, and α, α', and τ_0 are constants.

The second flow law presented above has been shown by Heard to fit the data for creep of marble better than the first.[16] For the only other rock material on which there are data (quartz and quartzite), either of the two laws fits within the scatter of the experiments. The second law is adopted in this paper but qualitatively similar results would be obtained using the first law. Weertman has shown that the exponent n has the value 2.5 for creep by dislocation climb.[17] Garofalo derives $n = 4.5$ for creep by dislocation climb.[18] Heard has found $n = 8.4$ for marble and $n = 4.5$ and 10 for "sub-basal" and basal slip, respectively, in quartz.[19] Griggs and Blacic found $n = 10$ for basal slip in quartz.[20] For this report two values of n were used: 4.5 and 9.0.

The activation energy is similarly uncertain in quartz: values from 10 to 110 kcal/mole have been found in the above experiments for different flow processes. In marble, Heard found that the activation energy was well defined and equal to that measured for self-diffusion (60 kcal/mole). In what follows, we consider the effects of different activation energies from 10 to 100 kcal/mole.

Since these constants are not known adequately, the flow law is normalized to Bridgman's measured values of stress and strain rate with the activation energy which has been assumed. The heat-flow equation cannot be integrated analytically, so numerical integration by the method described in Carslaw and Jaeger is used.[21] A further simplification is introduced: the calculations are done for the case of simple shear in a slab of infinite extent normal to the anvil face. The behavior of such a slab should be a good qualitative check on the conditions in Bridgman's actual experiment, since the instability should develop near the outer margin of the wafer and, once started, should propagate over the rest of the wafer readily due to the stress concentration. The usual rate of rotation in Bridgman's experiments was 0.2 radian/sec. The corresponding strain rates in a sample of typical thicknesses (0.012–0.024 cm) at two-thirds of the wafer radius are 3.55 and 1.78 sec^{-1}, respectively. The average shear stress prior to a snap is 15–20 kb in typical experiments. Since the force is observed by Bridgman to be nearly constant for a short interval just before a snap occurs, initial values for the strain rate $\dot{\gamma}_0$ and the stress τ_0 are chosen in the above ranges. The value of α' is determined from these and the activation energy. The plastic flow law and the heat-flow equation may then be written as follows:

$$\dot{\gamma} = \dot{\gamma}_0 (\tau/\tau_0)^n \exp(E/RT_0 - E/RT) \tag{1}$$

$$\dot{\gamma}_0 (\tau^{n+1}/\tau_0^n) \exp(E/RT_0 - E/RT) = \rho C_p \frac{\partial T}{\partial t} - k \frac{\partial^2 T}{\partial x^2} \tag{2}$$

where T_0 is the initial temperature in the calculation. The pistons and anvil have relatively high thermal conductivity and are massive, so that the temperature at the boundary of the wafer remains essentially constant at room temperature. In the calculations, the boundary is assumed to remain at 300°K. In the intervals between snaps, it was expected that the temperature within the sample would rapidly fall after the stress was released by shear melting and would remain near the boundary temperature during the period of stress increase preceding the next snap. This was verified by the calculations as will be shown later. For this reason, T_0 was assumed to be 300°K throughout the sample.

If the stress remains constant, this equation with these boundary conditions is identical in form to the equation for a slab of explosive whose reaction rate is governed by an activation energy. Robertson showed that such a slab will explode if its half-thickness (d) exceeds a certain value[22]:

$$d > [0.88 k RT_0^2 \exp(E/RT_0)/(E \cdot WA)]^{1/2}$$

where his term WA is the equivalent of our $\dot{\gamma}_0 (\tau^{n+1}/\tau_0^n) \exp(E/RT_0)$. The physical reason for this is readily seen. Any temperature perturbation will tend to grow because of the increase in reaction rate with temperature until the heat conduction just balances the increased heat generation, in the stable case where the thickness is less than the critical value. As the thickness is increased, the central temperature at equilibrium increases. The heat conduction is approximately proportional to the difference between the central and boundary temperature, while the rate of heat production increases exponentially with temperature. Hence, when the critical thickness is exceeded, heat production exceeds heat conduction in the center, the temperature rises very rapidly, and an explosion ensues.

For the case of the slab heated by plastic flow at constant stress, this "explosion" criterion may be rewritten as:

$$\psi = \frac{d^2}{k}(E/RT_0^2)\dot{\gamma}_0\tau_0 > 0.88 \qquad (3)$$

The Robertson criterion for the constant-stress case has been verified by machine solutions of Eq. (2). When the left-hand side of Eq. (3) is less than 0.88, the central temperature rises to a steady state value which depends on the value of this term, but never rises above about 310°K. When the critical value is exceeded, the central temperature rises at a rate which increases slowly at first and then exceedingly rapidly. Typical results of these calculations are shown in Fig. 4.

In the real experiment, however, the shear stress does not remain constant, but falls as the strain rate increases with temperature. Thus the heating rate is less as the temperature begins to rise than in the constant stress case.

We next consider the case of a constant displacement rate of the boundary of the slab, corresponding qualitatively to a constant rotation rate of the anvil relative to the pistons. The tangential displacement rate of the boundary of the material ($x = d$) relative to the center of the wafer is:

$$\dot{Z}_d = \int_0^d \dot{\gamma}\, dx$$

Since this is a constant given by the initial strain rate, from Eq. (1):

$$\dot{Z}_d = d\dot{\gamma}_0 = \dot{\gamma}_0(\tau/\tau_0)^n \int_0^d \exp(E/RT_0 - E/RT)\, dx \qquad (4)$$

Solving this for τ and substituting in Eq. (2):

$$\tau_0\dot{\gamma}_0 \exp(E/RT_0 - E/RT)\left[d\Big/\int_0^d \exp(E/RT_0 - E/RT)\, dx\right]^{(n+1)/n}$$
$$= \rho C_p \frac{\partial T}{\partial t} - k\frac{\partial^2 T}{\partial x^2} \qquad (5)$$

This case does not exhibit the explosive instability observed in the constant-stress case. The rate of temperature rise decreases with time, and although very high central temperatures may be reached, there is no behavior like the autocatalytic explosion of the previous case. Typical plots of central temperature versus time are shown in Fig. 5.

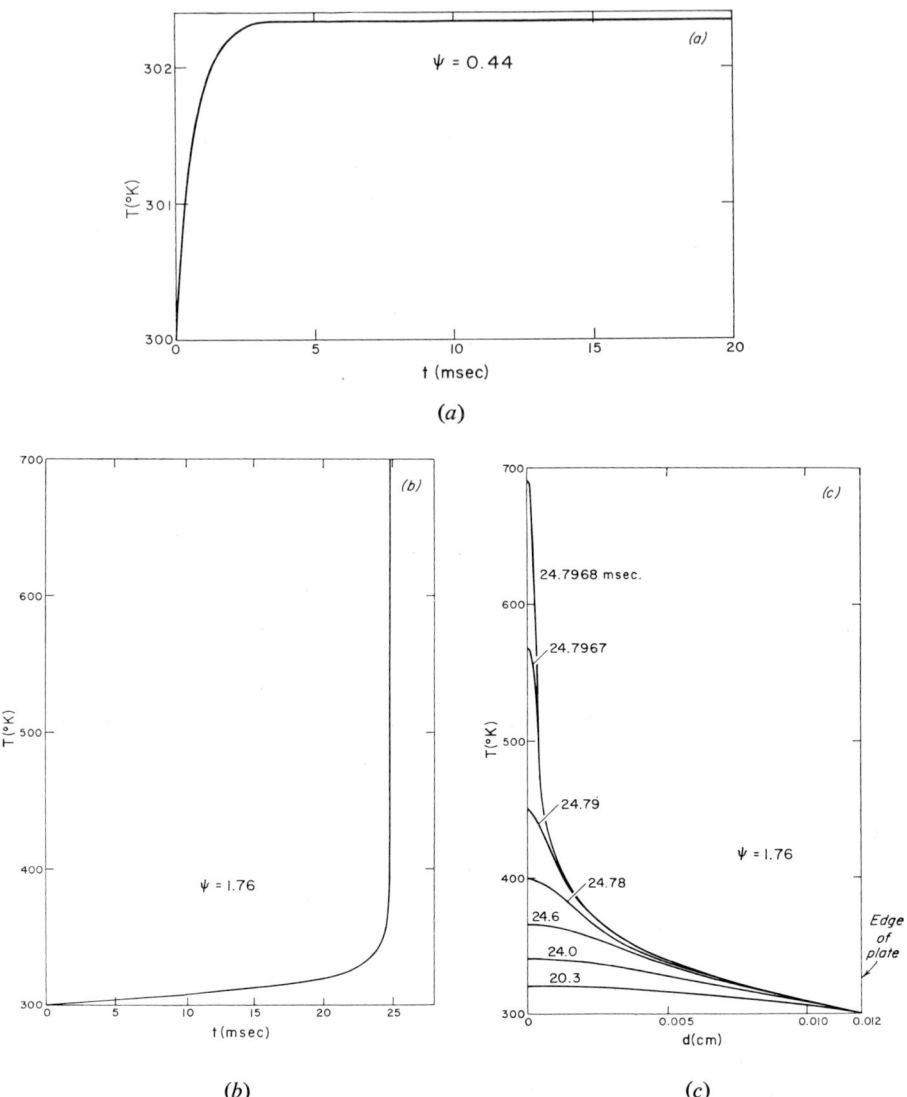

Fig. 4. Constant stress calculations from Eq. (2). (a) Rise of central temperature for $\psi = 0.44$ (half the Robertson explosion criterion). (b) Central temperature vs. time for $\psi = 1.76$. (c) Temperature profile across sample at different times with $\psi = 1.76$.

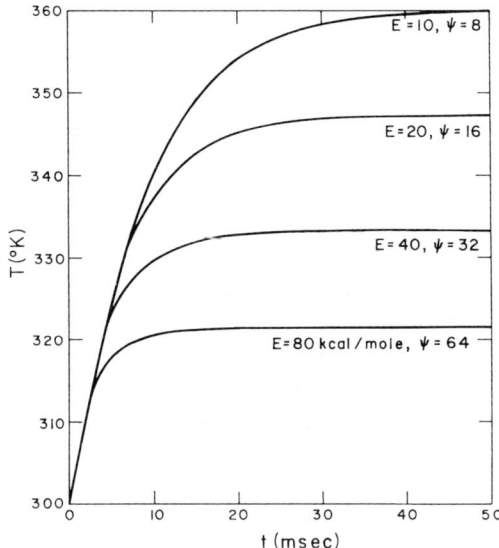

Fig. 5. Constant displacement rate $\dot{Z}_d$ at sample boundary. Central temperature vs. time for different activation energies. $\tau_0 = 30$ kb, $\dot{\gamma}_0 = 7.1$ sec^{-1}.

The Bridgman experiment is intermediate between these two cases. The rotation of the anvil was accomplished by applying force to a spring torquemeter, which deflected about 0.1 radian when a shear stress of 15 kb was applied to the sample. This elastically stored energy acts to keep the stress from falling as rapidly as in the constant-displacement rate case. If the spring constant were very low, the behavior would approach that in the constant-stress case, and if the spring constant were very high, it would approach the constant-displacement rate case. Since this same concept will later be applied to the earthquake source, the Bridgman experiment will be treated as though it were a thin slab the thickness of the wafer, subject to simple shear by an elastic element which is taken to be a pair of parallel elastic slabs of thickness L (Fig. 6).

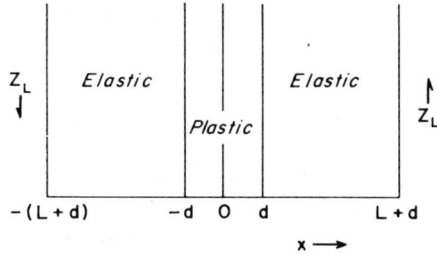

Fig. 6. Elastic–plastic slab model.

The shear strain of any element is the sum of the plastic and elastic components:
$$\gamma = \gamma_p + \gamma_E$$
The displacement rate (relative to the center of the plastic slab) at the outer boundary of the elastic slab is:

$$\dot{Z}_L = \int_0^d \dot{\gamma}_p \, dx + \int_0^d \dot{\gamma}_E \, dx + \int_0^L \dot{\gamma}_E \, dx \tag{6}$$

The change in elastic shear strain in the wafer is negligible and the plastic strain rate is given by Eq. (4). It is assumed (typical of many of Bridgman's snapping experiments) that the displacement rate at the point of application of force is constant, corresponding to steady-state rotation at the end of the torquemeter. Hence Eq. (6) may be rewritten as follows:

$$\dot{Z}_L = d\dot{\gamma}_0 = \dot{\gamma}_0(\tau/\tau_0)^n \int_0^d \exp(E/RT_0 - E/RT) \, dx + \frac{L}{\mu}\frac{d\tau}{dt} \tag{7}$$

where μ is the modulus of rigidity of the elastic slab.

Equations (2) and (7) define the behavior of this system. Explosive instability is possible in this case, as in the constant-stress case. If the Robertson criterion is sufficiently exceeded by the initial conditions, $\partial T/\partial t$ will increase with time. As the temperature increases, the strain rate in the plastic region will increase and the stress will drop. If the stored elastic energy is sufficient, this will develop into a very sudden rise in the central temperature. The temperature will not rise indefinitely as it does in the idealized constant-stress case, but will only rise until the increasing heat flow rate balances the rate of release of elastic energy. The Robertson criterion must be exceeded by a factor of about 4 for instability to develop in the Bridgman experiment. If the elastic compliance were less (stiffer torquemeter), the factor would be greater.

Numerical calculations were used to explore instabilities in this elastic–plastic model. The values of $\dot{\gamma}_0$, τ_0, T_0, ρC_p, k, d, and n were taken as described above. The value of L/μ is derived from the description of Bridgman's torquemeter. Preliminary calculations showed that, within the range of Bridgman's reported values, the model was stable at 10 kcal activation energy and unstable at 20 kcal and higher activation energies. Because of the exponential terms, Eqs. (2) and (7) exhibit calculational instability which becomes severe as the activation energy is increased beyond 20 kcal/mole. Although this subject has received much attention from von Neumann, Hartree, Crank, Nicholson, and many others, the authors do not know any time-saving foolproof prescription for avoiding these artificial instabilities. In the present calculation, the approach was to reduce the time step until no further changes occurred in the numerical solution. Double precision calculation is necessary (IBM 360/75) because the roundoff error becomes significant in these many-step calculations. A typical calculation requires 100,000 time steps.

When instability develops, the temperature rises exceedingly rapidly, so that even microsecond time steps cannot catch the details of the high temperature behavior. When the melting point is exceeded in the central cell of the space array of the calculation, the force is set equal to zero and

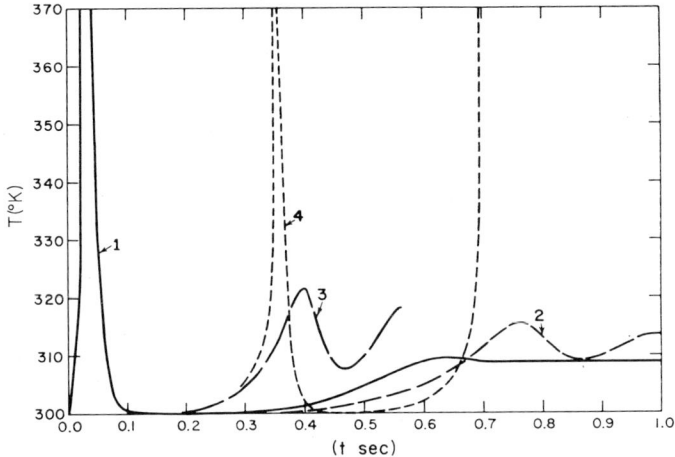

Fig. 7. Calculations of central temperature vs. time for elastic–plastic model subject to constant displacement rate $\dot{Z}_L$ at boundary of elastic plate.

	τ_0, kb	$\dot{\gamma}_0$, sec^{-1}	E, kcal/mole
(1)	15	1.78	20
(2)	20	1.78	20
(3)	15	2.67	20
(4)	15	2.67	25

half of the remaining elastic energy is added to the central cell. This corresponds to release of elastic strain by shear melting and seismic radiation at high efficiency.

The initial conditions assumed are artificial, since the past history effects the temperature and stress history. It was anticipated that after the first "snap," however, the behavior would repeat. Figure 7 shows the results of a series of calculations at the threshold of repetitive instability. All show initial instability due to the artificial starting conditions. In calculation (1), although the initial stability developed quickly, after this the model appeared to approach a steady state equilibrium in which no further snaps would occur. Calculation (2) shows the effect of increasing the initial stress by 33%. Here the temperature rose somewhat higher after the first snap, but resulted in damped oscillations which would apparently die out to a steady state equilibrium. Calculation (3) shows the effect of increasing the initial strain rate by 50%. Again, damped oscillations result, although the maximum temperature reached is somewhat higher, and the model is very close to developing explosive instability. In calculation (4) the activation energy is increased 25% while the other

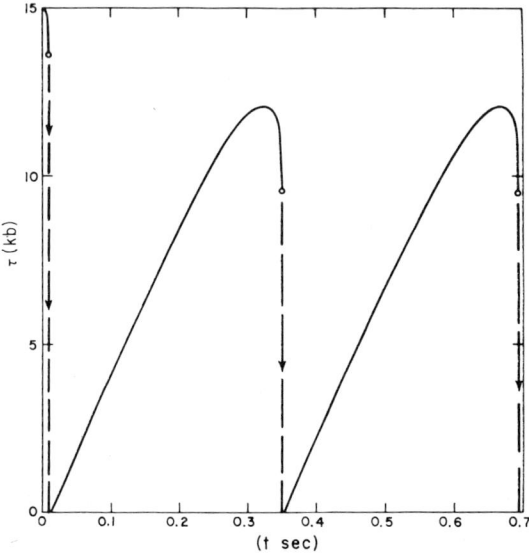

Fig. 8. Stress vs. time for calculation (4) in Fig. 7.

parameters are the same as in (3). Here repetitive snapping occurs which it is believed would continue indefinitely at a rate of about three snaps per second. The stress history of this last calculation is shown in Fig. 8. The stress, temperature, and strain-rate histories of the two snaps after the initial instability are not identical but agree to one part in a thousand, so that it seems pretty certain that this is the beginning of a sequence of regular periodic snaps.

Bridgman did two classes of experiments: constant force and constant speed. His "constant-force" experiment could not maintain the stress constant in the case of explosive instability because of the inertia of the apparatus. Nevertheless, the relative behavior is similar to that in our constant stress and constant displacement rate elastic–plastic case. For example, in the case of mica, he reported[23]:

> Mica shows phenomena considerably more complicated then the metals. Experiments made *at constant force* showed in the first place a much smaller

ORIGIN OF DEEP-FOCUS EARTHQUAKES 37

range of force within which the speed varied from too small to measure to too large to measure. Furthermore, the sharpness of the limits of force increases with increase of pressure. Thus at a mean pressure of 20,000 barely perceptible creep began at a mean shearing stress of 8300 kg/cm^2, and at 10,000 kg/cm^2 shearing stress the creep had become a running away of the lever. At a mean pressure of 40,000 there was no perceptible creep at a shearing stress of 19,300 kg/cm^2, and at 19,700 it ran away. Notice incidentally that these shearing strengths are very high. In the range of measureable motion just below the force producing instability the phenomena were quite capricious; creep was not at all regular, but there might be creep for awhile, then a little jump, then sticking for awhile, until the cycle of creeping and jumping repeated itself. The final instability and running away happened in the same way; for the critical force there was at first slow creep and then after a few minutes the system suddenly became unstable and ran away. At the highest pressure, 40,000, however, no creep was observed before the instability, but it immediately ran away on increasing the stress from 19,300 to 19,700.

The runs *at constant speed* showed different phenomena. For speeds corresponding to that of ordinary manual operation there was in the first place a perceptible variation of force with speed, and in the second place the rotation was punctuated with frequent snapping. This snapping means an internal rupture of the mica; because the apparatus is not infinitely stiff this rupture is accompanied by elastic yielding in parts of the apparatus and a drop of the rotating force. As rotation proceeds after a snap, force rapidly builds up again to its former value; rotation then proceeds quietly for awhile at a constant force, until there is another snap, and the cycle repeats. One surmizes that the snapping of the constant speed experiments corresponds to the running away of the constant force experiments. At lower speeds of rotation, however, under 0.04 radians per minute, the snapping entirely ceases, and rotation through the entire 60° is accomplished quietly.

It would seem that two phenomena coexist here. Mica has the capacity for continuous smooth plastic flow, but if the speed of deformation becomes too high, then the second phenomenon of rupture appears. This, as I have already emphasized in my geological paper, I believe to be of considerable significance for geology. What the exact mechanisms of rupture is cannot at present be stated. It must be that plastic flow is accompanied by some process of spontaneous internal readjustment, which proceeds at a characteristic speed. If the speed of deformation is so great that the spontaneous readjustment cannot keep the body in such a condition that it is adapted to flow, then it breaks. Whatever this process of internal adjustment is, it need not necessarily be accompanied by any important rise in the shearing strength; in the above experiments after each snap the force rapidly climbed back to its normal value and remained there without a sensible change until the next snap occurred.

These observations seem consistent with the shear-melting-instability hypothesis. In the constant-stress case, as stress is increased from a value at which imperceptible creep occurs, the strain rate will increase not only due to the stress effect in the flow law but also because the temperature of

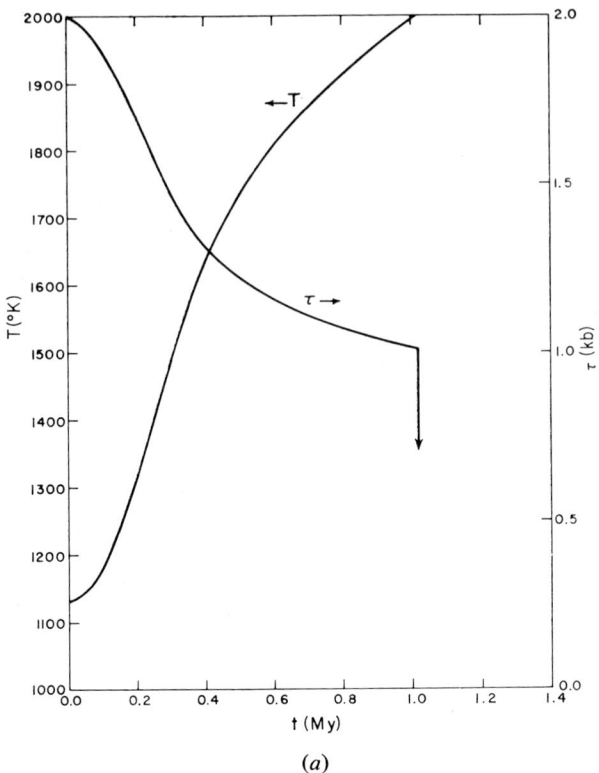

(a)

the sample increases and "running away" will occur suddenly as the Robertson criterion is exceeded.

The phenomena noted in the runs at constant speed are similar to those of our calculations in the elastic constant-displacement rate case, with one exception—Bridgman's observation of a period of constant force between snaps. This kind of behavior is possible in shear-melting instability, however, as will be shown below (see Fig. 9).

We thus believe that this shear-melting instability explains Bridgman's snapping phenomenon. To test this hypothesis, one should repeat Bridgman's experiment with varying elastic compliance of the torque system. Further, Bridgman-type experiments at various temperatures could perhaps establish the activation energy of the flow process and hence obtain quantitative verification or disproof of the hypothesis.

The calculational techniques should also be extended to include the effect of shear varying with radius, and the lenticular shape of the Bridgman wafer. Because of the great diameter-to-thickness ratio, it is believed

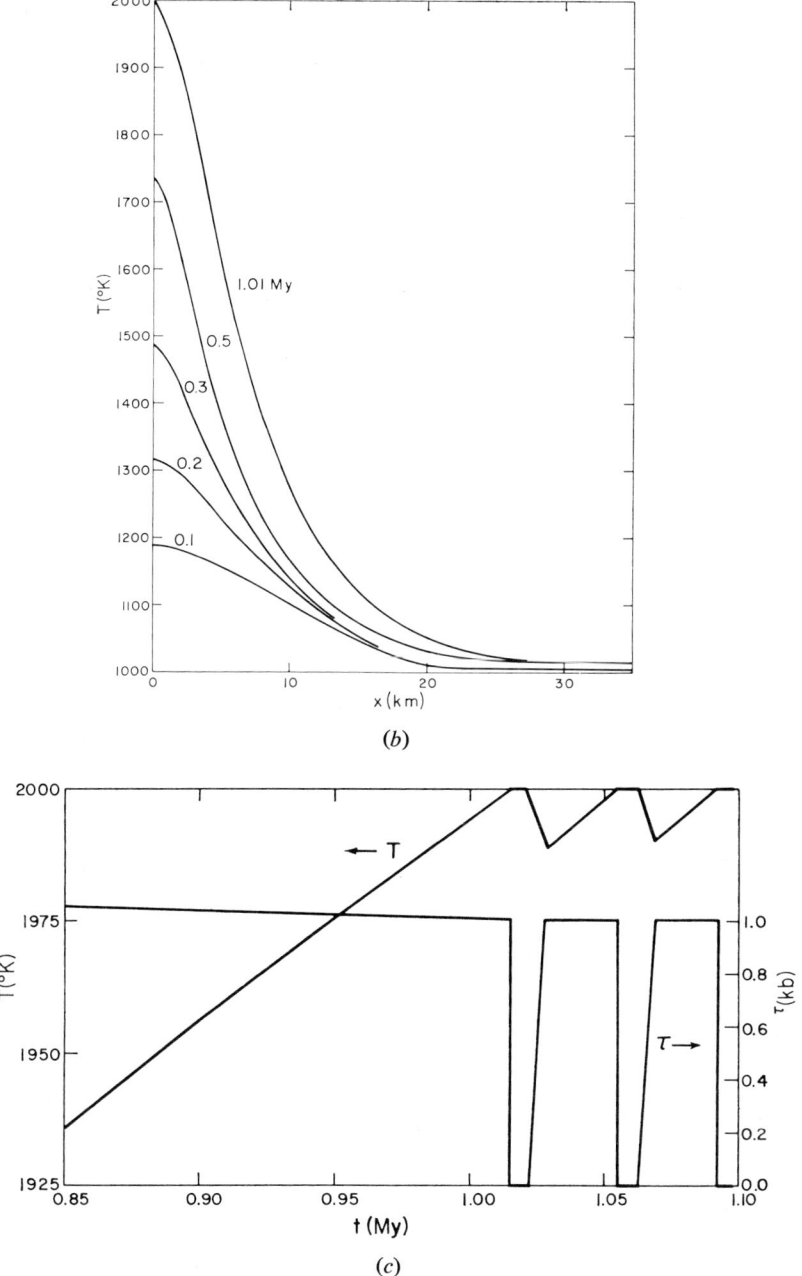

Fig. 9. Elastic shear melting calculations for 200 km thick plastic slab. (a) Central temperature and stress vs. time from initial conditions to first shear melting. (b) Temperature profiles during initial heating. (c) Central temperature and stress vs. time during first three cycles of recurrent shear melting. My, millions of years.

Shear-Melting Instability as a Source of Earthquakes

The Robertson criterion can be exceeded in large-scale mantle currents. Typical values in the mantle for the parameters in Eq. (3) are: $k = 0.01$ cal sec^{-1} cm^{-1} deg^{-1}, $E = 40$ kcal mole^{-1}, $T_0 = 1000°K$, $\dot{\gamma} = 3 \times 10^{-14}$ sec^{-1}, $\tau = 0.3$ kb. The half-width d corresponding to these values is 14 km. Thus, in the constant-stress case, any positive temperature perturbation of width greater than 30 km would be explosively unstable.

Because of the high elastic constants and presumed low stress in the mantle, however, the elastic stored energy is relatively far less than in the Bridgman experiment. Accordingly, the Robertson criterion must be exceeded by a much larger factor for instability to develop. The behavior of such elastic–plastic models has been explored by machine calculations of Eqs. (2) and (7). Referring to Fig. 6, the calculational zone (0–d) is chosen large enough (usually 100 km) so that the temperature perturbation does not appreciably affect the temperature at d, and a condition of zero temperature gradient across this boundary is usually imposed. The elastic constants of both regions are assumed to be the same, so that the distance L in Eq. (7) is taken from $x = 0$.

The numerical instability in the model is aggravated by the low elastic compliance, so that even smaller relative time steps than those used in the Bridgman model are necessary to avoid artificial instability.

We have not yet completed enough calculations to be sure that we have explored the range of applicability of this model to possible situations in the earth. Those calculations which have exhibited large temperature increases and have the property of recurrent "earthquakes" have rather improbably high values of L/μ and require a fairly high displacement rate. The behavior of a typical model is shown in Fig. 9.

The values of the parameters in this calculation are: $k = 0.01$, $E = 80$, $n = 9.0$, $T_0 = 1000°K$, $\dot{\gamma}_0 = 5 \times 10^{-14}$, $\tau_0 = 2.0$ kb, $L/\mu = 1.67 \times 10^{-4}$. An initial temperature perturbation of 130°C was applied with an appropriate width to grow uniformly under the initial conditions (about 20 km half-width). As shown in Fig. 9a, the temperature initially rises at an increasing rate. Then as the stress falls, the temperature continues to rise, but at a decreasing rate until the melting point (here taken as 2000°K) is reached. The temperature profiles at different times are given in Fig. 9b. When the melting point is reached, the stress is programmed to fall to zero, and half of the remaining elastic energy is placed in the central cell. There follows a short interval (Fig. 9c) as the melt refreezes by conduction and then a period of stress buildup which is essentially linear with time, until the integrated plastic strain rate is equal to the applied displacement rate. After this the stress remains essentially constant and the central region is again heated up to the melting point. This process recurs at regular intervals until the temperature perturbation has spread out by conduction, which is a relatively slow process.

At the present stage in our investigations we cannot say with any assurance that this shear-melting instability will apply to expectable mantle conditions. It does appear to have the characteristics required for deep-focus earthquakes, however.

Conclusion

It seems unlikely that deep-focus earthquakes can be caused by any of the known types of rupture, yet the seismic waves have the characteristics of a shear-fracture source. It has been shown that this type of behavior can be produced by shear-melting instability due to thermally activated flow in materials subject to shear strain at high pressure and temperatures. It is believed that this process was the cause of the "snapping" observed in Bridgman's high pressure shearing experiments. In this case, the hypothesis can be quantitatively checked by appropriate experiments. Application to deep-focus earthquakes is fraught with more uncertainty, but the fact that this process can qualitatively produce the effects of these seismic events warrants more thorough study of this phenomenon.

Acknowledgments

This research has been supported by NSF Grant GA 277. The authors gratefully acknowledge helpful discussions with F. Busse, R. Latter, C. Longmire, W. V. R. Malkus, A. C. McLaren, J. F. Nye, A. Smith, and E. Teller, many of whom spent long hours of study on the complexities of the heat-flow equation which is the basis of the hypothesis here presented.

References

1. L. R. Sykes, *J. Geophys. Res.*, **71,** 2981 (1966).
2. C. B. Raleigh and M. S. Paterson, *J. Geophys. Res.*, **70,** 3965 (1965).

3. C. B. Raleigh, *Geophys. J. Roy. Astron. Soc.*, **14**, 45, 113 (1967).
4. F. J. Vine and D. H. Matthews, *Nature*, **199**, 947 (1963).
5. J. T. Wilson, *Science*, **150**, 482 (1965).
6. F. J. Vine and J. T. Wilson, *Science*, **150**, 485 (1965).
7. W. C. Pitman III and J. R. Heirtzler, *Science*, **154**, 1164 (1966).
8. F. J. Vine, *Science*, **154**, 1405 (1966).
9. H. H. Hess, in *Submarine Geology and Geophysics*, W. F. Whittard and R. Bradshaw, Eds., Butterworths, London, 1965.
10. P. W. Bridgman, *Phys. Rev.*, **48**, 825 (1935).
11. P. W. Bridgman, *J. Geol.*, **44**, 653 (1936).
12. P. W. Bridgman, *Proc. Am. Acad. Arts Sci.*, **71**, 387 (1937).
13. E. S. Larsen and P. W. Bridgman, *Am. J. Sci.*, **36**, 81 (1938).
14. D. T. Griggs, in *Modern Physics for the Engineer*, L. N. Ridenour, Ed., McGraw-Hill, New York, 1954, p. 292.
15. D. T. Griggs and J. Handin, *Mem. Geol. Soc. Am.*, **79**, 347 (1960).
16. H. C. Heard, Final Report, Contract AF 19(628)-2784, Project No. 8652 Task No. 865211, Air Force Cambridge Research Laboratories, 1967, pp. 42–65.
17. J. Weertman, *J. Appl. Phys.*, **28**, 1185 (1957).
18. F. Garofalo, *Fundamentals of Creep and Creep Rupture in Metals*, Macmillan, New York, 1965, p. 187.
19. H. C. Heard and N. L. Carter, *Am. J. Sci.*, **266**, 1 (1968).
20. D. T. Griggs and J. D. Blacic, unpublished work.
21. H. S. Carslaw and J. C. Jaeger, *Conduction of Heat in Solids*, 2nd ed., Oxford University Press, London, 1960, p. 470.
22. A. J. B. Robertson, *Symp. Combust. 3rd, Madison, Wis. 1948* (Pub. 1949), p. 545.
23. P. W. Bridgman, *Proc. Am. Acad. Arts Sci.*, **71**, 410 (1937).

Response of Rocks to Stress

GARY H. HIGGINS

*Lawrence Radiation Laboratory, University of California,
Livermore, California*

Introduction

A number of important phenomena involve the response or behavior of rocks subjected to stress. A few obvious cases are earthquakes; meteor impacts; and explosions—either natural, as from volcanos, or man-made, as in mining or quarrying applications or the underground detonation of a nuclear explosive. Particularly, studies of nuclear explosions have yielded considerable detailed information of both experimental and theoretical nature concerning the subject matter of this chapter.

These studies have been largely motivated by Edward Teller. As early as 1944 he and others at the Los Alamos Scientific Laboratory foresaw the importance of nuclear explosives as new energy sources for other than weapons purposes. Many of the ideas then proposed involved underground detonations, which were expected to lead to the formation of large stable cavities and retention of energy within specified bounds or other expectations of rock behavior. The investigations which resulted from the critical analysis of these proposals led to most of the work summarized in the following pages. Teller has continued to be intimately involved in critically reviewing proposed models, identifying or sorting the "big" problems from the trivia, and encouraging many of us who have participated in the detailed efforts.

The information gained from studies of nuclear explosions has been added to the wealth of knowledge generated by those who have been concerned with geophysics and geochemistry, as well as to the body of information available from metallurgists and solid state physicists. It is not the intent, in this chapter, to review all of this tremendous body of information, but to summarize the more important publications covering perhaps the past five years, furnishing sufficient reference material so that the reader interested in the detailed pursuit of a given subject will have entrée to several different doors to the literature. Probably no one discipline or publication series provides sufficient bibliographies for a person seriously interested in the response of rocks to stress. This chapter draws heavily from U.S. Government report sources, AEC-sponsored programs, and NASA-sponsored cratering research, as well as from the published literature.

Whenever possible, the relevant data on rocks are included with a description of the methods of measurement and the analytical equations relating observed properties. No attempt has been made to criticize the methods or data, resolve differences between observers, or derive the mathematical relationships. The objective has been to achieve uniformity of symbolism; however, in a few cases, notably the Gruneisen gamma, Γ, bulk and linear compressibility, β, and bulk and linear coefficients of expansion, α, differences exist in different references. The distinctions between them have been defined.

For the purpose of this discussion a rock is defined as an assemblage of minerals as they might be expected to be found in the natural state; however, occasional inclusion of data from pure minerals is used to demonstrate a specific point.

1. Response to Very High Pressure Transients

In some instances (for example, explosions and meteor impacts), rocks may be subjected to almost instantaneous, very large increases in pressure or stress. We consider very high pressure as that ranging upward from 0.2 Mb. The response of matter under such conditions has often been discussed[1-3]; however, in order to introduce, later, one or two important differences between rock materials and homogeneous isotropic materials usually treated, the Rankine-Hugoniot equations of state will be presented. For the case of a plane wave front, they are:

$$P_1 - P_0 = U_s u_p / V_0 \tag{1}$$

$$V_1/V_0 = 1 - u_p/U_s \tag{2}$$

and

$$E_1 - E_0 = \tfrac{1}{2}(P_1 + P_0)(V_0 - V_1) \tag{3}$$

which, for very high pressure, can be approximated by

$$E_1 = \tfrac{1}{2}P_1(V_0 - V_1) \tag{3'}$$

where P is pressure, V is specific volume, E is the specific internal energy, U_s is the shock velocity and u_p is the particle velocity. Subscripts 0 and 1 refer to the initial and shocked states, respectively. Experimental shock compression curves, as shown in Fig. 1, can be constructed from laboratory specimens of rocks or rock-forming minerals. When plotted, as in Fig. 1, the resultant curve is not a true thermodynamic "equation of state," but rather the locus of P, V states which can be reached by single shocks from the $P_0 V_0$ state. Multiple or scattered shocks from the $P_0 V_0$ state to the same final pressure will generally result in less internal energy and, of course, larger final volumes. Also shown in Fig. 1 is a schematic unloading

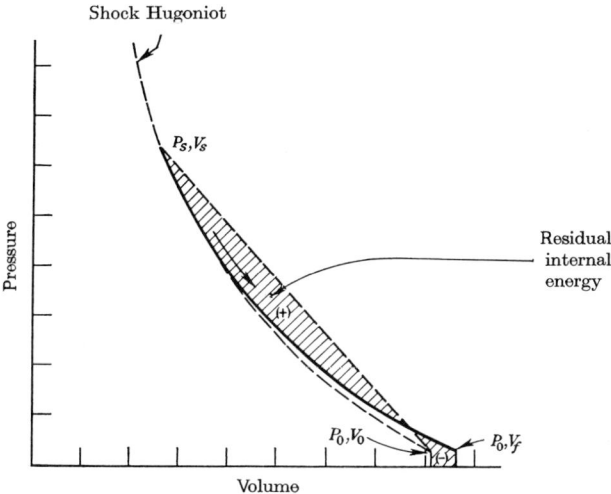

Fig. 1. Idealized shock Hugoniot, without polymorphic transitions or elastic limit. The transition from P_0, V_0 to P_s, V_s takes place discontinuously along the straight dashed line, and from P_s, V_s to P_0, V_f through the equilibrium states connected by the solid curve.

path or adiabatic through $P_s V_s$ to $P_0 V_f$. This unloading path is independent of the route by which the material arrived at $P_s V_s$, and, if multiple shocks were involved, the residual energy will be generally less than for single shocks; that is, the more "steps" there are in the compression, the more nearly reversible it will be.

As can be seen from Eqs. (1)–(3), the complete Rankine-Hugoniot relationship can be described if U_s and u_p are known. An empirical relationship has been observed between U_s and u_p in a large number of materials and is usually expressed in the following form:

$$U_s = C + S u_p \qquad (4)$$

The constants C and S are found empirically from graphs of U_s vs. u_p for each material. Since both C and S will be valid, in general, only for limited pressure ranges and for a single phase of material, tables of these empirical constants, such as Table I, generally include the pressure range applicable. Table I is a partial compilation from Ahrens,[4] Bass, Hawk, and Chabai,[5] and Bass.[6] A much more extensive tabulation is presented in the *Handbook of Physical Constants*.[7]

TABLE I. C and S Values for Use in the Relation $V_s = C + Su_p$

Material	Ref.	Density (g/cm³)	C (cm/μsec)	Sonic velocity (cm/μsec)	S	Pressure range (kb)
Limestone	6	2.66	0.40	0.58	1.140	$P > 200$
Andesite	6	2.59	0.35	0.44[b]	1.11	$P > 50$
Shoal granite	6	2.65	−0.111	0.554[b]	2.647	$P > 545$
			0.498		0.619	$130 < P < 545$
Volcanic breccia	6	1.82	0.157	0.28[b]	1.364	$P > 50$
Laramie oil shale	6	2.36	0.33	—	1.250	$P > 100$
Mahogany ledge oil shale (a)[a]	6	2.66	0.37	—	1.10	$P > 90$
Mahogany ledge oil shale (b)[a]	6	2.66	0.36	—	1.07	$P > 90$
Pahute mesa tuff (dry)	6	1.97	0.23	—	1.13	$P > 130$
Pahute mesa tuff (saturated)	6	2.19	0.34	—	0.9	$P > 100$
Alluvial material (dry)	6	1.54	−0.246	0.033[b]	2.33	$P > 330$
Alluvial material (dry)	6	1.54	0.04	—	1.53	$330 > P > 187$
Alluvial material (dry)	6	1.54	0.20	—	0.93	$187 > P > 98$
Alluvial material (dry)	6	1.54	0.04	—	1.83	$P < 98$
Playa[c]	4	1.41	0.4	0.4	2.0	$P < 87$
Playa[c]	4	1.41	2.5	—	0.88	$P > 87$

[a] Mahogany ledge oil shale: (a) Shock along bedding plane; (b) shock across bedding plane.
[b] *In situ* sonic velocity measurements.
[c] Nevada Test Site playa 6% H_2O.

Table I includes both rocks and minerals. Ahrens[4] shows that the Hugoniot for Vacaville basalt can be successfully constructed by adding, at each pressure level, the specific volume of each of the component phases. The line in Fig. 2 is the result of the addition of the experimental points

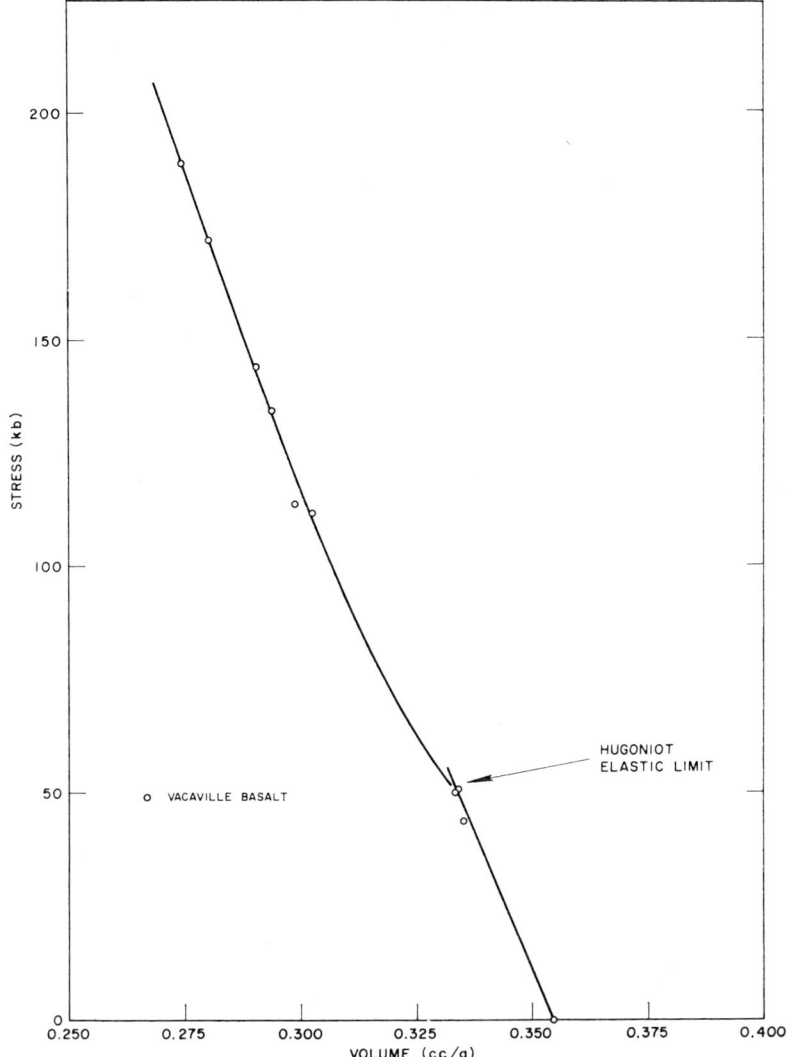

Fig. 2. Hugoniot for Vacaville basalt.

shown; however, for porous materials or water-containing silica sand as shown in Figs. 3 and 4, the procedure is obviously not valid.

A great deal of additional experimental data on Hugoniots for rocks and minerals has been collected and compiled. At pressures well in excess of 500 kb, the linear U_s–u_p relationship is often extrapolated to construct Hugoniot points beyond the range of experimental data. This procedure can be tested favorably with one of the theoretical or semiempirical equations of state, such as the Thomas-Fermi equation of state (Kompaneyets, 1962[8]) or the Mie-Gruneisen equation:

$$P - P_0 = \frac{\alpha V_0^2}{V C_p}(E - E_0) \tag{5}$$

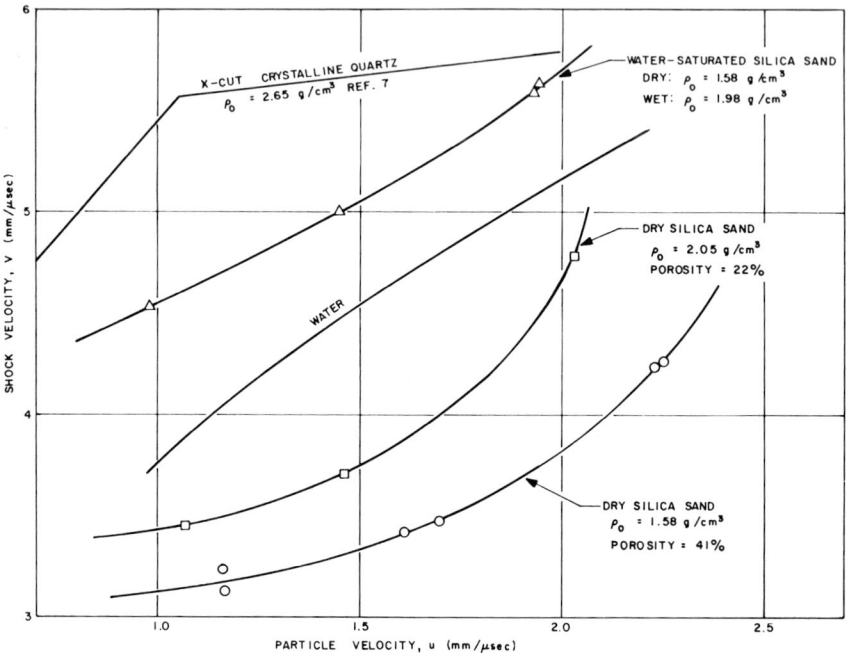

Fig. 3. Silica sand Hugoniot, shock velocity versus particle velocity.

where α is the volume coefficient of thermal expansion, V_0 the sound speed, and C_p the heat capacity.

Equation (5) is known to be invalid for a great many rock materials.[6] There is speculation that porosity causes difficulty in defining α, V_0, and C_p correctly.

The term in front of the energy in the right-hand side of Eq. (5) is called the Gruneisen coefficient, Γ_g, and is therefore defined as

$$\Gamma_g = \frac{\alpha V_0^2}{V C_p} \tag{6}$$

Γ_g should be equal[9] to $2S - 1$ [S from Eq. (4)], but actually, for dense rocks, Γ_g from shock data differs from Γ_g from Eq. (6) by twofold and for porous rocks it differs by as much as an order of magnitude.

It is possible to extend experimental measurements of the shock–pressure–volume–energy relationship to several megabars,[3] and this is sufficient for application to most problems. Extrapolation beyond 1 Mb

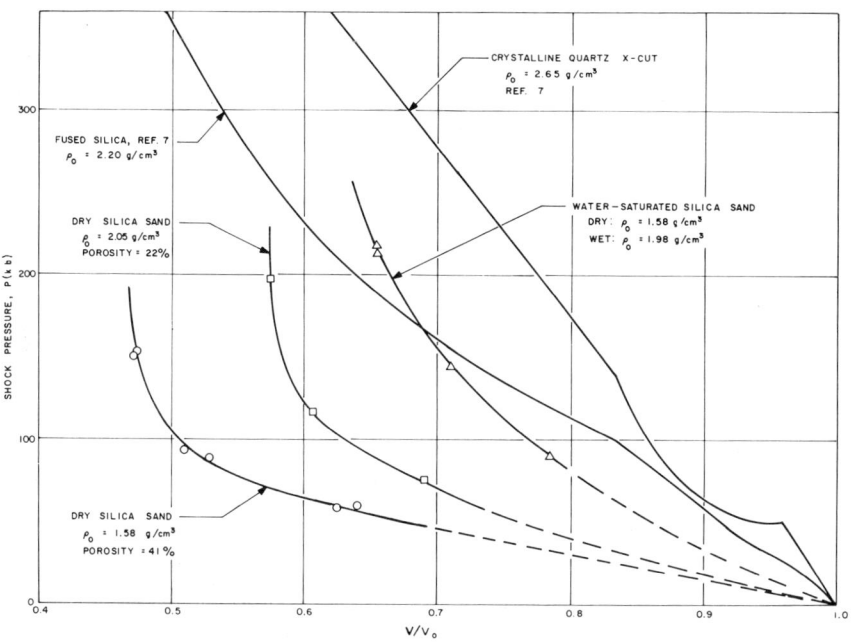

Fig. 4. Silica sand Hugoniot, shock pressure versus relative volume.

is considerably less subject to error than from pressures of about 200 kb, since material is generally in its most dense (metallic) form in the higher pressure range.

The foregoing discussion treats the irreversible compression of rock materials under conditions of extreme loading. Following compression, expansion occurs against some pressure less than the local shock loading pressure. This process is also shown schematically in Fig. 1, but may be treated more exactly by several different methods.

To get additional points not on the Hugoniot curve but in the PV plane, Rogers[10] recommends a semiempirical equation of state similar in form and development to Eq. (5), but modified to account for the phase changes which occur as energy is added to solids and for their large cohesive energies.

$$(P_1 + P_i)V_1^{(\gamma_g+\gamma_a)} = (P_2 + P_i)V_2^{(\gamma_g+\gamma_a)} \tag{7}$$

where

$$\gamma_g = \frac{3\alpha V}{C_p \beta_t} \tag{8}$$

and

$$\gamma_a = C_p/C_v \tag{9}$$

P_i is an "internal" or cohesive pressure which appears to be equal to the enthalpy of sublimation at absolute zero, or more approximately the

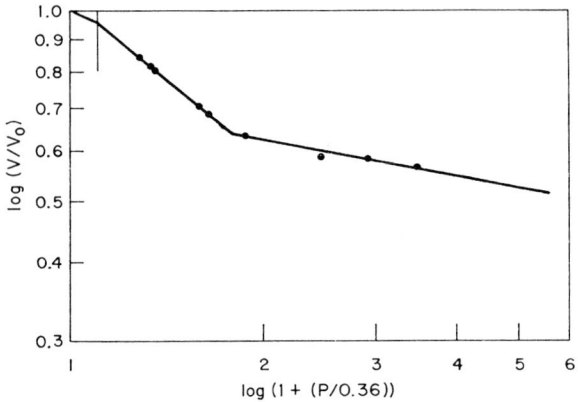

Fig. 5. Log (V/V_0) versus log $(1 + P/P_i)$ for shock compression of NTS basalt.

enthalpy of vaporization from 298°K; α is the linear coefficient of thermal expansion; β_t is the isothermal compressibility; and C_p and C_v are the heat capacities at constant pressure and volume, respectively. It should be noted that Eqs. (6) and (8) are identical under certain restrictive conditions (i.e., the material is gaseous) and that for solids $C_p \approx C_v$, so that γ_a is very nearly 1. Figure 5 demonstrates a plot of the basalt Hugoniot points according to Eq. (7), where P_i is 360 kb, $\gamma_g \approx 0.3$ with $P < 280$, $\gamma_g \approx 5.2$, with $P > 280$, and $\gamma_a = 1$. The utility of such a presentation in identifying phase changes is obvious and, once a fit for γ_a, γ_g, and P_i is selected, points lying off the Hugoniot can be computed.

A second method, which, while considerably more tedious, is not subject to extrapolative errors and provides a more complete description of a material, has been described in detail by Butkovich.[11] The method consists of computing, for various internal energies and densities, the pressure and temperature of a given material. For very high internal energies (above 3 eV) and low densities (less than 1 g/cm³), properties are calculated as perfect-gas equilibrium products, consisting of varying numbers of ions, molecules, and electrons. A computational version of the Saha equation, IEEOS,[12] is used to compute the number of particles, density, and temperature. Mixtures of up to ten elements can be accommodated if atomic numbers, weight, and ionization potentials are specified. At lower temperatures, but still at densities below the liquid–solid density region, thermochemical properties of the gaseous materials are used to compute state points. Under these conditions, it is presumed that there will be a few free electrons, or, stated another way, dissociated molecules will yield no more than one free electron. The perfect gas law is applied to each species calculated at a given temperature to compute pressures and total internal energies. A computer routine, HUG,[13] which can also include one solid or liquid phase (of constant volume) for each species is used to do the bookkeeping for mixtures.

Graphs are drawn through the state points from the computations specified above, including Hugoniot points and any experimental data available at higher densities and lower pressures. The result is a self-consistent graph describing some specified rock or pure substance. Figures 6 and 7 show results of the tabulation for NaCl and $SiO_2 + 1\% \ H_2O$.

Calculations leading to Figs. 6 and 7 can further be used to described the expansion of the gas phase once sufficient energy is imparted to the rock to cause vaporization. The amount required for vaporization, W_v, is obtained from thermochemical data, where the vapor is assumed to expand to about 100 bars. Equation (3) relates a shock pressure to each vaporization energy. This shock pressure, just sufficient to cause vaporization, is called P_v. For example, following a nuclear explosion, the spherically expanding shock wave reaches P_v at a specific radius, R_v, which depends only on the Hugoniot of the rock at its initial density. This radius is equivalent to a specific mass, M_v. Table II is a summary of the relevant calculated values for several rock types from Butkovich.[11]

TABLE II

	ρ_0 (g/cm³)	W (cal/g)	P_v (Mb)	$R_v^{1\,kt}$ (cm)	$M_v^{1\,kt}$ (g × 10⁶)
Granite	2.67	2800	1.80	183	68.6
Saturated tuff	1.97	2800	1.11	206	72.1
Dry tuff	1.76	2800	0.865	215	73.2
Alluvium	1.60	2800	0.703	220	71.4
Salt	2.24	1185	0.920	225	106.9
Water	1.00	620	0.196	330	150.5

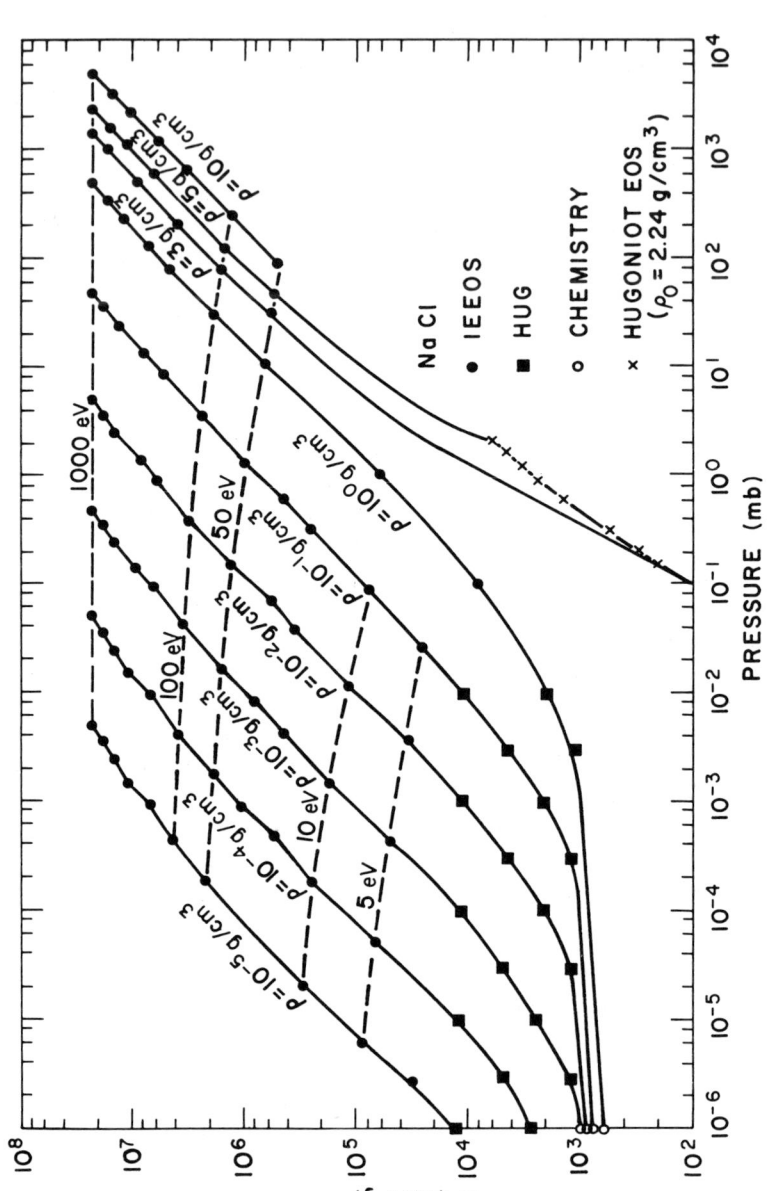

Fig. 6. Equilibrium equation of state for NaCl.

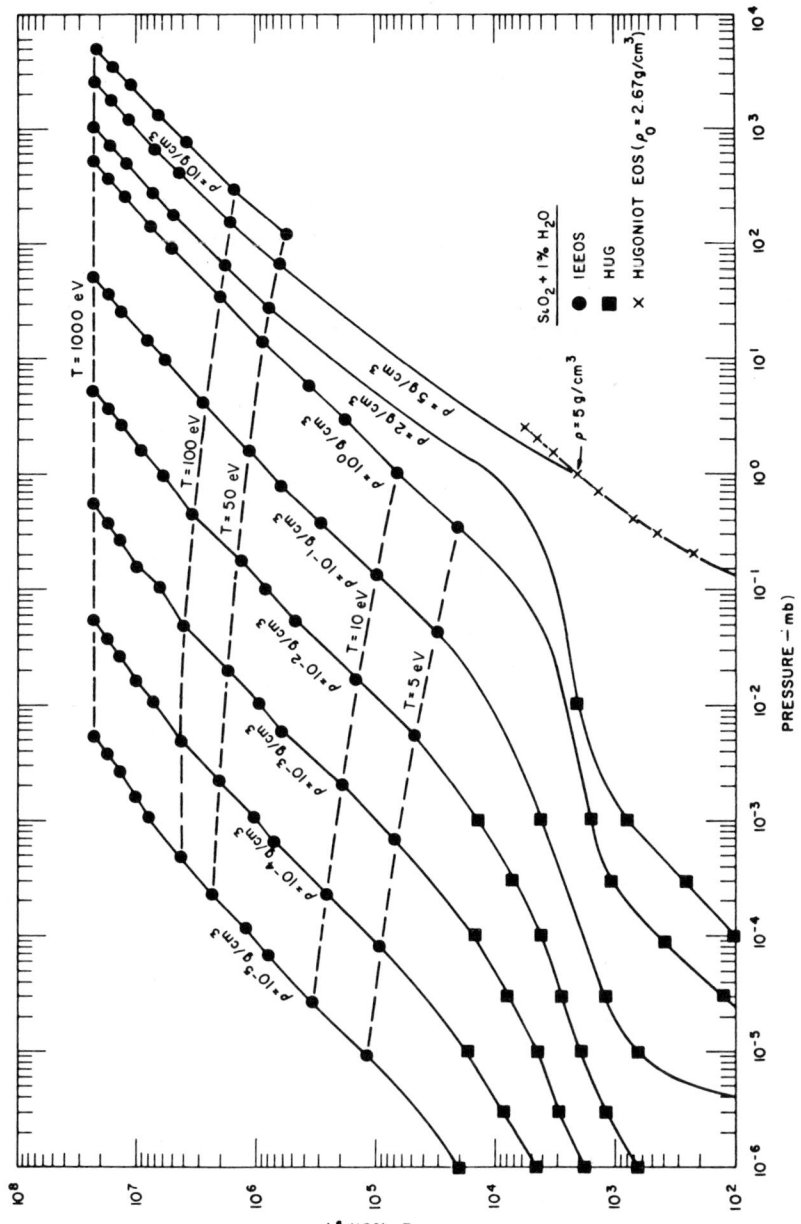

Fig. 7. Equilibrium equation of state for $SiO_2 + 1\%$ H_2O by weight.

Following vaporization, the gas will expand until its pressure is approximately equal to that of the surroundings. The tabular form of gas state points can be considerably simplified for limited regions of pressure by using the following equation:

$$PV = (\bar{\gamma} - 1)E \tag{10}$$

where $(\bar{\gamma} - 1)$ is a slowly varying quantity. Figure 8 is Butkovich's[11] tabulation of $(\bar{\gamma} - 1)$ as a function of P for three materials.

It is obvious from Table II that, for most phenomena, the equation of state and behavior of rocks at pressures in excess of a few hundred kilobars is important for relatively short distances, and then only to

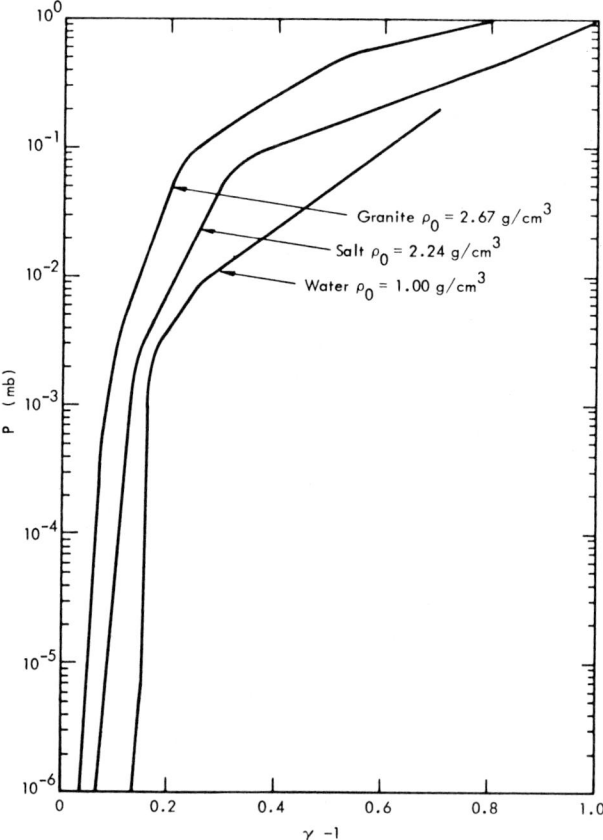

Fig. 8. Calculated relationship between $(\gamma - 1)$ and pressure for several natural materials.

RESPONSE OF ROCKS TO STRESS

determine the amount of vapor, or working fluid, which can later cause fracturing and yielding in the surrounding media, or in the case of a meteor impact, expand into the region above the surface which was struck by the meteor.

2. Behavior of Rocks Stressed by Transients from 10 to 500 kb

As the stress level decreases, the strength of rocks becomes more important in determining their overall behavior. Instead of responding as hydrodynamic* fluids, rocks subjected to these pressure levels display various amounts and kinds of resistance to failure. When a rock is subjected to a rapid increase in pressure just slightly greater than its dynamic yield point or strength, an elastic wave precedes the main shock front. This wave, called an elastic precursor, has an amplitude equal to the dynamic compressive strength of the material.

These dynamic yield strengths vary considerably from specimen to specimen of the same rock, and Ahrens[4] suggests that the observed amplitude of the elastic precursor is not independent of the specimen thickness. Thicker samples showed progressively lower elastic yield limits in his experiments. Results of one set of his measurements of this phenomenon are shown in Fig. 9.

Duvall[14] has suggested a theory which might account for this phenomenon. He assumes that the shocked material will momentarily support a higher-than-equilibrium value of shear stress. Then, after the sonic shock front has passed, the shear stress is reduced toward the static value by either plastic flow or fracturing. These processes would create forward-traveling rarefaction waves which would overtake the shock front and reduce its amplitude.

Another possibility is that as longer and longer paths are included, more and more chances for imperfections occur within samples. These imperfections would decrease the elastic wave amplitude by serving as minor reflection–rarefaction discontinuities. These discontinuities will have the effect of multiple shocks or "spreading" the shock front from a single discontinuity to several successive pressure increases.

Table III is a compilation of dynamic elastic strengths of several rock materials for ~ 1 cm sample lengths.

It is quite apparent from Table III that different rocks of the same name have quite widely different reported dynamic yield or "Hugoniot

---

* Or hydroelastic. The Hugoniot relationships are derived on the assumption that forces between molecules or particles are negligible. They are thus applicable only to perfect gases. At low pressures these are not valid assumptions.

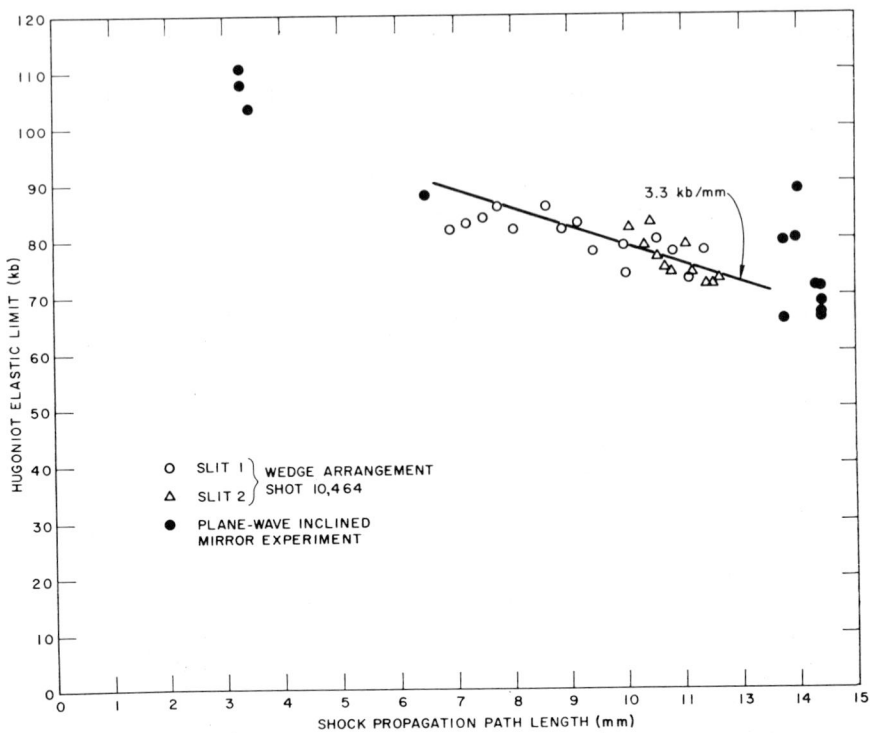

Fig. 9. Hugoniot elastic (limit) shock state versus shock propagation path length. Shot 10,464 (wedge) and plane-wave inclined mirror experiments.

elastic limits." It should also be apparent that, if Ahrens'[4] evaluation is correct, elastic waves resulting from large-scale natural phenomena—earthquakes and meteor impacts—or from large explosions will not be observed at the high pressure levels suggested in Table III, since the "length" of specimen will be several meters instead of about 1 cm, the commonly used laboratory sample size. The interpretation and constancy of the "elastic limit" or dynamic elastic strength is, therefore, still open to some speculation.

3. Lower Pressure Failure

Yielding of rocks at still lower pressure levels can be either brittle or ductile (plastic), and failure can be caused by tensile stress or by shear developed by compression, tension, or sliding. Laboratory measurements of failure of rocks in these different modes cannot generally be related to each other, and measurements with varying confining isostatic pressures

TABLE III

Material	Initial density	Dynamic shock compressive strength (kb)	Ref.
Novaculite	2.628	66–140	4
Quartzite,			
Eureka	2.624	68–84	4
Sioux	2.626	48–68	4
Augite	3.474	36–47	4
Diopside	3.283	69	4
Cracked	3.166	42	4
Calcite	2.71	24	4
Handcar dolomite		21	15
Yule marble		20	4
Solenhoffen limestone		10	4
Spergen limestone		5	4
Banded Mt. limestone		5	6
Buckboard basalt		40	5
Vacaville basalt	1.355	50	4
Shoal granite		60	5
Hardhat granite		40	16
Cowboy slat		1.2	17
NTS alluvium, tuff, and quartz sand		<1	

cannot be satisfactorily extrapolated to very low values of confining pressure. The approach used by Handin, Heard, and Magourik[18] is to plot the "octahedral stress," τ_{0c}, versus confining pressures at failure to display a material's failure responses. τ_{0c} is the stress resultant normal to the (111) plane. Figure 10 shows their results for dolomite at a strain rate of 10^{-4} sec^{-1} for failure in compression, extension, and torsion; and three distinctly separate curves should be noted.

Cherry, Larsen, and Rapp[19] have made a significant invention by analysis of failure tests of Handin et al.[18] and consideration of the resolution of stresses resulting from spherically symmetric waves originating from explosions. They define a yielding function, Y, in terms of the mean stress, P_m, and principal stresses T_{11}, T_{22}, and T_{33} (positive for tension) at failure of the test specimen. The relationship of the principal stresses in compression and extension, and the kinds of failure, are shown in Table IV.

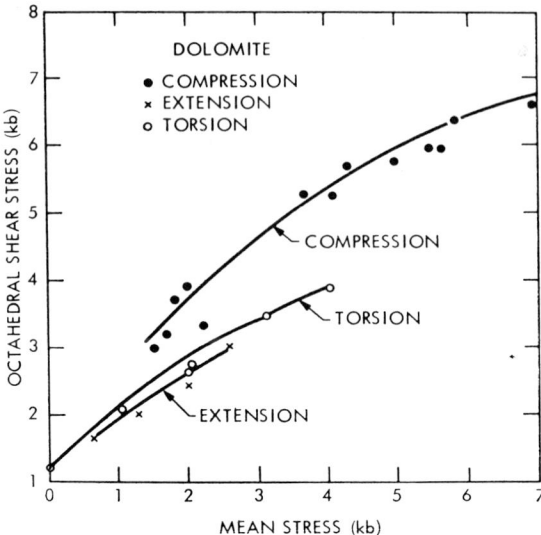

Fig. 10. Octahedral shear stress versus mean stress for dolomite.

TABLE IV

Compression	Extension (shear)	Extension (tension)
T_{11} ↓	T_{11} ↓	T_{11} ↓
← T_{22}	← T_{22}	← T_{22}
↗ T_{33}	↗ T_{33}	↗ T_{33}
$T_{11} < T_{22} = T_{33} < 0$	$T_{22} = T_{33} < T_{11} < 0$	$T_{22} = T_{33} < 0 < T_{11}$

The function Y is expressed in terms of the principal stresses, the space invariants of those stresses, or the stress deviators. These are defined as follows:

First invariant: $\quad I_1 = T_{11} + T_{22} + T_{33}$

Stress deviators:
$$T_1 = P_m + T_{11}$$
$$T_2 = P_m + T_{22}$$
$$T_3 = P_m + T_{33}$$

Second invariant of stress deviators:
$$I_{2D} = \tfrac{1}{2}(T_1^2 + T_2^2 + T_3^2)$$

Third invariant:
$$I_3 = T_{11}T_{22}T_{33}$$

Third deviatoric invariant:
$$T_{3D} = T_1 T_2 T_3$$

Finally,
$$Y = \tfrac{3}{4}\left[(3I_{2D})^{1/2} + \left(\frac{I_3}{|I_3|}\right)\left(\frac{|I_{3D}|}{2}\right)^{1/3}\right] \tag{11}$$

[It should be noted also that τ_{0c} is $(\tfrac{2}{3}I_{2D})^{1/2}$ and $P_m = -\tfrac{1}{3}I_1$.]

Values of Y deduced from data can be extrapolated smoothly to zero mean stress, as shown in Fig. 11, where the data for Fig. 10 were used.

The curve defined by the function Y and P_m is the boundary for failure. If, at a point in a material, P_m reaches specific value P, but the quantity on the right side of Eq. (11) is not as great as the value of Y_p

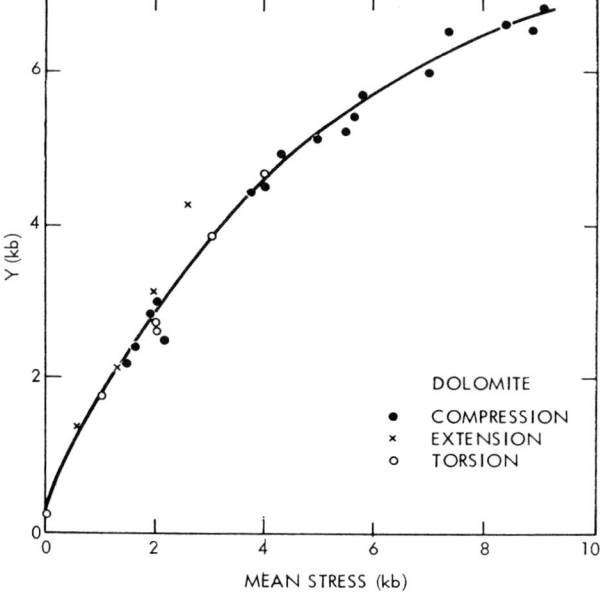

Fig. 11. Y versus mean stress for dolomite.

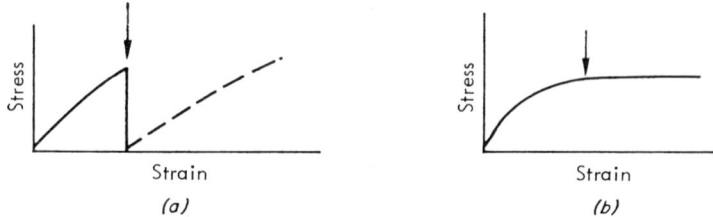

Fig. 12. Qualitative relationship between stress and strain (*a*) for brittle failure and (*b*) for plastic failure. The arrow indicates the point of stress identified with failure.

associated with P from the curve, the material will not fail but will respond elastically. If stresses develop which would tend to exceed Y_p, failure occurs and P increases; i.e., the forces become more isotropically distributed.

Test specimen behavior is used to determine whether failure along the Y–P_m curve is brittle or ductile. For each point on a plot (Fig. 11) a curve such as those shown in Fig. 12 is obtained, where the arrow is the chosen value indicating failure, which, in turn, is the point plotted in Fig. 11. Figure 12*a* indicates typical behavior of a brittle failure, and 12*b* ductile failure.

Further, it is presently assumed that the value of Y is limited by the dynamic elastic limit discussed in the previous section. Typical Y–P_m curves are shown for granite and salt in Fig. 13, along with the extended dolomite curve from Fig. 11.

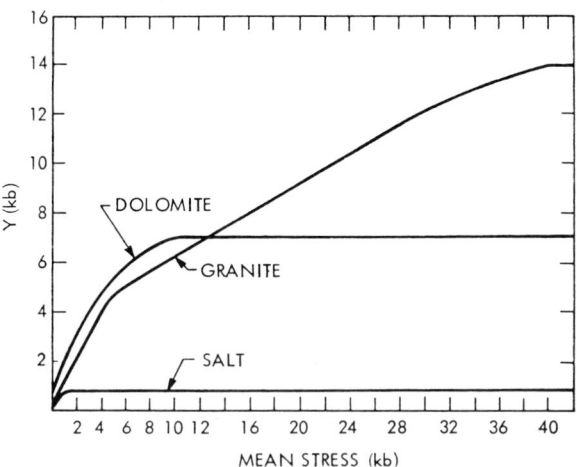

Fig. 13. Y versus mean stress for dolomite, granite, and salt.

4. Compression of Fractured and Unfractured Rocks

The elastic response of rocks has been measured by two commonly used methods. Stephens and Lilley[20] use the isostatic piston–cylinder compression equipment similar to that described by Kennedy and Newton.[21] A sample to be measured is placed in an apparatus which consists of a press and a cylindrical die. The sample is encased in a fluid or plastic material and the volume decrease is measured as pressure is increased by moving close-fitting pistons into each end of the cylindrical die. For low pressures, below the yield point of the piston and cylindrical die, the piston displacement is a good measure of volume decrease. The slope of the $P-V$ curve at constant temperature is the isothermal compressibility [β_t from Eq. (8)]. By using the same apparatus with a heater and a thermocouple, the volume coefficient of expansion can be obtained at several constant pressures [α of Eq. (5) but *not* of Eq. (8)] from the slope of the volume–temperature function.

Data of this kind are available for a number of pure minerals and condensed uniform rocks. Brase[22] and Stephens[23] have listed current bibliographies of these measurements, in addition to some recent additional data.

The elastic response of the rock materials, very much like the Hugoniot equation of state properties, depends rather strongly on whether or not the material is initially well compacted or porous and, much more importantly, on the previous stress history of the sample. In the study of many natural phenomena, the failure of a given segment of rock does not signify a complete loss of interest on the part of the experimenter in that group of fragments and their further response to pressure. Particularly in the case of brittle failure illustrated by Fig. 12a, if the stress is not relieved by the initial failure, the response of the segment may well follow a path similar to the dotted line, or, as confining pressure falls, a material which was flowing ductily may fail by brittle fracture. Stephens and Lilley[20] have attempted to describe the behavior of failed material by cracking specimens in equipment as described earlier in this section, but with a faulty piston. The sudden release of pressure when the piston fails breaks the sample by tensil fracturing. The broken piston is removed from the die, a new sound piston inserted, and the $P-V$ data are taken with the cracked specimen.

Results of these tests have been reported by Stephens for a variety of materials. These fall into three categories: (*1*) dense, consolidated materials, (*2*) consolidated porous materials, and (*3*) unconsolidated porous materials. Figure 14 shows compressions of two basalts, one initially porous and the other initially dense, and the compressions of both after fracturing. These exemplify all three categories.

The initially porous consolidated basalt has a smaller specific volume after failure than before, indicating fracturing has caused collapse of the porous structure. Felsites, tuffs, trachytes, and alluvium all behaved similarly. After fracture the initially dense consolidated material, on the

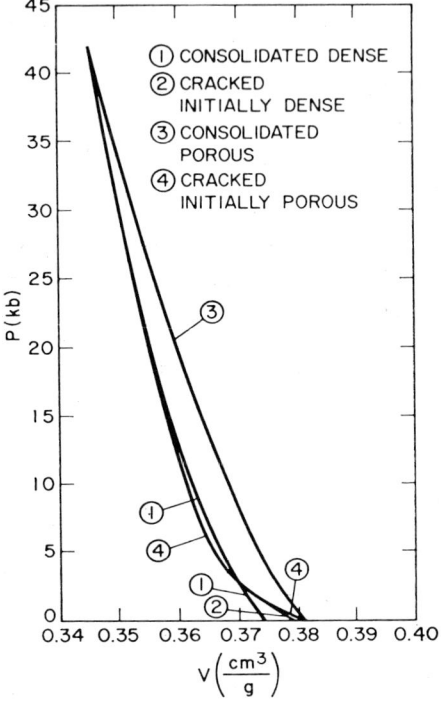

Fig. 14. Comparison of loading curves of dense and porous basalt.

other hand, had a larger specific volume at low pressure, indicating a "disorganization" of the grains. Granite and dolomite behaved more like the dense consolidated materials.

Response of any of the fractured materials to compression above 3–5 kb is essentially the same as the initially dense rocks. Initially porous rocks, however, apparently do not behave as dense consolidated materials until surprisingly large confining pressures are reached—on the order of 40 kb and higher. This phenomenon—the support of surprisingly high pressures by pores in rocks—has been the subject of discussion by Stephens and Lilley[20] and Walsh.[24] The purpose of their arguments is to attempt to obtain the compressibility of rock materials by the appropriate weighted average of the compressibilities of the mineral constituents and voids of various kinds. The effective compressibility, β_{eff}, for material with spherical voids is[24]

$$\beta_{eff} = \beta(1 + aN_p) \qquad (12)$$

for cracks,

$$\beta_{eff} = \beta(1 + \alpha \bar{C}^3) \qquad (13)$$

and ellipses,

$$P_c = E\alpha' \qquad (14)$$

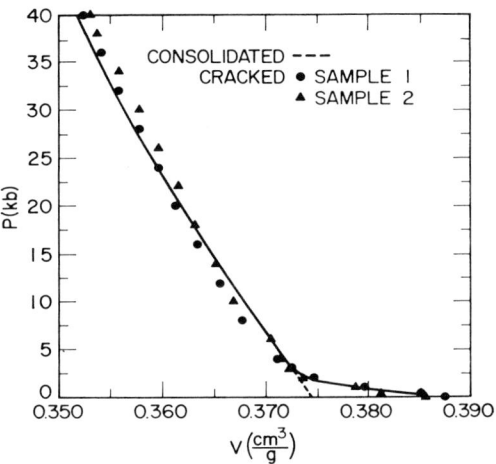

Fig. 15. Compression of NTS granodiorite.

where β is the dense compressibility; a, a constant; N_p, the porosity due to spherical pores; $\bar{C}$, the average crack length per unit volume; E, Young's modulus; and α', the ratio of major to minor axes of the ellipses.

It was found that for low porosity such a treatment was valid, but for fractured samples and for the porous basalt the observed compressibilities were much too large, so that pores of a variety of shapes and lengths must be present in concert.

Figures 15 through 19 show the P–V behavior of Hardhat granodiorite, Handcar dolomite, dry Nevada Test Site tuff, and Nevada Test Site alluvium.

The second well-known standard method for determining elastic response of rocks is by determining their elastic moduli from *in situ* sound wave propagation velocities.* If V_p is the longitudinal or dilatational

---

* For an extensive discussion, see Jaeger.[25]

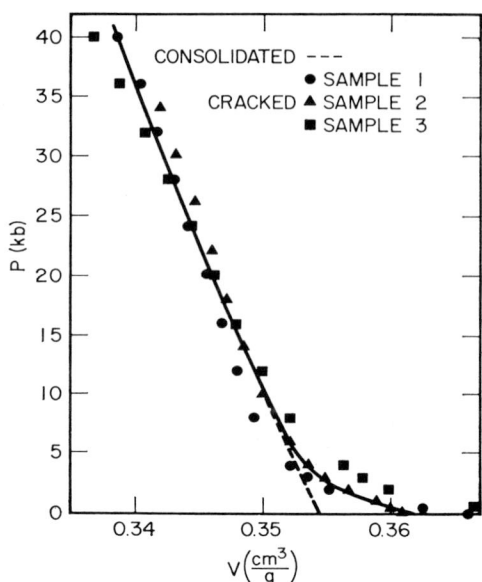

Fig. 16. Compression of Handcar dolomite.

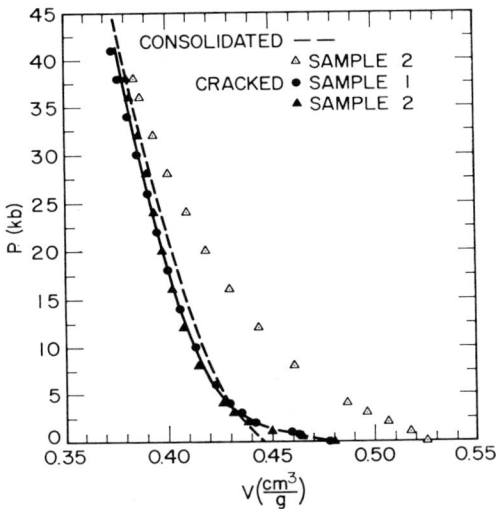

Fig. 17. Compression of NTS dry tuff.

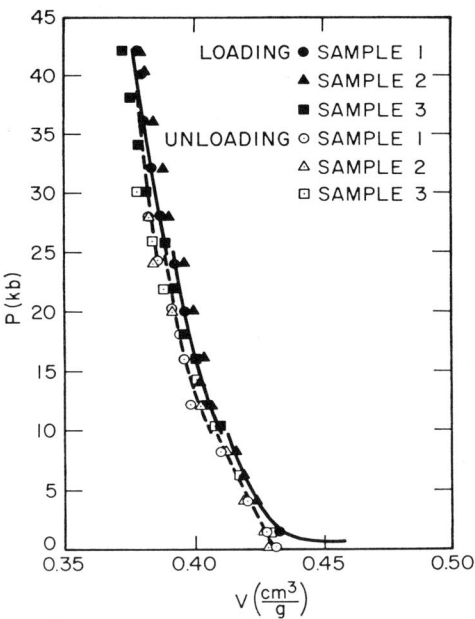

Fig. 18. Compression of NTS alluvium (with 10% water).

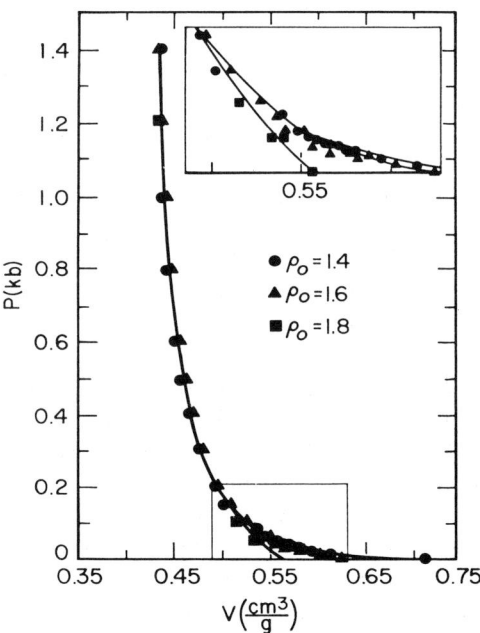

Fig. 19. Loading P–V curves of alluvium, 10% water, broken material, low pressure detail.

sound velocity and V_s the transverse or shear sound velocity, the elastic equations are:

$$\rho V_p^2 = k(\tfrac{4}{3}\mu) \tag{15}$$

and

$$\rho V_s^2 = \mu \tag{16}$$

where k is the bulk modulus of compressibility and μ the shear modulus. Bulk compressibility, α, is defined as

$$\alpha = -\frac{1(\partial V)}{V(\partial P)_t}$$

and the bulk modulus is

$$k = -V\left(\frac{\partial P}{\partial V}\right)_t$$

its reciprocal.

Stephens[23] has shown that there is excellent agreement between measurements taken by the two methods. Porous materials, as mentioned in Section 2, have considerably greater compressibilities than nonporous. They also obviously have lower densities. Thus, examination of Eqs. (15) and (16) and the definition of μ and k show that the sound velocity must decrease notably as porosity increases. Porous materials containing water have less compressibility, higher sound velocities, and greater moduli.[26] Thus, measurements of *in situ* density and sonic velocity coupled with dry samples analyzed for sonic velocity may accurately yield a measure of *in situ* water content—a property very difficult to measure in partially saturated rocks.

Table V lists the values of V_p, V_s, k, μ, and ρ for several rocks measured both by the static compressibility measurement and by *in situ* sound velocity measurements. Reference 7 contains values for a great many more rocks.

The values in Table V are valid for static measurements and for very low strain rates or for frequencies in the low "sonic" range. As has often been noted, larger values should be observed at higher strain rates. Cherry and Hurdlow[27] have attempted to correct for this in their seismic source calculations by treating the material as "slightly plastic" using the Voigt solid model[26] in which k and μ are modified into time-independent and time derivative-dependent portions. This has the effect of attenuating low frequencies in proportion to the square of frequency, and high frequencies in proportion to the square root of frequency.

If we define the ratio of compressional sound velocity, V_p, to shear

TABLE V

Material	V_p (km/sec)	V_s (km/sec)	k (kb)	μ (kb)	ρ_0 (g/cm³)
Hardhat granite[1]	5.44	3.05	361	315	2.67
Salmon salt[1]	4.55	2.16	324	104	2.24
Salmon salt[17]	4.30–4.49	2.35–2.63	245–324	104–122	2.24
Handcar dolomite[1]	5.50	3.00	473	224	2.59
NTS tuff[1]	2.0–2.42	n.m.	66.5	33.2	1.89
NTS alluvium[1]	0.5–1.0	n.m.	5.71	3.43	1.522
Cowboy salt[17]	4.375	2.555	226	141	2.16
Gnome salt[17]	4.08	2.88	224	134	2.1–2.0
Salmon anhydrite[17]			518	275	2.77
Salmon limestone[17]			186	87.3	2.47
Salmon sediments[17]			56.8	11.6	2.0

wave velocity, V_s, as R, then Poisson's ratio,* σ, is defined by the equation

$$\sigma = \frac{R^2 - 2}{2(R^2 - 1)} = \frac{3K - 2\mu}{2(3K + \mu)} \tag{17}$$

When insufficient data are available for the determination of both μ and k, a value of σ is usually assumed (in the range 0.20–0.33 for brittle rocks); then, depending on the available measurements, Eqs. (15)–(17) are used to arrive at "best guess" values of μ and k. Francis Birch, in the revised edition of *Handbook of Physical Constants*[7] lists values of elastic constants for about 150 rocks. His discussion of elasticity in rocks and minerals is extensive; Table VI, reproduced from his work, is helpful in transposing the various "constants" and ratios reported in the literature.

The values listed in Table V apply to rocks near atmospheric pressure. Isentropic moduli defined by Eq. (15) can be found for higher pressure and temperature materials by examining the velocity of sound and particle velocities behind shock fronts in rocks[4] or by a method so far only applied to metals,[3] where rarefactions from the wall of a cylindrical sample interact and retard the shock front. Examples of isentropic expansion of several silicate rocks are listed in reference 3; however, work in this interesting area of materials behavior has only recently begun, and much more remains to be learned.

* Poisson's ratio, σ, is the ratio of lateral contraction to longitudinal extension for a cylinder which is under axial tension and laterally unrestricted.

TABLE VI. Connecting Identities for Elastic Constants of Isotropic Bodies

[K = bulk modulus; E = Young's modulus; μ = shear modulus; β = compressibility = $1/K$; λ = Lamé's constant; σ = Poisson's ratio; ρ = density; $R_1 = V_p/V_s$; $R_2^2 = K/(\rho V_s^2)$; $R_{32} = K/(\rho V_p^2)$]

K	E	λ	σ	ρV_p^2	$\rho V_s^2 = \mu$
$\lambda + 2\mu/3$	$\mu\dfrac{3\lambda + 2\mu}{\lambda + \mu}$	—	$\dfrac{\lambda}{2(\lambda+\mu)}$	$\lambda + 2\mu$	—
—	$9K\dfrac{K-\lambda}{3K-\lambda}$	—	$\dfrac{\lambda}{3K-\lambda}$	$3K - 2\lambda$	$3(K-\lambda)/2$
—	$\dfrac{9K\mu}{3K+\mu}$	$K - \dfrac{2\mu}{3}$	$\dfrac{3K-2\mu}{2(3K+\mu)}$	$K + 4\mu/3$	—
$\dfrac{E\mu}{3(3\mu - E)}$	—	$\mu\dfrac{E-2\mu}{3\mu - E}$	$E/(2\mu) - 1$	$\mu\dfrac{4\mu - E}{3\mu - E}$	—
—	—	$3K\dfrac{3K-E}{9K-E}$	$\dfrac{3K-E}{6K}$	$3K\dfrac{3K+E}{9K-E}$	$\dfrac{3KE}{9K-E}$
$\lambda\dfrac{1+\sigma}{3\sigma}$	$\lambda\dfrac{(1+\sigma)(1-2\sigma)}{\sigma}$	—	—	$\lambda\dfrac{1-\sigma}{\sigma}$	$\lambda\dfrac{1-2\sigma}{2\sigma}$
$\dfrac{2\mu(1+\sigma)}{3(1-2\sigma)}$	$2\mu(1+\mu)$	$\mu\dfrac{2\sigma}{1-2\sigma}$	—	$\mu\dfrac{2-2\sigma}{1-2\sigma}$	—
—	$3K(1-2\sigma)$	$3K\dfrac{\sigma}{1+\sigma}$	—	$3K\dfrac{1-\sigma}{1+\sigma}$	$3K\dfrac{1-2\sigma}{2+2\sigma}$
$\dfrac{E}{3(1-2\sigma)}$	—	$\dfrac{E\sigma}{(1+\sigma)(1-2\sigma)}$	See below.	$\dfrac{E(1-\sigma)}{(1+\sigma)(1-2\sigma)}$	$\dfrac{E}{2+2\sigma}$
$\rho(V_p^2 - \tfrac{4}{3}V_s^2)$	$\dfrac{9\rho V_s^2 R_2^2}{3R_2^2 + 1}$	$\rho(C_p^2 - 2V_s^2)$	—	—	—

$2\sigma = (R_1^2 - 2)/(R_1^2 - 1) = (3R_2^2 - 2)/(3R_2^2 + 1) = 2(3R_3^2 - 1)/(3R_3^2 + 1)$.

5. Conclusion and Speculation

Even though the foregoing discussion treats the equation of state of rocks in separate sections—high pressure shock, transition from shock to plastic, fractured, and finally elastic—it is obvious that rock response must be continuous as stresses change.

It has been noted that the pressure–volume relationships obtained for rocks with static and dynamic methods compare favorably. Thus, the data of Figs. 2 and 14 for the compression of basalt by static and shock techniques are comparable. This implies that for isotropic and one-dimensional stresses, strain-rate effects cannot be very significant. It does not imply, however, that there may not be large strain-rate effects in adjustment of the principal stresses when nonisotropic strains exist.

It should be noted that, while we can discuss mean confining pressure and principal stresses independently for theoretical or laboratory experimental treatments, the environment can place rather great restrictions on the modes of possible response of rocks. For example, using Cherry's yield criteria and Stephens measurement on halite, it can be concluded from consideration of the mean pressure that brittle failure cannot occur in salt below depths in the earth of about 15,000 ft (5 km) unless, somehow, large tensile stress is developed. This is not surprising, but the implication from the same kind of analysis that brittle failure cannot occur in dolomite below 40 km or in granite below 70 km might cause some comment. It would be interesting, indeed, to have the analysis of similar data on the rocks which comprise the earth's mantle to be able to evaluate deep earthquake sources which are known at depths greater than 150 km.

Since it is almost impossible to instrument the pressure history in rocks very near earthquake foci* or meteor impact points, *in situ* tests of the combined rock properties are presently unavailable. The only phenomena which have permitted such tests are buried chemical and nuclear explosions. Figure 20 shows the consequence of a calculation taking into account the various equations of state for salt, anhydrite, limestone, and sediments over a 5-kt† nuclear explosion, using entirely conventional equations of motion and equations of state in all the pressure ranges described above. Figure 20 warrants detailed study, but for purposes of this chapter it should be sufficient to note that calculations of peak stress and particle velocity are within 20–50% of measured values, and displacement of the ground surface is within about 20%. The reduced displacement potential (seismic source potential function[28]) in the region of elastic

---

* A few tens of meters or, perhaps, a few hundred meters, are considered "very near."

† $5 \text{ kt} = 5 \times 10^{12}$ calories released energy.

response of the rocks is within about 50% of measurements. This is especially noteworthy because the energy reaching the elastic region is only about 1% of the original explosive energy—the remainder, of course, going ultimately to heat the rock which is fractured or plastically deformed.

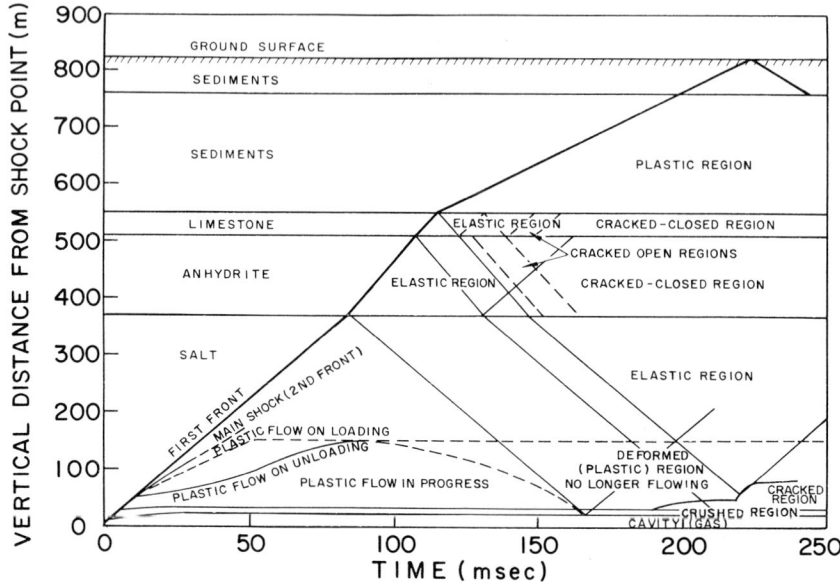

Fig. 20. Time–distance plot of arrival times and various regions of rock alteration from the preshot computer calculation.

Finally, at several points in the text, specific questions have been raised and no answers furnished. There is obviously much yet to be learned about the responses of rock to stress. To mention a few in summary, there is the matter of the variable dynamic compressive strength of rock; the dependence of bulk and shear moduli on strain rate; dependence of failure type, brittle or ductile, on strain rate; the effect of porosity on the bulk and shear moduli as well as the failure criteria; and the inapplicability of property addition for some rocks in construction of equations of state. There are also several omissions in this discussion, the most notable of which is the plastic failure mode.

References

1. F. Holzer, *Proc. Roy. Soc. (London)*, Ser. A, **290**, 408–429 (1966).
2. E. Teller, W. Talley, G. Higgins, and G. Johnson, *Constructive Uses of Nuclear Explosives*, McGraw-Hill, New York, 1967.

3. L. V. Al'tshuler, *Soviet Phys.—Usp.*, **8**, No. 1 52.91 (1965) (*Usp. Fiz. Nauk*, **85**, 197 (1965)).
4. T. J. Ahrens, "Dynamic Properties of Rocks," Proc. Ground Shock Calculations Review Meeting, Rand Corp., Santa Monica, Oct. 1965.
5. R. C. Bass, H. L. Hawk, and A. J. Chabai, "Hugoniot Data for Some Geologic Materials," SC-4903(RR), June 1963.
6. R. C. Bass, "Additional Hugoniot Data for Geologic Materials," SC-RR-66-548.
7. S. P. Clark, Jr., Ed., *Handbook of Physical Constants*, Geological Society of America. New York, 1966.
8. A. S. Kompaneyets, *Theoretical Physics*, Dover Press, New York, 1962, p. 487.
9. I. S. Dougdale and D. McDonald, *Phys. Rev.*, **89**, 832 (1953).
10. L. A. Rogers, "Shock Compression of Several Rock Types," UCRL-12027, 1964.
11. T. R. Butkovich, "The Gas Equation of State for Natural Materials," UCRL-14729, 1967.
12. C. A. Rouse, "Twenty-Two Element Mixtures at Stellar Temperatures and Densities," *Astrophys. J.*, **137**, 1286 (1963).
13. P. F. Bird, R. E. Duff, and G. L. Schott, "HUG, a Fortran-Fap Code for Computing Normal Shock and Detonation Wave Parameters in Gases," Los Alamos Sci. Lab. Rept., LA-2980, 1963.
14. G. E. Duvall, "Propagation of Plane Shock Waves in A Stress Relaxing Medium," in *Stress Waves in Anelastic Solids*, H. Kolsky and W. Prager, Eds., Springer-Verlag, Berlin, 1964.
15. R. E. Marks and L. A. Rogers, "Shock Front Studies in the Handcar Event," PNE 801-F, April 1966.
16. D. R. Grine, "Equation of State of Granite and Salt," Stanford Research Inst. Final Report, Project PGD-3244-1961.
17. L. A. Rogers, "Free-Field Motions Near a Nuclear Explosion in Salt, Project Salmon," UCRL-14463, Rev. 1, Feb. 1966.
18. J. H. Handin, C. Heard, and J. N. Magourik, "Effects of Intermediate Principal Stress on the Failure of Limestone, Dolomite and Glass at Different Temperatures and Strain Rates," *J. Geophys. Res.*, **72**, 611 (1967).
19. J. T. Cherry, D. Larsen, and E. Rapp, "Brittle Failure and Extent of Fracturing from Explosions in Rocks," UCRL-70617, Sept. 1967.
20. D. R. Stephens and E. M. Lilley, "Static P-V curves of Cracked and Consolidated Earth Materials to 40 Kilobars," Proc. Conf. Shock Metamorphism of Natural Materials, Greenbelt, Md., April 1966; UCRL-14711.
21. G. C. Kennedy and R. G. Newton, *Solids Under Pressure*, McGraw-Hill, New York, 1963, pp. 163–168.
22. W. F. Brase, "Some New Measurements of Linear Compressibility of Rocks," *J. Geophys. Res.*, **70**, 391 (1965).
23. D. R. Stephens, "The Hydrostatic Compression of Eight Rocks," *J. Geophys. Res.*, **69**, 2967 (1964).
24. J. G. Walsh, "The Effects of Cracks on the Compressibility of Rock," *J. Geophys. Res.*, **70**, 381 (1965).
25. J. C. Jaeger, *Elasticity, Fracture and Flow*, Wiley, New York, 1964.
26. W. M. Ewing, W. S. Jardetzky, and F. Press, *Elastic Waves in Layered Media*, McGraw-Hill, New York, 1957.
27. J. T. Cherry and W. R. Hurdlow, "Numerical Simulation of Seismic Disturbances," *Geophysics*, **31**(1), 33 (1966).
28. G. C. Werth, R. Herbst, and D. Springer, *J. Geophys. Res.*, **67**, 1587 (1962).

Material Properties At High Pressure

RUSSELL E. DUFF

*Lawrence Radiation Laboratory, University of California,
Livermore, California*

This part of our birthday present to Edward Teller is rather different from the others. It is not a discussion in depth of some particular subject; rather it is a collection of brief reports of work now in process under Dr. Teller's general direction. I have tried to connect these reports into a meaningful story of research activities which will improve our understanding of materials at extreme conditions. A discussion of several theoretical equation-of-state programs is followed by brief accounts of several experimental investigations of the properties of materials being made by dynamic, high pressure techniques.

Of course, most of the work to be discussed here has been and is being done by others. In most instances, the references cited will serve to identify the originators. However, it is appropriate to make special acknowledgment to several of my colleagues who helped in the preparation of this manuscript. R. Grover, H. C. Graboske, M. Ross, E. B. Royce, and R. N. Keeler deserve special mention.

All of us together with our colleagues join in extending our congratulations and best wishes to Edward Teller on his sixtieth birthday.

Equation of State

A central question in every discussion of phenomena at extremely high temperatures and pressures is, what is the equation of state of the materials involved? This question must be answered if one is to understand the interior of the earth, the structure of the sun, or the evolution of a star. It must also be answered if one is to understand the effects of a nuclear explosion.

For the general atomic system of ionic nuclei and compensating electrons, the calculation of thermodynamic equilibrium requires the eigenstates of the many-body Schrödinger equation, the construction of the partition function, and the minimization of the free energy, or other state variables, to define the equilibrium state. This complete procedure, in practice, is not feasible due to the complex nature of the quantum-mechanical many-body problem and the difficulty of obtaining solutions of the appropriate wave equation.

Many theoretical efforts have been made to solve this problem. It would be pointless to try to review all of them here. However, many of them fall into several broad categories defined by the fundamental point of view adopted. Three of these categories are of particular interest to us as we try to answer better the central question.

Statistical Atom

When the temperature or density is sufficiently high to introduce a large mixture of electronic quantum states, a simple statistical approximation to the equation of state can be made based on independent free electrons and nuclei. In the Thomas-Fermi theory[1] each atom of the material occupies an independent spherical cell, and the electron distribution is determined to a first approximation about a nucleus fixed in the center of the cell. The electrons are assumed to be free Fermi-Dirac particles, and all other details of the quantum mechanics of atoms are ignored. Thus, the distribution of the cloud of partially degenerate, nonrelativistic electrons is related to the electrostatic potential by Poisson's equation. In this manner the main effects of coulombic interactions are included self-consistently to all orders in the electric charge.

The Thomas-Fermi equation is obtained directly from Poisson's equation by a suitable choice of variables.

$$\phi''(x) = ax I_{1/2}\left[\frac{\phi(x)}{x}\right]$$

where

$$\phi(x) = \beta[\mu + eV(r)]r/r_0$$

$$x = r/r_0, \qquad \beta = 1/kT$$

μ is the chemical potential, $V(r)$ is the electrostatic potential, and

$$a = \left(\frac{r_0 4\pi e(2m)^{3/4}}{h^{3/2}\beta^{1/4}}\right)^2$$

The Fermi-Dirac function is

$$I_n[x] = \int_0^\infty \frac{y^n\, dy}{e^{(y-x)} + 1}$$

The applicable boundary conditions are that the electrostatic potential and its gradient are zero at the cell boundary r_0, and the potential has a coulombic singularity at the origin. This equation has been solved by Latter[2] for a wide range of conditions.

Once the electron distribution is known, the pressure and energy can be calculated. The former is just the momentum transport across the outer boundary of the cell, and the energy is determined by the volume integral of the kinetic and potential energies of the electrons.

The nucleus was assumed fixed in the above derivation, but, of course, it is in motion in a real system. Except for the light atoms, the contribution of this nuclear motion to the thermodynamics of the system is small and for most purposes may be crudely approximated.

Of course, many improvements have been proposed to better represent the quantum-mechanical reality of the system. In particular, statistical corrections for electron exchange and correlation effects have been added by many workers.

This simple picture of an atom has provided a most useful approximation to the equation of state of matter between the completely degenerate region at high density and low temperature and the classical ideal gas region at high temperature and modest density. Surprisingly, the theory is also useful in many types of low temperature, normal density applications.

From its statistical basis, the Thomas-Fermi theory would not be expected to apply to atomic problems at normal densities and low temperatures. For instance, an interesting theorem of Teller's[3] shows that at normal conditions molecular binding cannot occur in the Thomas-Fermi approximation even when exchange corrections are made. A related problem for the theory is its well-known inability to distinguish between the ground states of the atom and the solid. However, Brillouin[4] showed long ago that the Thomas-Fermi equation corresponding to zero temperature is in part a W-K-B approximation to Hartree theory for the atomic ground state. Actual calculations show the average electron density distribution[5] and energy eigenvalues in the atom as calculated from the electrostatic potential in Thomas-Fermi theory[2] to be in close agreement with Hartree values. Another remarkable success of the theory is its ability to predict the value of the atomic number, Z, at which new angular momentum components appear in the atom.[5]

An interesting development in this connection has been the demonstration of the Thomas-Fermi equation as a first step in an asymptotic expansion of the Hartree-Fock equations in powers of Planck's constant, $\hbar$. The next higher order terms in this expansion were shown to include quantum corrections as well as exchange. These have been investigated by Kirzhnits[6] and others in the Soviet Union. In their method the electron density at a point about the atom is expanded in higher commutators of the individual electron momentum operators by a standard but formal method which has in the past been used, mainly in high temperature approximations. Only the leading term of the density expansion is used to derive the Thomas-Fermi equation, and the remaining terms represent quantum corrections which contain various local derivatives of the electrostatic potential at the point. It was pointed out that the first quantum correction term in the density and the exchange density correction are both of order $\hbar^2$ and can be treated more consistently as perturbation corrections in the Thomas-Fermi equation rather than in the self-consistent manner used in Dirac's modification of Thomas-Fermi theory.[1]

However, Grover[7] has shown that there is a more important class of quantum corrections which are overlooked by the Kirzhnits method, and which can be similarly expanded in an asymptotic series in $\hbar$. These corrections were derived for the particle density of a bound system of particles in a one-dimensional potential by extending the Brillouin method of summing directly over individual particle states represented by W-K-B eigenfunctions. A more complicated asymptotic expansion in powers of $\hbar$ results which include the Kirzhnits corrections but in which the leading correction term is of order $\hbar$. This term has an oscillating magnitude arising from individual particle levels. In addition, it is nonlocal in nature, since it depends on the spacing of the eigenvalues of the highest occupied states as given by the well-known W-K-B integral condition for bound-state eigenvalues. This is no more than a complication, since the level spacing may be consistently evaluated from the zero-order Thomas-Fermi potential. The improved density can then be used in the Poisson equation to obtain the electrostatic potential through the first order in $\hbar$, etc.

Such an iterative expansion in powers of $\hbar$ appears to provide, for the first time, a formal justification of the common usage of the Thomas-Fermi potential in atomic problems as a zero-order approximation. The success of the calculation of atomic energy levels and angular momentum thresholds in atoms, which were mentioned previously, and the apparent accuracy of the new first-order density corrections indicate that this iteration scheme converges rapidly. Higher order corrections will introduce great complications into the scheme, but it should at least be possible to estimate truncation errors from them.

A problem of particular interest to which this method appears well suited is the study of the manner in which the high-pressure equation of state of solids blends into the Thomas-Fermi limit at high compression. The extension of this density expansion to the spherical Hartree atom presents some analytical problems which have as yet not been completely worked out. It is clear, however, that in the spherical atom the additional correction terms are of the order $\hbar^2$, the same as the Kirzhnits corrections, and that there are both oscillating and slowly varying corrections of this order.

In view of the as yet incomplete theoretical picture, it is interesting to note that high pressure experimental data already illustrate some of the expected properties of this expansion. The very regular and large periodic influence of shell structure is evident in plots of the atomic volume of the solid elements at fixed pressure (Fig. 1). All the Thomas-Fermi predictions, with and without the Kirzhnits correction as well as with the Dirac modification, parallel the average increase of atomic volumes with Z throughout the periodic table. The Kirzhnits correction is somewhat less than the amplitude of the periodic variations in all shells. It is expected that the inclusion of the oscillating correction terms mentioned above will bring the modified Thomas-Fermi theory into even better agreement with observations.

Band Structure

The central question in the determination of the equation of state of materials at extreme conditions is the calculation of the electronic configuration of the system. The previous section described one approach to the problem; however, a severe limitation of the Thomas-Fermi theory arises from its use of classical electrostatics to describe the electron density distribution. It does not predict the forbidden electron energy bands which are known to be important in solids. Instead, it allows electrons to be excited into a continuous set of energy levels immediately above the ground state. Consequently, the statistical model may correctly predict the averaged thermodynamic properties but will fail to show some of the most subtle behavior which is a consequence of quantum mechanics, such as the effects of stripping off shells of electrons when the temperature becomes comparable to the gap in energy between filled and empty bands. A more realistic approach to the problem is therefore achieved by the Hartree-Fock band structure calculation or even by the less sophisticated Wigner-Seitz model.

Equation-of-state information can, in principle, be derived from the band calculations in a straightforward way. At $0°K$ it is necessary first to calculate the total energy of the system, a difficult task to say the least. The pressure is then the volume derivative of this energy. At finite temperature the band calculation can provide the electronic part of an

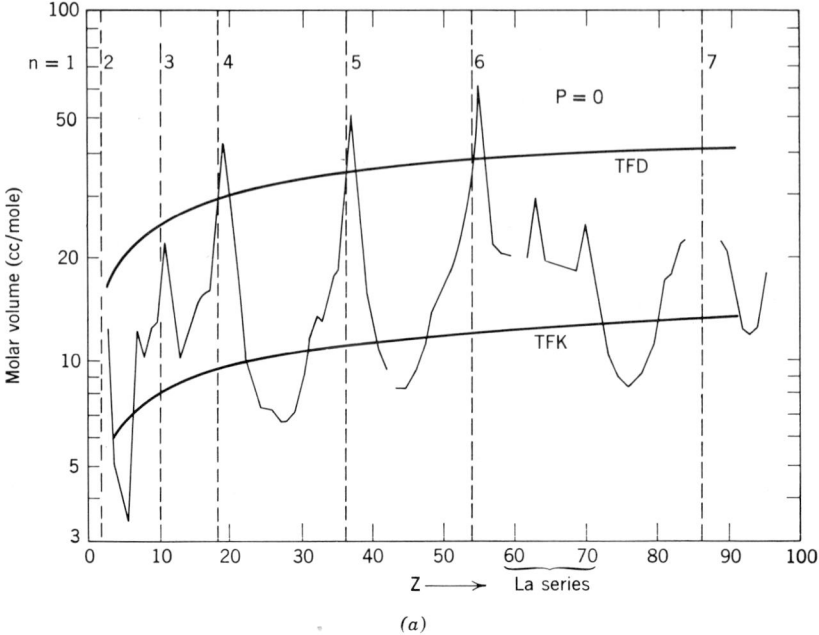

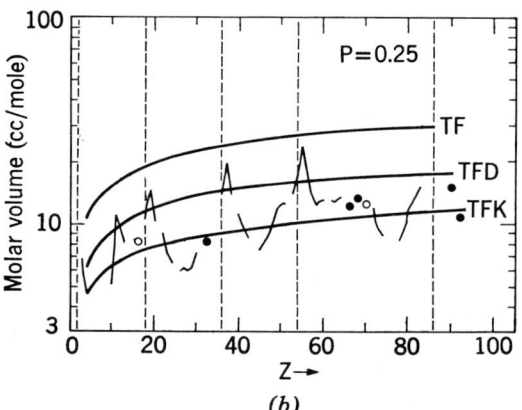

Fig. 1. Comparison of the atomic number (Z) dependence of the zero-temperature, molar volume of the solid elements with statistical theories. (a) Normal data at zero pressure. (b) At a pressure of 0.25 Mb as determined by shock wave compression data. TF, Thomas-Fermi theory. TFD, Thomas-Fermi theory with Dirac modification. TFK, Thomas-Fermi theory as modified by Kirzhnits. Note that at zero pressure the TF molar volume is infinite for all Z. Vertical dashed lines indicate completion of successive shells of p electrons.

expression for the free energy of the system. The pressure and other thermodynamic variables follow by differentiation of the free energy.

The latter technique was recently used by Ross to explain a surprisingly high compression in strongly shocked liquid xenon. But first we need the background to this story.

The inert gases, argon and xenon, have been shock-compressed to two and three times their normal liquid density[8] as part of a program for the study of equations of state at high pressures. These experiments were theoretically analyzed using the Monte Carlo method, and a repulsive intermolecular potential for argon was determined. This potential was then shown to be in good agreement with results obtained from molecular beam methods.[9] If one applies the law of corresponding states to the Hugoniots and scales the argon measurements to the xenon results, it is found that below 200 kbar the two sets of data are in very good agreement. However, above this pressure the xenon points lie significantly below those of argon in the P–V plane. These results are shown in Fig. 2, where solid curve A is the averaged argon experimental results scaled up to xenon, and the dashed extension is an extrapolation of the theoretical curve which fits the entire Hugoniot.

The disagreement between the two Hugoniots at high pressure was assumed related to differences in electronic excitation in the two cases. Because of the high temperatures generated and the fact that the first excited state of xenon (8.4 eV) is significantly lower than that in argon (11.5 eV), xenon should undergo considerably more electronic excitation than argon, and its Hugoniot should be softer. In order to place these qualitative ideas on a more quantitative basis, energy band calculations have been made for both materials as a function of compression. The band calculations show that at the highest xenon pressures the energy gap between the valence and conduction band has narrowed by about 2 V, and the $5d$-like conduction bands now lie below the $6s$, as is the case in compressed cesium. Using the results of these band calculations, a theoretical Hugoniot curve, B in Fig. 2, has been calculated and is in agreement with the xenon experimental curve.

This is a conspicuous success for the method. The band model correctly predicts the equation of state, whereas the Thomas-Fermi-Dirac statistical model was unable to predict the electronic excitation across the forbidden gap, and it could not explain the observations.

At the highest point on the xenon Hugoniot the temperature is near 2 eV, and the energy gap is about 6 eV. Therefore, electrons from one atom in five are promoted into the conduction band. Under these conditions xenon becomes metal-like, even though a large band gap still exists. However, consideration of the electron excitation energy alone is not sufficient to explain the softening of the xenon Hugoniot. More important is the compression-induced shift of the energy levels which leads to a large negative contribution to the total pressure when the free energy is differentiated with respect to volume.

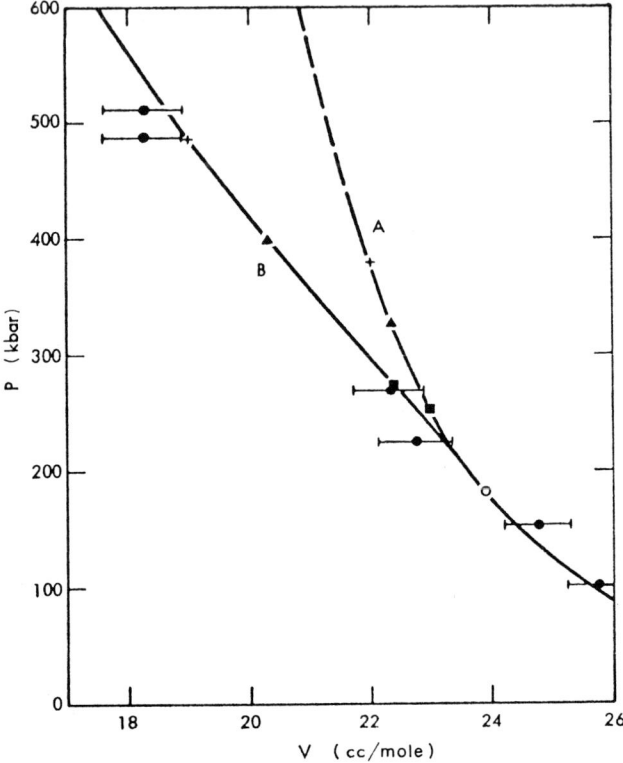

Fig. 2. Experimental measurements and theoretical calculations of Hugoniots for argon and xenon. The law of corresponding states was used to scale the argon results up to xenon. Curve A is the scaled argon experimental curve extended by theoretical calculations. Curve B is based on the xenon band calculations. Calculated Hugoniot temperatures on the two curves are (●) 8000°K, (■) 12,000°K, (▲) 16,000°K, and (+) 18,000°K.

Hartree-Fock band calculations are also being used in an effort to delineate the nature of electronic phase transitions in solids. Such transitions occur when an unfilled band is forced to overlap a filled band by compression of the material. At present the best known of these transitions, the 45 kbar transition in cesium, is being investigated.

Chemical Thermodynamics

One can also adopt a view derived from low density, high temperature considerations in which the actual concentrations of various ions, atoms, and molecules are determined. This requires that the thermodynamic equilibrium state and, in particular, the equilibrium concentrations must first be calculated. The extension of this model to conditions of interest requires a consideration of the plasma effects of the long-range coulomb interactions between ions as well as their effect on the bound levels of the un-ionized components. The full quantum-mechanical description of this non-ideal, charged chemical mixture presents a difficult problem in manybody physics, which has not yet been solved except in certain limiting cases.

Research into the behavior of such systems has been of continuing concern to us. Work has proceeded along two fronts. Approximate physical models have been developed which can be readily evaluated, but they are of limited applicability. Also rigorous quantum-statistical theories have been proposed which are very difficult to apply, but which are generally valid.

In the case of the ideal classical gas with Maxwell-Boltzmann statistics and negligible coulomb interactions, the thermodynamic equilibrium is solvable analytically and is determined by solution of the Saha equations for the various components. In the limit of high density, this ideal classical equilibrium causes a shift toward atomic recombination and molecular association, contrary to the expected physical behavior. Since high density implies electron degeneracy, Chandrasekhar[10] demonstrated that in the limit of complete degeneracy, the inclusion of Fermi-Dirac statistics in the equilibrium equations did not remove the spurious density effects.

Recently, Harwood has extended this analysis to cover the entire range of partial degeneracy for the ideal quantum gas by including Fermi-Dirac statistics for the electron component of the mixture. The ideal quantum gas equilibrium has been solved analytically as in the ideal classical case. The effects of the electron degeneracy are to shift the equilibrium at a given density toward slightly less ionization than is obtained for the classical equations. This effect is to be expected since the region of phase space available to the degenerate free electrons is smaller than in the nondegenerate case. The consistent treatment of the thermodynamic equilibrium and of the equation of state for an ideal quantum gas indicates that the classical density effect is slightly increased for the degenerate gas. Thus, the generalization of the model to provide pressure ionization at high densities is even more necessary in the quantum-gas case. It seems obvious that the required generalization will include the coulomb interactions between the various constituents.

Harris[11] and Harris and Trulio[12] developed a thermodynamic method for computing the equilibrium composition and thermodynamic properties of a weakly interacting, degenerate gas. This method consisted of using various approximations to describe the interactions of the charged particles and the coulomb perturbations of the bound states of atoms and molecules. Their model included quantum-statistical behavior for the free electrons, treated the free particle interaction in the Debye approximation, and used the analytical solution of the constrained hydrogen atom as a basis for the description of the perturbed bound states. The electronic orbitals of the "caged" hydrogen atom were used to construct atomic and molecular models of density-perturbed, multielectron systems for all components present in the mixture. The various contributions to the free energy were combined, the free energy minimized with respect to the concentrations, and the equilibrium determined. From the equilibrium free energy and concentrations, the equation of state and thermodynamic properties were determined by numerical differentiation. Although it was limited by the differences in the treatment of the perturbations of the bound states and of the plasma, this method was an interesting attempt to fully describe a real gas. The model is valid in the region of negligible degeneracy and weak interactions, where the Debye-Hückel correction is accurate.

More recently, Rouse has developed two approaches to the problem which have been used in several astrophysical applications. The first model,[13] the ionization equilibrium equation of state, is valid for a weakly interacting classical gas. It is an application of the Debye-Hückel approximation to both the plasma interaction and the bound state perturbation. A generalized Saha equation is derived by including a Debye potential term which describes the influence of the plasma on the atomic levels, in particular, a lowering of the ionization potential. This gives a better estimate of the equilibrium concentration than the classical Saha equation. Then the Debye-Hückel corrections to the ideal gas pressure and internal energy are added to ideal gas values to take into account the plasma interaction.

A second method[14] employed by Rouse applies to degenerate gases. A density-dependent factor is added to the equilibrium constant in the Saha equation. The factor can be related to the influence of density on the wave functions of an isolated atom. This normalization introduces two free parameters which can be adjusted to provide an empirical description of pressure ionization for monatomic systems. This method is to be used in regions of moderate to strong interactions where no experimental data or rigorous theoretical models are available. The resultant pressure–ionization curves for H and Fe show qualitative behavior that seems realistic. However, at high density, where the normalization term is important, the calculated thermodynamic variables fail to satisfy thermodynamic consistency requirements, for example,

$$\left(\frac{\partial E}{\partial V}\right)_T \neq T\left(\frac{\partial P}{\partial T}\right)_V - P$$

The major source of this inconsistency is thought to be the neglect of the contribution arising from the shift of the energy levels with volume. As demonstrated by Harris,[15] the change in free energy with volume due to the shift in energy levels as the system is compressed can be quite large, producing significant contributions to the pressure.

The rigorous analytical theory of the completely ionized quantum-mechanical coulomb plasma has been developed by De Witt[16,17] and others. De Witt has derived the grand partition function for a quantum plasma of arbitrary degeneracy by generalizing the Mayer cluster expansion for non-ideal gases. The resulting perturbation expansion is solved by use of quantum-statistical diagram techniques, yielding a partition function valid for the completely ionized, multicomponent plasma. This theory reduces to the Debye–Hückel classical result at low density, and reproduces the Gell-Mann–Brueckner high density correlation energy for the electron gas. De Witt also demonstrated that wave-mechanical effects in the coulomb interaction are important in the high temperature region, even though quantum statistics are not necessary here. This theory thus provides a rigorous, analytical form for the equation of state and other thermodynamic properties of an ionized, multicomponent plasma of arbitrary chemical composition.

The corresponding theory for partially ionized gases is still far from complete. Recent work by De Witt[18] has indicated a possible extension of the diagrammatic expansion of the two-component partition function to include the effect of bound states. A preliminary analysis of this problem has indicated that in the quantum-mechanical perturbation treatment, the low-lying bound states are not shifted significantly, a result which differs significantly from the conclusions of the near-classical treatment.

A general program combining the results of several of the above studies is currently being developed by H. Graboske and D. Harwood. The object of the study is to produce a flexible, generalized method for computing all the thermodynamic properties of interest for a multicomponent coulomb quantum gas which may be easily modified to test various physical approximations to the real situation.

Since the more accurate physical theories of the plasma interaction and the bound level perturbations are usually nonanalytic and must be solved by computer, the inclusion of such models requires that the entire thermodynamic equilibrium must be numerically evaluated. It is possible to start with the energy eigenstates of the free and bound constituents of a multicomponent gas and to proceed to calculate the entire partition function and the Helmholtz free energy. This state variable is minimized with respect to the composition variables to determine the equilibrium state and the corresponding concentrations. Then the equation of state is computed by appropriate differentiation of the free energy. This numerical thermodynamic equilibrium method thus proceeds in a consistent manner from the statistical description of the microstates of the system to a complete evaluation of all the thermodynamic properties.

The basic building blocks for the method are the partition functions for translational motions, Z_0, coulombic interactions, Z_I, and bound levels, Z_{int}. Then

$$Z(N_i, V, T) = Z_0(N_i, T) Z_I(N_i, V, T) Z_{\text{int}}(N_i, V, T)$$

The translational partition function is given for a general Fermi-Dirac gas so that both classical and quantum statistics can be used. The interaction partition function can be unity (no interaction) which yields the ideal gas approximation, or the Debye-Hückel approximation, or the more rigorous quantum-statistical results of De Witt. The internal partition function can also be treated in a variety of ways, from the unperturbed (ideal gas) approximation to the, as yet incomplete, quantum-statistical bound-state perturbation theory.

The results of this program of research should provide some insight into the degree of validity and the range of applicability of the various models for the partition function which have been discussed above. We hope it will also point the way toward further progress in the understanding of the equation of state of ionized gas mixtures.

Experimental Investigations of Properties of Materials

The equation of state of a substance is fundamental to a discussion of its thermodynamics and hydrodynamic behavior. Some equation-of-state information can be determined by experimental techniques which have been developed at several laboratories. Simultaneous measurements of pressure, density, and specific internal energy at pressures up to 1–10 million atmospheres can be made by dynamic measurements. When any two properties of a shock wave are measured as the wave engulfs a sample of the material of interest at known initial conditions, pressure, density, and specific internal energy are then calculable from these measured properties through the algebraic relations which embody the conservation relations for mass, momentum, and energy. The variables most often measured are the shock velocity and the material velocity generated by the wave.

It should be stressed that only P, ρ, and E can be determined by these dynamic techniques. If one is interested in a more complete equation of state, additional assumptions must be made. For instance, if one wishes to calculate the temperature, an assumption must be made about the specific heat.

Ingenious schemes have been developed to generate the required shock waves by high explosives or by guns. The latter are particularly useful in the relatively high and low pressure regimes. These schemes have been adequately described in the literature,[19-21] so there is little point in discussing them here.

Instead, I assert that there is more to physics and chemistry than thermodynamics, and I support the assertion with a discussion of several investigations which use dynamically generated high pressures to determine other properties of interest. Many things are of interest because the high pressures, densities, and temperatures induced in shock experiments permit one to investigate physical phenomena in a domain of parameter space unattainable by more familiar laboratory techniques. It is feasible to measure saturation magnetization, electrical conductivity, and certain optical interactions at extreme conditions in times comparable to a microsecond, the time required for a shock wave to pass through a specimen. Examples of such investigations are given below.

Magnetic Studies

Ferro- and ferrimagnetism arise from the interaction of electrons of neighboring atoms. It seems obvious that this interaction must depend on atomic spacing. Dynamic compression can be used to investigate how magnetic properties change with lattice spacing and, therefore, to learn more about the nature of the magnetic interaction.

The short time available for an experiment imposes a constraint. The sample must not trap the magnetic field and prevent the changed magnetic state from being observed. This means that the samples must be electrical insulators or, if metals, quite thin.

Many experiments have been made by Royce[22,23] and by Graham[24] at another laboratory. In the work of Royce, measurements have been made of the shock-induced transverse demagnetization of polycrystalline nickel ferrite, $NiFe_2O_4$; yittrium iron garnet (YIG), $Y_3Fe_5O_{12}$; and iron. In these experiments, a sample of material is biased nearly to saturation by a magnetic field transverse to the shock propagation direction. As the shock wave propagates through the sample, measurements are made of the change in the component of the magnetization along the applied field.

The experimental geometry is shown in Fig. 3. The high-explosive assembly produces a plane shock wave in the metal base plate, from which the shock progresses into the thin sample of the material being studied. The element C is a permanent magnet which produces the bias field for the sample. The elements B are ferrite bars or arms which complete the magnetic circuit between the sample and bias magnet. The shock-induced

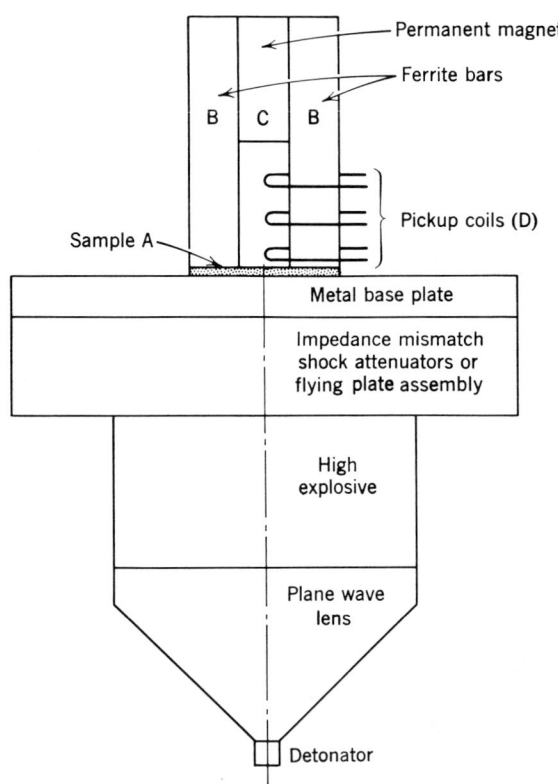

Fig. 3. Assembly for shock-induced demagnetization experiments (not drawn to scale). (A) Sample of material to be studied; (B) ferrite bars or arms completing a magnetic circuit; (C) permanent magnet; (D) single-turn pickup coils on printed circuit boards. The magnetic circuit was normally surrounded by epoxy.

reduction in the magnetic flux through the sample was detected by the pickup coils, D, wound on one arm of the magnetic circuit. The demagnetization of the ceramic materials takes place as the shock wave progresses through the sample. The magnetic circuit $CBABC$ is interrupted, and the magnetic flux through the sample before compression redistributes itself across the gap between the arms. Signals are generated as this flux cuts the single-turn pickup coils, and the flux change is determined by integrating these signals. In measurements on the ceramic materials, the change in the magnetization of the sample was measured to an accuracy of $\pm 5\%$; the measurements on iron are probably less good.

Note that in this system, only the component of the magnetization transverse to the shock propagation direction is measured. The symmetry of the apparatus requires that if there is a longitudinal component, the sample must be divided into a domain structure, with the longitudinal component either parallel or antiparallel to the shock propagation direction in different domains.

The demagnetization of the ceramic materials is observed to be completed as soon as the shock wave passes from the sample. Thus, the distortion and destruction of the magnetic circuit presents no problem. For iron, on the other hand, the demagnetization rate is determined by eddy current diffusion in the iron and is longer than the shock transit time through the sample.

Demagnetization experiments were performed on nickel ferrite at a bias field of 300 Oe over a pressure range from 43 to 410 kbar. The results are shown in Fig. 4, where the transverse demagnetization is seen to be independent of the pressure.

The observed change in the value of the transverse component of the magnetization is due to two effects: the change in the magnitude of the magnetization itself caused by shock compression and heating, and the rotation of the direction of the magnetization resulting from a shock-induced uniaxial magnetic anisotropy. The latter effect is an anisotropy in which the shock propagation direction becomes an easy direction of magnetization. The orientation of the magnetization is then determined by the competition between this induced anisotropy and the applied bias field since the latter is perpendicular to the easy direction of the anisotropy in these experiments.

More specifically, the induced magnetic anisotropy is the result of the anisotropic compression of individual volume elements which is associated with the anisotropic stress existing behind the shock front. This stress anisotropy, of course, results from the nonzero yield strength of the material. In the simple elastic–plastic model of material distortion, the difference between the longitudinal and transverse components of the stress

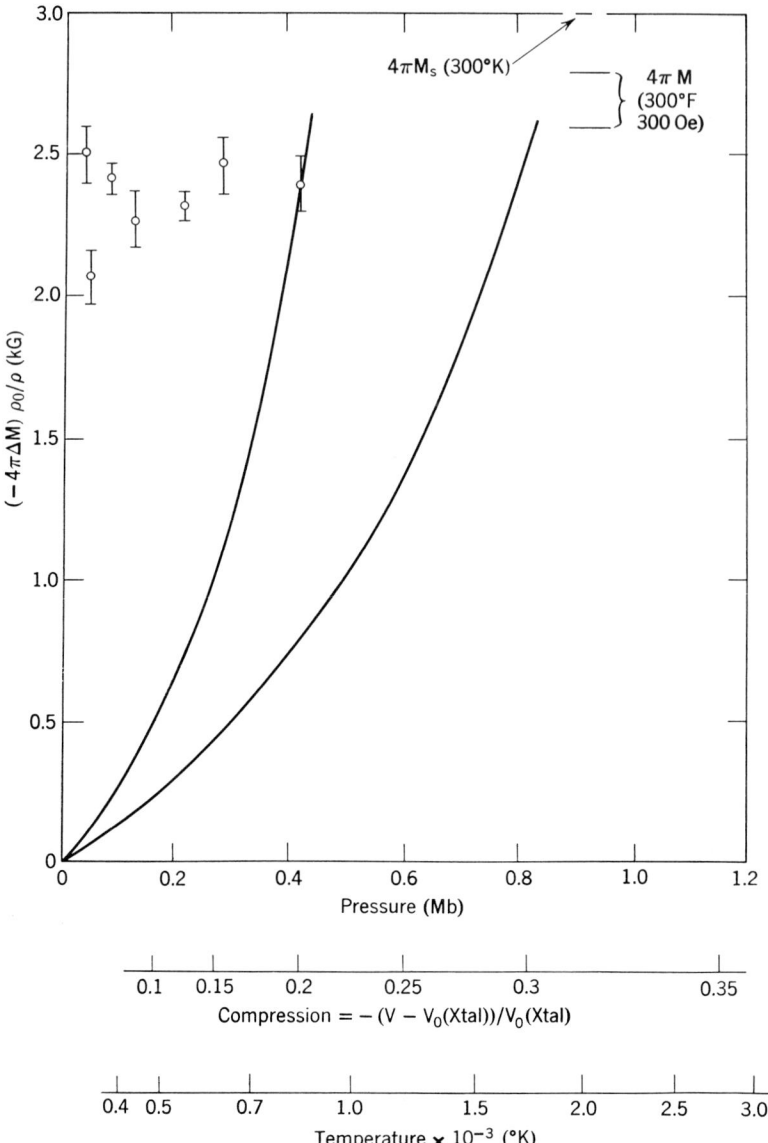

Fig. 4. Shock-induced demagnetization $(-4\pi\Delta M)\rho_0/\rho$ of nickel ferrite. Points shown by circles are measurements in a bias field of 300 Oe. The more reliable points are average results from two or more shots. The curves show the theoretical demagnetization predicted from shock compression and heating alone. The upper curve is based on the assumption that the Néel temperature is constant at 590°C; the lower curve is based on an increase in the Néel temperature at 1.16°C/kbar. Also indicated are the saturation magnetization $(4\pi M_s)$ at 20°C and the estimated magnetization at 300 Oe, both for the unshocked material.

is proportional to the yield strength of the material and is assumed to be constant at all pressures above the yield point.

Additional experiments at a high bias field were performed in an effort to prove the consistency of this model of the phenomenon and to actually measure the strength of the anisotropy field. Experiments at 950 Oe showed somewhat less demagnetization below 150 kbar. Measured values of demagnetization at two known bias fields together with the relation between magnetization and applied field permit one to determine the magnitude of the anisotropy field and the magnitude of the magnetization of nickel ferrite. They also led to an estimated value of $\mu' - 1 = 0.5$ for the transverse susceptibility of the shocked material.

The effective anisotropy field was calculated to be 5500 Oe, a value which is of the same order of magnitude as the value 10,200 Oe estimated from the ordinary magnetostriction effect. The uncertainty in both of these values is probably at least a factor of 2, and within these limits they should be considered to be in agreement. If one accepts an anisotropy field of 5500 Oe, the magnitude of the magnetization is calculated to be 2510 ± 150 G, a value to be compared with the value 2700 ± 100 estimated from the published magnetization curves of Pauthenet[25] at 300 Oe bias field and 20°C. These values appear to be in agreement, as they should be in this pressure range.

Since the anisotropy field is large compared with the applied field, it is difficult to study the magnetic state in detail. This large magnetic anisotropy in nickel ferrite arises from the interaction between the distorted cage of oxygen ions surrounding the magnetic ion and the nonspherical charge distribution of the nickel ion. This interaction reduces the orbital degeneracy of optically excited states and leads to an alignment of the spin system through the spin-orbit coupling. On the other hand, iron ions have a spherical charge distribution. Therefore, to first order there should be no interaction between the crystal lattice distortion and the magnetic ion for the iron ions in nickel ferrite or in YIG. Since the only magnetic ions in YIG are iron, one should expect the induced magnetic anisotropy field to be much weaker in YIG than in nickel ferrite. Measurements were made of the shock-induced transverse demagnetization of YIG at bias fields of 140, 280, and 550 Oe and at several pressures.

The induced magnetic anisotropy energy may be written in the form

$$W_{\text{anis}} = K_1' \sin^2 \theta + K_2' \sin^4 \theta$$

where θ is the angle between the magnetization and the shock propagation direction, or easy direction of the magnetization. The K_1' term is obtained from the normal, linear magnetoelastic theory; the magnetostriction mentioned earlier is related to K_1'. The K_2' is a higher order term not normally observable. The energy resulting from the applied field is

$$W_{\text{app}} = -MH_{\text{app}} \sin \theta$$

where the applied field H_{app} is taken in the direction transverse to the shock propagation direction. If one minimizes the total magnetic energy, a relation between the anisotropy constants K_1' and K_2' and experimental

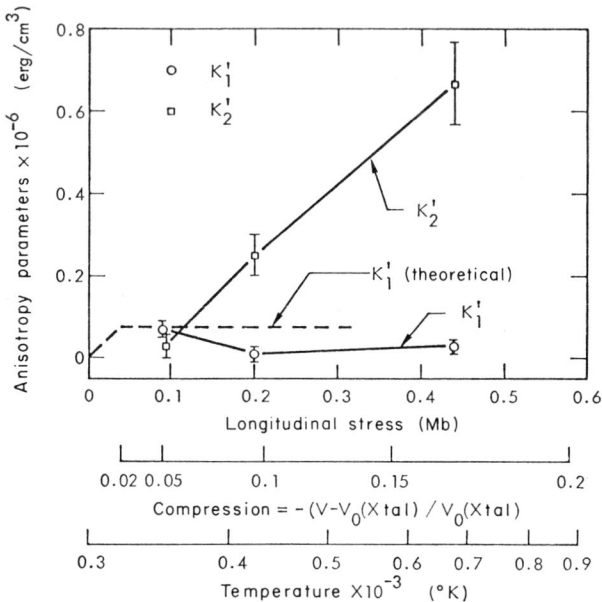

Fig. 5. Parameters of the shock-induced magnetic anisotropy of YIG as deduced from the field dependence of the transverse magnetization. The broken line is the value of K_1' calculated under the assumption of a Hugoniot elastic limit of 50 kbar, a constant yield strength, and constant magnetoelastic interaction coefficients.

observables is obtained. Resulting values of these constants as a function of total stress or pressure are presented in Fig. 5.

At 90 kbar, the value of K_1' is comparable to that predicted from the linear magnetoelastic theory, and the value of K_2' is small, as expected. At 200 and 440 kbar, however, the K_2' term shows a severalfold increase. This increase is thought to be the result of a breakdown of internal scaling of the compression and a differing compression of the tetrahedral and octahedral magnetic sites. The large effect arises because the induced anisotropies on the two sites are oppositely directed and almost equal. The increase is probably not the result of a rapid increase in the yield strength with pressure.

This work on YIG has demonstrated the existence of the induced anisotropy presumably arising from an anisotropic strain field in ceramic materials. It may be noted, however, that for nickel ferrite one should expect the induced anisotropy to be relatively independent of pressure because only the nickel sublattice contributes to the anisotropy. This expectation appears to be consistent with the data on nickel ferrite since

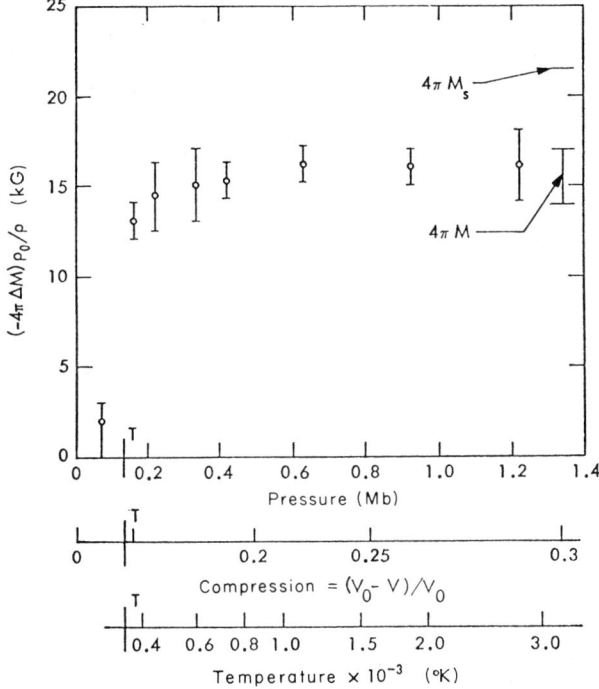

Fig. 6. Shock-induced demagnetization $(-4\pi\Delta M)\rho_0/\rho$ of iron. Also shown are the values of the measured initial magnetization of the iron samples $(4\pi M_s)$ and the accepted value of the saturation magnetization $(4\pi M_s)$ at 20°C.

the transverse magnetization of the shocked material appears to be independent of pressure, within experimental error.

The yield strength of metals is normally much lower than that of ceramic materials. Thus, the stress associated with shock compression is more nearly isotropic, the induced anisotropy should be small, and one should observe nearly the true demagnetization in experiments on iron. Iron exhibits a polymorphic phase transition near 130 kbar, and from static high pressure experiments, the high pressure ε phase is known to be HCP and nonferromagnetic.[26] Thus, one should observe virtually complete demagnetization above 130 kbar and essentially none below. Since the duration of the demagnetization signal is governed by the decay of eddy currents, it provides a simultaneous estimation of the electrical conductivity of the iron.

The shock-induced demagnetization of pure iron is shown in Fig. 6. Most of the points shown are averages of the results from samples of different thicknesses taken on different shots. The data show no significant effect of varying sample thickness. Bias fields between 220 and 300 Oe were employed, and the bracket on the right indicates the measured initial magnetization of the samples in this field. This measurement was made by quickly withdrawing the sample from one of the magnetic assemblies shown in Fig. 3. The demagnetization obtained by the integration of the signal from the pickup coil is then just the magnetization of the iron before the sample was withdrawn.

At high pressures, the experiments show a complete shock-induced demagnetization of iron. As the final pressure is lowered toward the 0.13 Mbar phase transition, somewhat less demagnetization is observed. Pressure–volume data indicate that the phase transition may not be complete in this region in shock experiments, and these magnetic results indicate the same result. They also indicate that the untransformed material

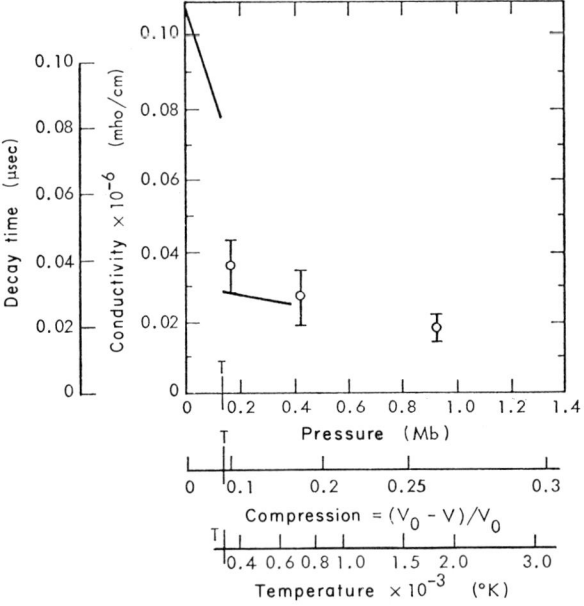

Fig. 7. Pulse decay time for the demagnetization signal from samples of iron 0.36 mm thick. The conductivity inferred for the shocked material is also shown. The solid lines present the results of the conductivity measurements of Fuller and Price.

extends over significant regions, since the observed ferromagnetic ordering is a cooperative phenomenon extending over many crystalline unit cells. Below 0.13 Mbar the experiments show a small demagnetization where none should be expected. One interpretation of the Hugoniot compression data indicate that there may be a small amount of phase transformation below 0.13 Mbar in iron. This would explain the magnetic results. Graham[24] has observed both of these effects in the demagnetization of Silectron cores (grain oriented 3% Si–97% Fe).

It can be shown that the decay of demagnetization pulses after the passage of the shock wave through the iron sample is exponential if the process is governed by the decay of eddy currents. The time constant of the decay is proportional to the conductivity of the shocked material and the square of the sample thickness. It thus provides a means of determining the conductivity. At pressures of 0.168, 0.42, and 0.92 Mbar, several experiments were performed in which the sample thickness was varied. The observation of the predicted quadratic variation with thickness verified that the observed decay is in fact due to the eddy currents. Figure 7 shows the observed decay time and the conductivity inferred. The conductivity measured directly by Fuller and Price on shocked iron wires is also shown.[27] The conductivity shown in Fig. 7 is for material which has been shock heated. In Fig. 8, this conductivity is corrected to 300°K and compared with static measurements of Balchan and Drickamer.[28] In making this correction it was assumed that the resistivity would be associated with phonon scattering and would, hence, be directly proportional to the

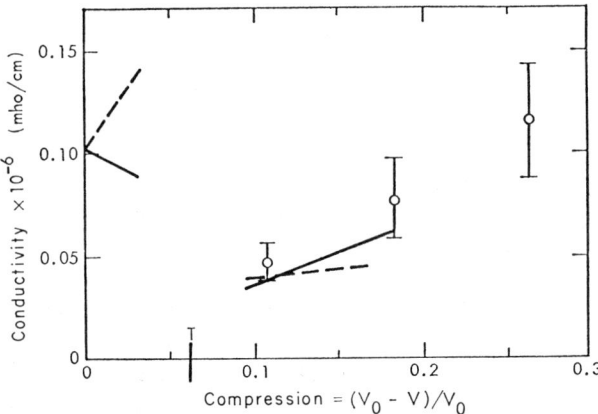

Fig. 8. Conductivity of iron at 20°C calculated from measurements on the shocked material. The solid line shows the results of Fuller and Price and the broken line the results of Balchan and Drickamer.

temperature. The three types of measurements appear to be in agreement in the range where the data overlap.

Electrical Conductivity

The material property which was thought to be the easiest to measure in a dynamic experiment is the electrical conductivity. It varies over an extremely wide range, and it is readily related to the physical state of

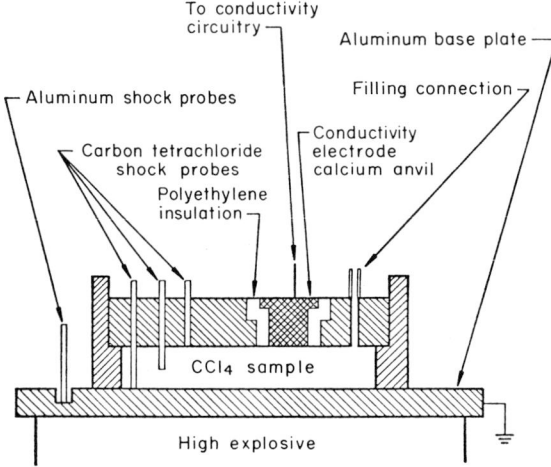

Fig. 9. Sketch of a typical experimental assembly used to measure the electrical conductivity of shock-compressed fluids. Not to scale.

the material. The importance of such conductivity measurements was recognized long ago, and many experimental investigations from many laboratories have been reported.[27-31] Unfortunately, since the electrical conductivity is sensitively dependent on the physical state, it is also very sensitive to small departures from experimental ideality—fancy words to say that the scatter in the data reported is unfortunately large. Scatter of a factor of 2 or an order of magnitude is not uncommon.

Without doubt, such relatively crude information is useful to suggest a qualitative picture of a situation, but as one moves toward a detailed investigation of some case, more accurate data are needed. To obtain such data, the techniques for measuring electrical conductivity must be improved. Mitchell and Keeler[32] have recently completed this task for fluids. The extension of much of what has been learned to solid insulators is obvious and straightforward.

MATERIAL PROPERTIES AT HIGH PRESSURE 95

Mitchell and Keeler carefully considered the experimental problems of making conductivity measurements in fluids and decided that the geometry shown in Fig. 9 was most suitable. It enables one to understand all details of the hydrodynamics of the experiment. Lateral rarefactions, electrode

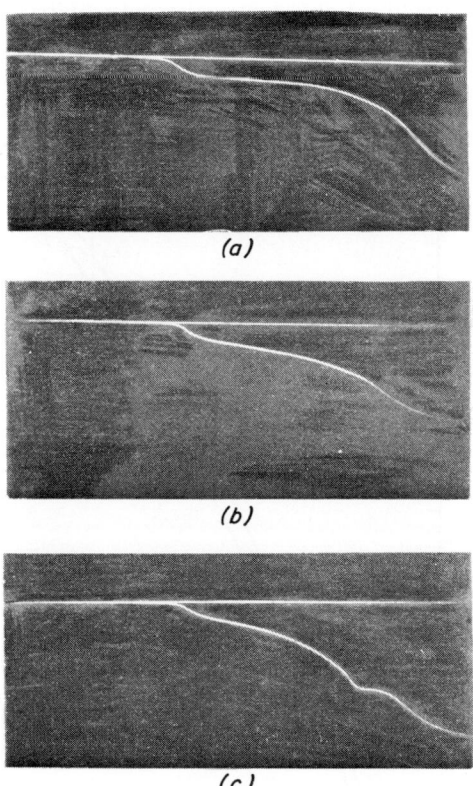

Fig. 10. The effect of shock wave reflection from a metal anvil on electrical conductivity records. (a) Calcium, (b) magnesium, (c) aluminum. All three records have a sweep duration of 2 μsec.

motion and distortion, and material nonuniformity will not complicate the data. Note that plane parallel electrodes are used, and that the electrode spacing is small compared to the diameter. This made it possible to determine conductivity from the measured resistance of the sample by making a straightforward correction for the fringing electric field. Note also that shock velocity was measured in the shocked fluid and in the base plate by carefully positioned probes or pins. The sample-loading apparatus was arranged to prevent contamination of the fluid by any foreign materials.

Resistances in the range from 1 ohm to 500 kohms were determined by a simple circuit in which the voltage was measured across a known resistance in series with the sample and a charged capacitor. For resistances between 500 kohms and 10 megohms a cathode follower with high input impedance and an amplifier were added to the circuit. By careful attention to system calibration, electrical conductivity for a fluid sample was determined with an estimated uncertainty of $\pm 10\%$ or less.

During the preliminary experiments, the second conductivity electrode or anvil was changed so as to investigate the influence of mechanical impedance on the measurement. Some of the results are shown in Fig. 10.

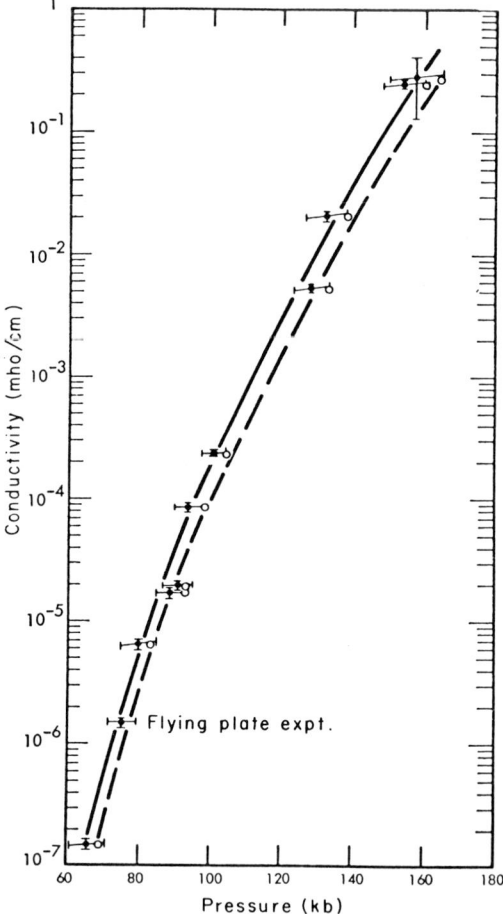

Fig. 11. Electrical conductivity of shock-compressed CCl_4 as a function of shock pressure. (, – – –) Without and (●, —) with shock wave attenuation correction behind the shock front.

Resistance records from carbon tetrachloride under anvils of calcium, magnesium, and aluminum are presented. The relatively slow rise of the signal is caused by the changing electrical capacity of the sample as the shock wave approaches the anvil. It is possible to calculate the rise exactly, and calculations agree very well with the observed shape. The further rise in the records in which relatively high impedance materials were used as anvils is caused by the lower resistance of the material behind the shock reflected back from the anvil into the sample. Clearly, it is very difficult to determine the resistance of a singly shocked sample precisely unless there is a good impedance match between sample and anvil as exists in the case of carbon tetrachloride and calcium.

Conductivity measurements were made of shocked carbon tetrachloride for several reasons. Pure material is readily available, and it is easy to handle. It was known to become a good conductor in a readily accessible pressure range. Finally, the CCl_4 molecule is spherical to a first approximation, so it may be possible to make an interesting theoretical interpretation of the results. Figure 11 shows the conductivity of singly shocked CCl_4 at pressures between 60 and 160 kbar. The data cover a range of more than six orders of magnitude. It is particularly significant to note that shock wave attenuation in experiments that use a 4-in. thick charge of high explosive lowers the apparent conductivity by more than a factor of 2.

These results clearly demonstrate that excellent measurements of electrical conductivity of shock-compressed fluids can be made if sufficient attention is paid to the many experimental details that can influence the measurement.

Electrical conductivity is usually examined as a function of temperature rather than pressure. The temperature of shocked CCl_4 can be estimated from the Lennard-Jones–Devonshire cell theory using an intermolecular potential derived from the Hugoniot measurements on this fluid. It is interesting to note that the temperatures thus calculated differed by only about 100°C from temperatures calculated by assuming the applicability of the Grüneisen equation of state and the adequacy of the Dugdale-MacDonald model giving γ as a function of specific volume.

Conductivity is plotted as a function of reciprocal temperature in Fig. 12. The data define an excellent straight line whose slope indicates an activation energy for the production of charge carriers of 4.3 eV.

The techniques described above for measuring the electrical conductivity of fluids are not suitable for metals which have high conductivity. The resistance of a sample is too low to be measured accurately in the short time available in a dynamic experiment. However, metallic conductivity can be investigated at high pressure by sandwiching thin foils

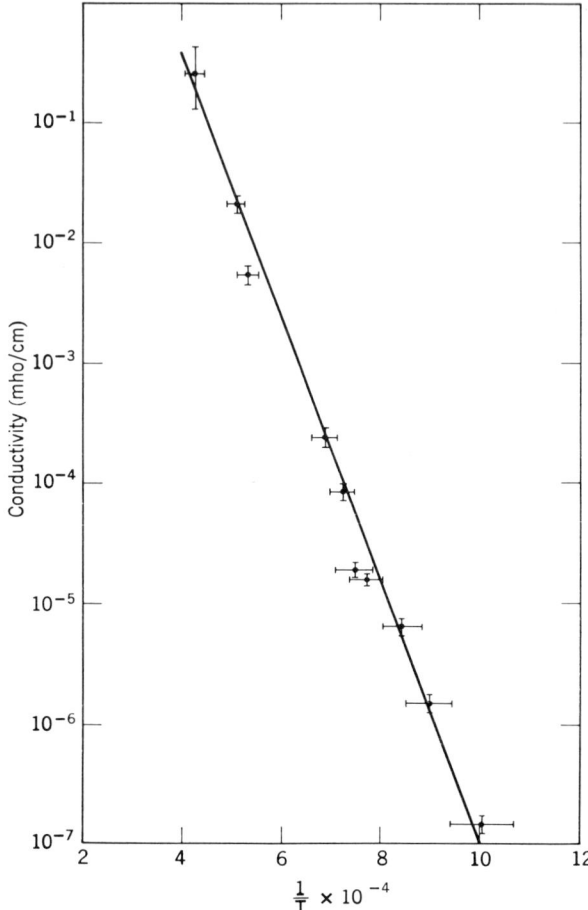

Fig. 12. Electrical conductivity for shock compressed CCl_4 as a function of $1/T$.

between two blocks of insulating material. Similar techniques were introduced by Fuller and Price.[27] Unfortunately, the thermodynamic state of the sample is not simply determined in such experiments because a thin sample is used. The pressure rapidly equilibrates to the pressure in the much thicker insulator. The temperature of the sample will equilibrate with its surroundings much more slowly. Some care must be used in comparing these measurements of conductivity with measurements made by other techniques. A similar situation existed in the magnetic studies of iron mentioned in the last section.

Preliminary experiments have been performed on 0.25 mm thick copper and iron foils. The copper was surrounded by Teflon plates, and alumina ceramic was used as the insulator for the iron sample. Oscilloscope records of the apparent resistance are shown in Fig. 13. The copper record is easy to interpret. A 0.26 Mbar shock arrives soon after the start of the trace, and the resistance increases markedly. This increase is approximately that expected if resistance values determined at ambient conditions are extrapolated to the shock condition. The shock-induced temperature rise contributes the major part of the resistance increase.

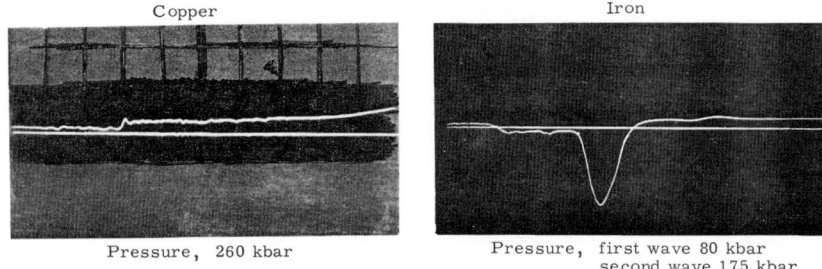

Fig. 13. Conductivity records for copper and iron. Total sweep duration of both records is 1 μsec.

The iron record is much more interesting. The resistance of the sample appears to go slightly negative shortly after the start of the trace. It stays negative for about 0.3 μsec and then goes strongly negative. Finally, the resistance increases to the expected value at the end of the trace. This complexity is not an experimental artifact. It is caused by the demagnetization of iron known to occur at high pressure.

Consider first the wave system in the alumina plate holding the iron foil. A 0.08 Mbar elastic wave precedes a second shock which raises the pressure to 0.175 Mbar. There is a well-known phase transition in iron at a pressure of 0.13 Mbar. This transition pressure is insensitive to temperature. As mentioned before, the high pressure phase has been shown to be a nonmagnetic, hexagonal, close-packed material.[26] The large negative pulse is associated with this phase transition in iron.

A simple sketch of the resistance-measuring circuitry is shown in Fig. 14. The inductance in the primary circuit ensures that the externally applied current through the sample will be constant during the experiments. Therefore, the voltage indicated by the relatively high impedance recording circuit is a direct measure of the resistance of the sample as long as no magnetically induced currents are present. There are no such currents in copper, but the situation is different in iron.

The externally applied current generates a magnetic flux, **B**, in the wire which is approximately one thousand times larger than the applied field, **H**. When the pressure in the wire exceeds 0.13 Mbar, the material rapidly transforms to the nonmagnetic phase, and the **B** field can no longer be supported by the resistance-measuring current. As the field collapses, it generates a voltage in the sample which tends to increase the current in the foil so as to maintain the field. The large inductor prevents this current from flowing in the primary circuit; it must flow backward through the metering circuit, thus causing the negative voltage spike and the apparent

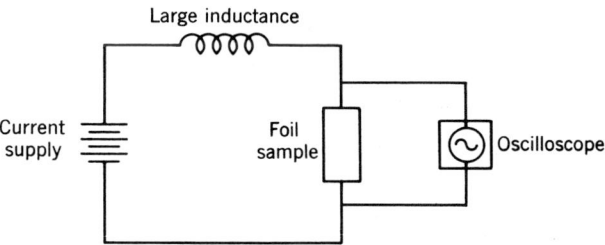

Fig. 14. Simplified diagram of the electrical circuit used to measure the electrical resistance of metal foils.

negative resistance in the sample. This pulse lasts until the magnetic field can diffuse out of the wire. The diffusion rate depends on the electrical conductivity of the iron, the quantity these experiments started out to measure. The conductivity indicated by the pulse decay and by the asymptotic signal level at late time are in substantial agreement.

This explains the main pulse, but the small signal associated with the elastic wave in alumina still must be considered. The iron sample seems to have a more-or-less constant negative resistance which lasts for a long time compared with magnetic field diffusion time. This requires the magnetic permeability to decrease at a roughly constant rate. Such a decrease could come from a relatively small amount or slow rate of phase transformation below 0.13 Mbar, or it could indicate a change of permeability of α-iron near 0.08 Mbar, an unlikely explanation. Either effect could indicate a gradually increasing pressure between the two main waves in alumina, a situation quite likely to exist. The experimental evidence is now inadequate to permit a choice between these alternatives.

The magnetic phenomenon can be described in a different way. If one shocks a current carrying iron wire above its phase-transition pressure, the wire acts as an autotransformer. The single wire acts simultaneously as primary, secondary, and magnetic core. The driving function for the circuit is not a change in the current or voltage applied to the element;

it is a change in the magnetic properties of the core. More conventional, two-winding, shock-driven transformers with separate iron cores have been described by Neilson and co-workers[33] and by Besancon et al.[34]

It is an interesting commentary on research that experiments designed to measure magnetic properties of iron also provided electrical conductivity data, and experiments intended to measure the electrical conductivity provided interesting magnetic information.

Laser Experiments

The development of the laser has opened up new possibilities for experimental investigations of the properties of materials by dynamic means. The most obvious and most widely developed of these possibilities uses the laser as a radar set to measure the velocity of surface motion as a function of time. The Doppler shift in the optical frequency of laser light reflected from a surface can be measured by beating the reflected signal against the incident beam,[35] by beating the reflected signal against a part of the reflected beam which has been delayed a known amount,[36] or by passing the light through an extremely sharp interference filter or Fabry-Perot interferometer.[37] These techniques will permit very precise measurments of the free surface velocity of a small region of a reflective surface. Such measurements should be particularly valuable in the investigation of surface motions caused by elastic–plastic wave propagation, spalling phenomena, and crystallographic phase transformations.

However, this application of the laser only permits better measurement of something which could be measured well by other techniques. A more exciting possibility is the use of nonlinear optical phenomena to measure material properties which could not be measured accurately or at all by other methods.

It is well known that intense laser light focused into a dielectric medium with a large electrostrictive coefficient, usually an organic fluid, may produce stimulated Brillouin scattering. The **E** field of the electromagnetic radiation acting through the electrostrictive effect sets up a spatially varying density field in phase with the light. This field acts as a Bragg grating and reflects back a fraction of the incident light. The reflected beam beats with the incident beam and sets up a standing wave which increases the density fluctuation. If the optical intensity is sufficiently large, internal loss mechanisms are overcome, and the amplitudes of the density fluctuation and that of the reflected wave grow exponentially. Stimulated Brillouin scattering occurs. The back-scattered beam is coherent, and its intensity may be more than 30% of that of the incident beam.

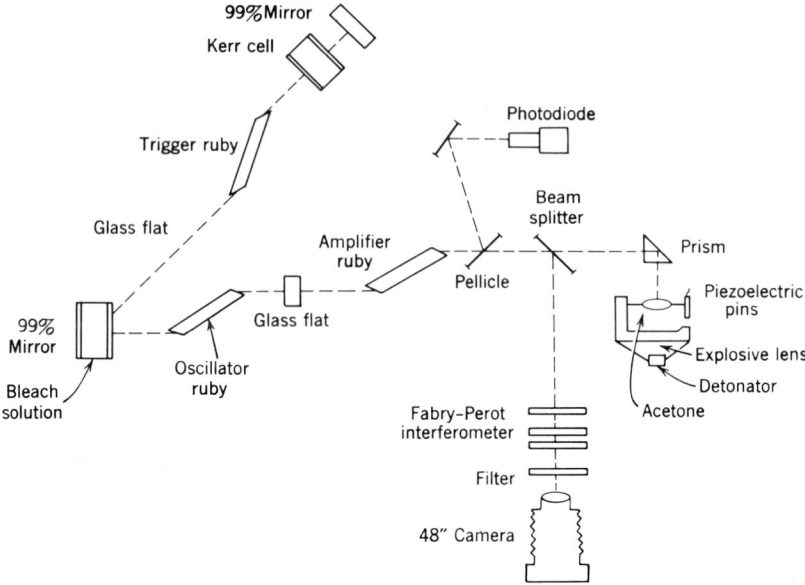

Fig. 15. Experimental apparatus for the observation of stimulated Brillouin scattering from shock-compressed liquids.

However, the spatially varying density field may also be recognized as a very high frequency phonon. It moves with the local sound speed in the fluid. Therefore, the scattered beam is Doppler shifted by an amount proportional to the ratio of the sound speed to the speed of light. The Doppler shift can be measured with a Fabry-Perot interferometer and the sound speed determined.

Keeler, Bloom, and Mitchell[37] have demonstrated that stimulated Brillouin scattering experiment can be conducted behind a shock wave in a fluid. Now there are three Doppler shifts to consider. Shifts are caused by the moving shock front, the moving fluid, and the moving phonons. The downshift in optical frequency of a light beam incident normal to the shock front is

$$\frac{\Delta \nu}{\nu} = \frac{2}{c}(n_2 V_s + (n_2 - n_1) U_s - n_2 U_p)$$

where n_1 and n_2 are the index of refraction of the unshocked and shocked fluids, respectively, V_s is the velocity of sound, U_s is the shock velocity, and U_p is the material velocity behind the shock. Obviously, the relative importance of these terms varies as the angle of incidence between the light beam and the shock front is changed.

A sketch of the apparatus used is shown in Fig. 15. This sketch should indicate that this is a hard experiment to do. One needs a single-mode, high intensity, Q-switched ruby laser which can be timed to a small fraction of a microsecond. Such a laser system was developed as shown, but it was difficult to keep operating. Nevertheless, the sound speed behind a 0.035 Mbar shock in acetone was measured and found to be approximately the value predicted by simple theory. It has not yet been possible to run a successful series of experiments to serve as a real test of a particular fluid model because the improved system is now being tested. It should be much more stable.

When this experimental technique has been perfected, precise measurements of sound speed can be made in conjuction with measurements of P, ρ, and E in a shocked state. The sound speed will determine the isentropic slope at a point on the Hugoniot. This will be new information.

Further in the future is the possibility of measuring transport properties of shocked fluids by laser techniques. The lifetime of a phonon is dependent on the viscosity. It should be possible to measure the decay of a phonon produced by stimulated Brillouin scattering by observing ordinary scattering from the phonon. Also it may be possible to do ordinary Brillouin scattering experiments behind a shock front. This will require more powerful lasers than now available, but if it can be done, the width of the scattered line would give a measure of the thermal diffusivity and the specific heat ratio. If such measurements can be made, the power of dynamic high pressure techniques will have been significantly increased, and additional material properties at extreme conditions will become known.

References

1. The Thomas-Fermi theory has been discussed in many places, but a particularly good treatment of the theory in a general equation of state context can be found in S. G. Brush "Equation of State. A Survey of the Thermodynamic Properties of Matter at High Pressures and Temperatures," UCRL 7437 (1964); also in *Progress in High Temperature Physics and Chemistry*, Vol. I, Carl A. Rouse, Ed., Pergamon Press, New York, 1967.
2. R. Latter, *Phys. Rev.*, **99**, 1854 (1955).
3. E. Teller, *Rev. Mod. Phys.*, **34**, 627 (1962).
4. L. Brillouin, *L'Atome de Thomas-Fermi*, Hermann, Paris, 1934.
5. P. Gombas, *Die Statistische Theorie des Atoms*, Springer-Verlag, Vienna, 1949.
6. P. A. Kirzhnits, *Zh. Eksperim. i Teor. Fiz.*, **32**, 115 (1957) [English transl.: *Soviet Phys.—JETP*, **5**, 64 (1957)].
7. R. Grover, *J. Math. Phys.*, **7**, 2178 (1966).
8. R. N. Keeler, M. van Thiel, and B. J. Alder, *Physica*, **31**, 1437 (1965).
9. M. Ross and B. J. Alder, *J. Chem. Phys.*, **46**, 4203 (1967).

10. S. Chandrasekhar, *Phil. Mag.*, **9**, 292 (1930).
11. G. M. Harris, *J. Chem. Phys.*, **31**, 1211 (1959).
12. G. M. Harris and J. Trulio, *J. Nucl. Energy, Part C: Plasma Phys.*, **2**, 224 (1961).
13. C. A. Rouse, *Astrophys. J.*, **136**, 665 (1962).
14. C. A. Rouse, *Astrophys. J.*, **139**, 339 (1964).
15. G. M. Harris, *Phys. Rev.*, **133**, A427 (1964).
16. H. E. De Witt, *J. Nucl. Energy, Part C: Plasma Phys.*, **2**, 27 (1961).
17. H. E. De Witt, *J. Math. Phys.*, **7**, 616 (1966).
18. H. E. De Witt, Proceedings of the Workshop Conference on Lowering of the Ionization Potential and Related Problems of the Equilibrium Plasma, Univ. Colorado, Nov. 1965.
19. R. G. McQueen, "Laboratory Techniques for Very High Pressures and the Behavior of Metals Under Dynamic Loading," in *Metallurgy at High Pressures and Temperatures*, K. A. Gschneider, M. T. Hepworth, and H. A. D. Parlee, Eds., Gordon and Breach, New York, 1964.
20. L. V. Al'tshuler, *Usp. Fiz. Nauk*, **85**, 197 (1965) [English transl.: *Soviet Phys.—Usp.*, **8**, 52 (1965)].
21. S. Thunborg, Jr., G. E. Ingram, and R. A. Graham, *Rev. Sci. Instr.*, **35**, 11 (1964).
22. E. B. Royce, *J. Appl. Phys.*, **37**, 4066 (1966).
23. E. B. Royce, "Shock-Induced Demagnetization of Nickel Ferrite, Yttrium Iron Garnet, and Iron," Proc. IUTAM Symp. Behavior Dense Media Under High Dynamic Pressures, Paris, France, Sept. 11–15, 1967.
24. R. A. Graham, "Pressure Dependence of the Magnetism of Invar and Silectron from 30 to 450 kbar," IUPAP Intern. Congr. Magnetism, Sept. 11–15, 1967. *J. Appl. Phys.*, to be published.
25. R. Pauthenet, *Ann. Phys. (Paris) (Ser* 12), **7**, 710 (1952).
26. J. C. Jamieson and A. W. Lawson, *J. Appl. Phys.*, **33**, 776 (1962); T. Takahashi and W. A. Bassett, *Science*, **145**, 483 (1964); R. L. Clendenen and H. G. Drickamer, *J. Phys. Chem. Solids*, **25**, 865 (1964); D. N. Pipkorn, C. K. Edge, P. Debrunner, G. de Pasquali, H. G. Drickamer, and H. Fraunfelder, *Phys. Rev.*, **135**, A1604 (1964).
27. P. J. A. Fuller and J. H. Price, *Nature*, **193**, 262 (1962).
28. A. I. Balchan and H. G. Drickamer, *Rev. Sci. Instr.*, **32**, 308 (1961).
29. B. J. Alder and R. H. Christian, *Discussions Faraday Soc.*, **22**, 44 (1956).
30. S. D. Hamann and M. Linton, *Trans. Faraday Soc.*, **62**, 1 (1966).
31. L. V. Al'tshuler, L. V. Kuleshova, and M. N. Pavlovskii, *Zh. Eksperim. i Teor. Fiz.*, **39**, 16 (1960) [English transl.: *Soviet Phys.—JETP*, **12**, 10 (1961)].
32. A. C. Mitchell and R. N. Keeler, *Rev. Sci. Instr.*, to be published.
33. R. W. Kulterman, F. W. Neilson, and W. B. Benedick, *J. Appl. Phys.*, **29**, 500 (1958).
34. J. E. Besancon, J. L. Champetier, Y. Leclonche, J. Vedel, and J. P. Plantevin, *Conference on Megagauss Magnetic Fields Generated by Explosives and Related Experiments*, H. Knoepfel and F. Herlach, Eds., EURATOM, Brussels, 1966, p. 331.
35. L. M. Barker and R. E. Hollenbach, *Rev. Sci. Instr.*, **36**, 1617 (1965).
36. L. M. Barker, "Fine Structure of Compressive and Release Wave Shapes in Aluminum Measured by the Velocity Interferometer Technique," Proc. IUTAM Symp. Behavior of Dense Media Under High Dynamic Pressure, Paris, France, Sept. 11–15, 1967.
37. T. Burgess, private communication.
38. R. N. Keeler, G. H. Bloom, and A. C. Mitchell, *Phys. Rev. Letters*, **17**, 852 (1966).

Polymeric Materials for Extreme Conditions

HERMAN F. MARK

Department of Chemistry, Polytechnic Institute of Brooklyn, Brooklyn, New York

I. Introduction

When I was an Associate Professor in the Chemistry Department at the Polytechnic Institute in Karlsruhe in 1929, I gave a course on "wave mechanics," which, at that time, was a very new and exciting discipline. Its mathematical concepts were somewhat unusual for a chemist in those days, and on some occasions, I was not able to clarify the situation to my audience. As often as that happened, however, I was extricated by a young student who sat in the first row, waited patiently for his time to come, and then would say in a soft, friendly voice with a Hungarian accent, "I think, Professor Mark, it must be quite clear to all of us what you have just now explained, namely ... " and then he went to the blackboard and gave a completely correct and lucid version of my somewhat obscure presentation. Then he sat down again and the course continued.

The following little report on polymeric materials for extreme conditions is, therefore, not only an enthusiastic tribute to the great and important discoveries with which Edward Teller has enriched modern physics, but also a cordial greeting to his 60th birthday and a modest expression of thanks for the most welcome services which he rendered me when I was a young teacher and gave a course on a topic which I did not really understand.

— — — — —

The carbon atom has a remarkable versatility in providing the backbone for molecules of an enormous variation in size and structure. This makes it stimulating to explore the possibility to design carbon-based polymers which can withstand extreme conditions in thermal and mechanical exposure.

Two natural occurrences of the element carbon are known: *diamond* and *graphite*.

The first type exists in well-developed octahedral crystals, the lattice of which is formed exclusively by single-bonded tetrahedral carbon atoms;

all bonds are identical and are provided by L electrons which are all of the sigma type. The dissociation energy of such bonds is 80 kcal per mole of bonds, the stretching modulus is 4.6×10^5 dynes cm^{-1}, and the bending modulus is 1.25×10^5 dynes cm^{-1}. The most important data on diamond are given in Table I.

TABLE I. A Few Relevant Data on Diamond

Length of the cubic cell at 25°C	3.5596 Å
Length of the carbon–carbon bond	1.54 Å
Specific gravity	3.51
Tensile modulus	
normal to the cubic plane	7.2×10^6 kg/cm^2
normal to the octahedral plane	10.3×10^6 kg/cm^2
normal to the dodecahedral plane	9.3×10^6 kg/cm^2
"Mohs" hardness	10
Compressive strength normal to the octahedral plane	8×10^5 kg/cm^2
Compressive modulus normal to the octahedral plane	6×10^6 kg/cm^2
Temperature of fusion	>3500°C
Temperature of sublimation	4200°C

It is possible today to synthesize diamond from simple carbon compounds,[1] but the resulting crystals are so small (usually below 0.1 in.) that they cannot be used as structural elements for mechanical, electrical, or optical devices although they are finding important commercial application as superior abrasives.

The other modification of the element carbon is graphite; it occurs in the form of polycrystalline aggregates from which small platelets have been picked which were single crystal, or at least represented mosaic crystals with very close geometrical arrangements of the individual components.[2] Graphite consists of a typical layer lattice; the hexagonal basis planes are formed by condensed six-membered rings; the bonds between the individual carbon atoms have a length of 1.42 Å and are of the aromatic type; they are all identical and have a dissociation energy of 125 kcal per mole of bonds, a stretching modulus of 6.5 dynes cm^{-1}, and a bending modulus of about 2 dynes cm^{-1}. These lattice planes are thermally and mechanically very stable, but their extraordinary rigidity and infusibility cannot be put to practical use for the construction of larger objects because the bonding *between* these planes is relatively weak. It is provided by dispersion forces between the aromatic rings of subsequent planes along the hexagonal axis which have a dissociation energy of less than 20 kcal per mole of bonds. This explains the large distance (3.41 Å) between the planes and the pronounced cleavability of graphite which led to its use as a solid lubricant. The most important data on graphite are given in Table II.

TABLE II. A Few Relevant Data on Graphite

Carbon–carbon distance in the hexagonal basis	1.42 Å
Distance between the hexagonal basis planes	3.41 Å
Specific gravity	2.1–2.5
Tensile modulus parallel to the basis plane	Very low because of slippage
Mohs' hardness	0.1–1.5
Temperature of fusion	Above 3500°C
Temperature of sublimation	4200°C

Neither diamond nor graphite melt; at sufficiently high temperatures, carbon atoms or C_2 molecules are broken off and form a vapor phase: both modifications *sublime*. If we consider a ring of six carbon atoms in the lattice, then we see that in diamond it is held together only by single bonds, with six single bonds for six carbon atoms; in graphite it is held together by a combination of single and double bonds with three single and three double bonds for six carbon atoms; the first design does not utilize the available bonding and stabilizing strength of the double bonds; the second uses so much of it inside the hexagonal basis plane that no strong bonding opportunities are left to tie those planes into a firm three-dimensional, macroscopic system. This situation obviously poses the question: would it be possible to produce other modifications of the element carbon with more double bonds than in diamond but with less than in graphite; for example, with two double bonds and four single bonds for six carbon atoms. Because of restrictions imposed by steric considerations, it might occur that in such structures not all valences could possibly be satisfied by a carbon–carbon linkage which might make it necessary to saturate such bonding facilities (impaired electrons, free valences) by hydrogen atoms. Instead of a new modification of the element carbon, one would then obtain a new hydrocarbon molecule of the formula C_nH, where n would be a number of 4 or 5 or even more. From the practical point of view this would be of no avail because these few hydrogen atoms would not weaken the overall structure either thermally or mechanically. It could also be possible to saturate the free-bonding facilities with oxygen, nitrogen (NH), or sulfur and provide for crosslinking between the individual aromatic layers.

II. Some Speculative Approaches

One way to prepare for a solution of the problem is to take a model of the diamond lattice, open a certain number of single bonds, convert them into double bonds, and consider the geometrical changes in the lattice which will be the consequence of such a progressive but limited "aromatization." Another way would be to look at a model of the graphite lattice, open a few double bonds, and connect the free valences with each other in order to get a firmer and stronger bonding between the individual layers. Considerations and speculations of this type are perfectly legitimate and advisable as long as one does not take them too seriously, because they are very inexpensive (a sheet of paper, a pencil, and a slide rule) and clearly indicate the difficulties which one has to face if one wants to put such enforced changes of natural structures into practice.

A different approach is to look at the actual armamentarium of synthetic polymer chemistry and to visualize practical polymerization processes which would not yet lead to the most desirable ultimate concepts but which still would, hopefully, result in structures which are not too different from them and eventually, by additional steps of minor difficulty, could be transformed into something which would be close to the desired design.

There exist, in fact, rather numerous ways to prepare linear macromolecules which consist essentially of aromatic units. Once they have been built up to a certain desired length, they can be dehydrogenated, aromatized, and eventually crosslinked in such a manner that the resulting network does not exactly and literally, but effectively and statistically, correspond to a highly resonance-stabilized system with an uninterrupted thoroughgoing network of covalent bonds which ought to have—and in fact has—unusual thermal and mechanical stability.

It is necessary to add that all presently existing efforts to prepare synthetic carbon analogs with superior property combinations are still in their infancy, but it would be unrealistic to ignore the fact that their existence proves for all practical purposes the possibility to arrive at a wide spectrum of synthetic polymers somewhere between the structure of diamond and graphite which will be distinctly superior to both extremes in their practical applicability and availability.

III. Some Practical Approaches

An interesting and promising starting material for such synthetic efforts is acetylene (HC≡CH) because it is available from vast natural resources, is inexpensive, and is already used as a commercial gas in considerable quantities. It has long been known that acetylene polymerizes spontaneously to a brown, insoluble, and infusible powder known as cuprene. It is apparently an irregular, partly crosslinked, network of carbon and hydrogen atoms which is so intractable that it has not found any practical uses. More recently it was possible with the aid of special catalysts to obtain long linear chains of *trans*-polyacetylene,[3] also known as

trans-polyvinylene, which form darkly colored powders, films, and fibers having high rigidity and high softening ranges. Careful dehydrogenation in the absence of oxygen leads progressively to black materials which are very hard, practically infusible, and contain only about one hydrogen atom for four to five carbon atoms. It is probable that an irregular, highly aromatized network is formed in the course of this process which is essentially amorphous and quite porous.

Another relatively early effort was based on the use of styrene and divinylbenzene which can readily be copolymerized into a rigid crosslinked network and later dehydrogenated (carbonized) to form a black, hard, and infusible powder.[4] Baker and Winslow have not only investigated the unusual mechanical and thermal properties—hardness, abrasion resistance, infusibility, etc.—but also the electrical conductivity and have established with the aid of the electron spin resonance method the presence of a considerable number of unpaired electrons which are produced by the removal of hydrogen from the carbonaceous skeleton.

Another attractive approach to penetrate into the field of highly carbonized synthetics starts with such simple aromatics as benzene, toluene, xylene, phenol, 2,5-dimethylphenol, and other related materials. A new, very general, and promising method is now developed to prepare long linear chains from these monomers, namely, the principle of oxidative coupling with the aid of heavy metal halide complexing agents in the presence of a hydrogen-abstracting component. In this manner long chains of the following representative structures have been made.[5]

Poly-phenylene

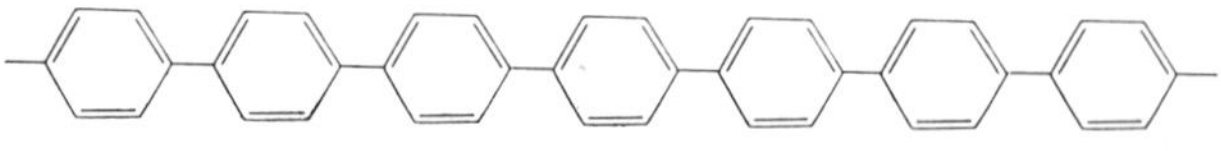

Poly-tolylene

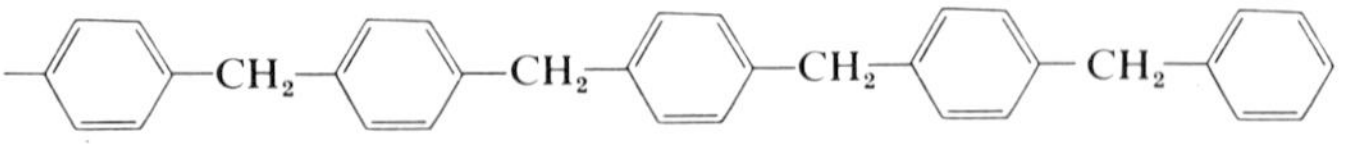

Poly-xylylene

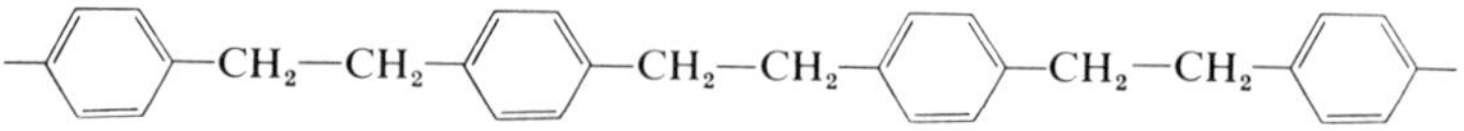

Poly-2,6-p-xylylene oxide

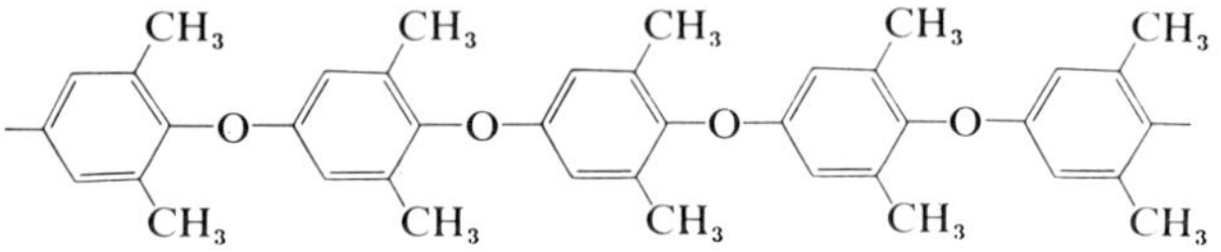

They all represent relatively rigid, high melting, and difficultly soluble materials which can be converted by additional dehydrogenation and crosslinking into dark-colored, hard, infusible, and insoluble systems of distinctly carbon-like character which contain only small proportions of hydrogen and oxygen, give only poorly organized x-ray diagrams and exhibit electronic semiconductance.

Still another way starts with readily polymerizable aliphatic compounds such as butadiene, hexatriene, acrylonitrile, and vinylidene cyanide, converts them into long chains of the form shown in Fig. 1, induces

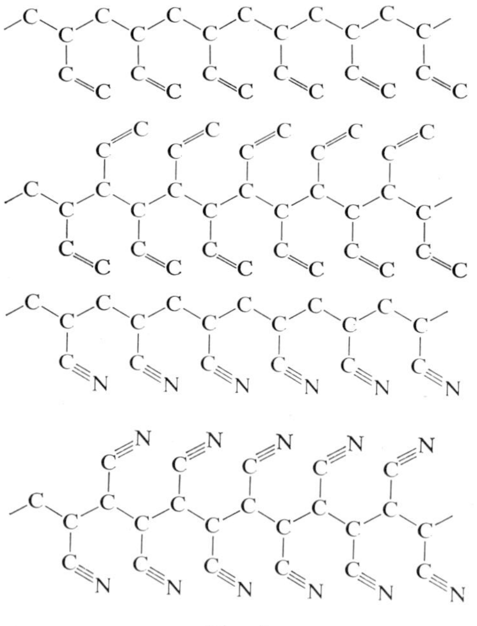

Fig. 1

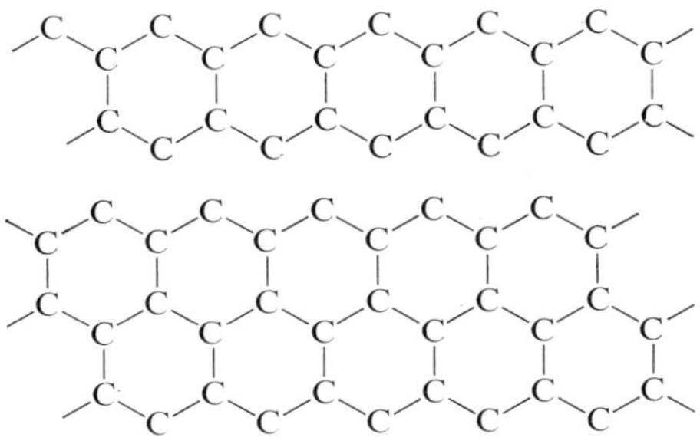

POLYMERIC MATERIALS 111

Fig. 2

catalytically additional ring closure along the length of the chains to obtain ladder-like structures of the type shown in Fig. 2 (together with the corresponding nitrogen containing analogs made of acrylonitrile and vinylidene cyanide), and aromatizes them *in vacuo*, at elevated temperatures, and preferably in the presence of an acidic catalyst by dehydrogenation into dark-colored, very hard, completely infusible and insoluble material having such structures as illustrated in Fig. 3.

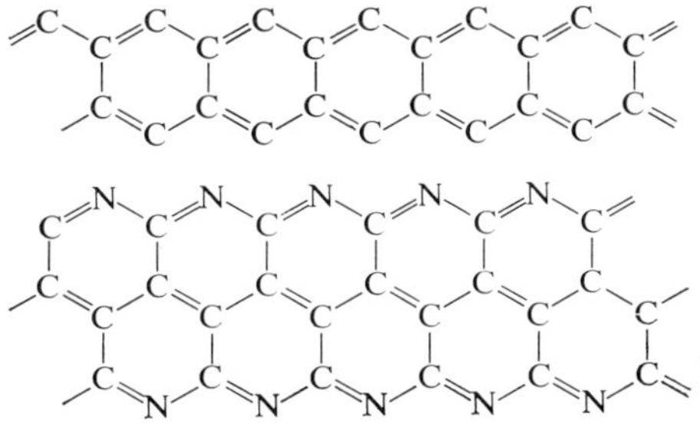

Fig. 3

It must be understood that the "ladderized" and "aromatized" segments in these molecules do not extend from one end of the chain to the other but are only of limited length—probably six to eight monomers—and are linked together by pairs of single bonds which, however, may provide for crosslinks between adjacent chains, and therefore can also contribute to the firmness of the entire structure.[6] An important advantage of this approach consists in the fact that the linear uncyclized and even the cyclo aliphatic ladder structures are still fusible and soluble and can be readily spun, cast, extruded, or molded into almost any desired form and subsequently converted into the final carbon analog which itself is obviously highly intractable and refractory. All materials of this type are based on well-known, readily available, and inexpensive monomers, the polymerization, ladderization, and carbonization of which take place under relatively mild conditions, all of which shows that materials of this type will be available in the near future in large quantities, with a wide variety of properties and at reasonable cost. It should be added that all structures of this type are not insulators but semiconductors and can, by the introduction of metal atoms through complex formation, even be converted into electronically conducting materials with specific resistivities less than that of many metals.

In view of these experiences it seemed interesting to investigate in a systematic manner the cautious carbonization of several natural and synthetic fibers. At this point, it is worthwhile to remember that Edison prepared his first carbon filaments through the carefully controlled carbonization of natural cellulosic fibers. Following this lead and improving the techniques involved, it is possible today to arrive at fibers, yarns, cords, and fabrics which have considerable strength and consist essentially of a carbon skeleton which usually contains a few other elements (H, N, O, B, P, Zn) in small quantities and has remarkable thermal stability. The guiding principle in the attainment of a favorable combination of rigidity, strength, and high melting point is to achieve a high degree of aromatic cyclization without causing too many bond breaks in the main chains of the polymeric system. One way to arrive at good results is to start with a fiber which has a high melting point, such as cellulose, aromatic polyesters, and polyamides, bring it to a state of high orientation and crystallinity, and heat it in the presence of condensing agents such as heavy metal salts (phosphates and silicates of zinc, molybdenum, tungsten, etc.) and oxides (Sb_2O_3) *in vacuo* to elevated temperatures in the range of 1000°C. If the process is carried out slowly, the system darkens, loses weight (hydrogen, water), and is eventually converted into a black, very rigid, strong infusible, and insoluble material which still possesses considerable strength and toughness. Table III lists a series of such new products which are beginning to find increased use in the technology of

TABLE III. A Few Pertinent Data on High Temperature Polymers

Material	Range of useful service life, °C	Softening or decomposition range, °C
Aromatic condensation polymers	200–300	400–500
Cyclized and aromatized polybutadiene	up to 600	800–1000
Cyclized and aromatized acrylonitrile	up to 650	800–1100
Highly carbonized aromatic yarns	up to 700	800–1200
Structural carbon fiber	up to 700	900–1500

jet and rocket engines, heat shields, plasma containers, and vessels and pipes in atomic energy plants.

Whenever one wants to get plates, rods, or tubes of these materials, it is advantageous to start with a powder, a strand of fibers, a felt, or a fabric, impregnate it with a thermosettable system, form it in the desired shape, thermoset the binder, and give the entire object another carbonizing and devolatilizing heat treatment. Even the classical thermosetting resins, such as phenol formaldehyde, lead to interesting results, but more sophisticated formulations, such as trivinylbenzene–acrylonitrile or triallylbenzene–phthalonitrile, are now being studied for superior performance of the ultimate products.

While the reported approach leads to a large number of obviously interesting synthetics which exhibit unusual stability under extreme conditions, it must be admitted that we have learned very little on the mechanism of the individual steps which are necessary to form long linear chains with recurring aromatic or aromatized units, the resonance stabilization of which causes their superior mechanical and thermal performance. The desire to arrive at a better understanding has led to another approach, which is parallel and simultaneous with the first and concentrates on the stepwise buildup of long chains from aromatic monomers with the aid of polycondensation. In the classical chemistry of polycondensation products, one operates with bifunctional units to penetrate into the macromolecular range. Using, for instance, an aromatic diamine and an aromatic acid, one obtains chains of the following type:

They contain the desirable aromatic constituents, because we have put them in, but between each two rings there are three consecutive single bonds which possess rotational mobility and therefore represent an undesirable element of molecular limpness. In order to avoid this it is obviously necessary to produce *new rings* in the course of the polycondensation process; this can be done by using the functional groups (COOH and NH_2) not in a 1:1 ratio but at 2:1 or 1:2 or 2:2. Using this idea, several authors[7] have very successfully prepared numerous new polymers which possess considerable thermal stability and which can be converted by additional cyclization into another class of carbonaceous structures having extreme properties.

One representative class of these materials is obtained by condensing an aromatic tetramine with an aromatic dicarboxylic acid to obtain a polybenz-imidazole of the form (1). Another is prepared from an aromatic tetracarboxylic acid and an aromatic diamine, which gives a polyimide such as (2). Still other designs are combinations of both types such as can be obtained from aromatic triamines and aromatic tricarboxylic acids (3).

It can readily be seen that there exist many different embodiments of this principle since organic chemistry offers a plurality of monomeric units which can be condensed with the formation of new rings.

A somewhat different way leads over intermediates which still contain too many single bonds but which can be stiffened by subsequent cyclization. It is, for instance, possible to prepare the hydrazid of terephthalic acid (4) and produce additional ring formation by dehydration of the hydrazidic bonds (5).

All these polycondensation products can be spun, cast, or molded in the form of incompletely cyclized precursors which are still soluble in certain liquids such as sulfuric, phosphoric, trifluoroacetic, and toluene sulfonic acid and can be brought into their final structure when the desired shape of the ultimate object is reached. They represent, in general, yellow to brown powders with softening ranges between 500 and 1000°C and considerable rigidities and are very strong and tough even at temperatures as low as 4°K.

IV. Considerations on Particle Size

Probably the oldest attempt to prepare a synthetic modification of the element carbon is the commercial production of carbon black by the controlled dehydrogenation of oil or gas in furnaces or in heated channels. The final product is obtained in the form of a very fine, substantially amorphous, dark powder which is the most widely used reinforcing agent for rubbers and plastics. A very important factor in the characterization of these materials is the magnitude of their specific surface, i.e., the accessible surface of one gram of the powder. This quantity is of general interest and significance for all highly subdivided systems, such as fine powders, thin filaments, foams, or other media which possess very small pores. Since the days when colloid chemistry came to prominence—between 1910 and 1920—many efforts were directed to the determination of the average particle size of colloidal systems, particularly of catalysts, but it is interesting to realize that one of them eventually survived all others because of clarity, simplicity, and reproducibility. This is the BET method which was worked out by Brunauer, Emmett, and Teller in 1938.[8]

The leading idea of this contribution is to generalize the classical treatment of unimolecular layer adsorption given by Langmuir and to extend it to multilayer formation. This is done by considering the evaporation–condensation equilibria for the various surface areas S_i ($i = 0, 1, 2, 3, \ldots$) which are bare ($i = 0$) and covered by a unimolecular ($i = 1$), bimolecular ($i = 2$), and i-molecular ($i = i$) layer, respectively. The resulting equation, which expresses the volume of adsorbed gas or vapor in terms of the total volume v of adsorbate used and the pressure p can be put in the form

$$\frac{p}{v(p_0 - p)} = \frac{1}{v_m c} + \left(\frac{c-1}{v_m c}\right)\left(\frac{p}{p_0}\right)$$

where $p_0 =$ vapor pressure of the liquid at the temperature of the test,

v_m = volume of gas required to give a complete monomolecular layer, and c = empirical constant.

A plot of $p/v(p_0 - p)$ versus p/p_0 gives a straight line with the intercept of $1/v_m c$ and a slope of $(c - 1)/v_m c$, which allows the experimental determination of v_m. Since the molecular volume of the adsorbate is in general known, one can readily compute the accessible surface area in square centimeters. This method enjoys a tremendous popularity because of its simplicity and reproducibility, and because of the very valuable information which it provides for many problems.

The BET equation has been, and still is, very useful in the study of reinforcing carbon blacks, catalysts, and chromatographic fillers. In all these cases it was found that the efficiency and activity of the finely divided materials (as they are actually used in large-scale operations) depends on a *combination* of the accessible surface area and the existence of certain active centers in that area. Depending on the preparation of the reinforcing filler or catalyst, the number and specific reactivity of these active centers can be varied within certain limits, and it is therefore of great importance to know the total accessible surface area of a given amount of the sample. The BET method relies on the *unspecific physical* adsorption of vapors of various chemical character and permits, therefore, the experimental determination of the *entire accessible* area without reference to the existence of active centers. Its results are then combined with the specific reactivity of the material and have in many cases permitted a successful separation of the two important conductivity factors: total surface area and concentration of active centers.

Conclusion

Although the designing and manufacturing of highly aromatized and eventually carbonized polymers is still in its infancy, there exist already a number of interesting and promising applications.

Wholly aromatic polyamides, polyesters, and polyhydrocarbons are spun into fibers which serve to make high temperature resistant cords and fabrics. The cords are used in tires which have exceptional strength and rigidity and for heavy aircraft and road building equipment; the fabrics are the only presently available materials for parachutes which have sufficient impact strength to be used for the landing of space capsules or other objects descending from orbit.

Aromatic polyamides and related substances are used as high temperature insulators in motors, dynamos, and electronic circuits serving cars, planes, space capsules, and computers at places where high thermal gradients exist and where mechanical strength is needed. Introduction of heavy metal ions in such structures has made it possible to produce materials with unusually high dielectric constants (up to 2000 in the range of megacycles), semiconductor characteristics, and even electronic conductivities in the range of aluminum and iron.

Carbonized ladder polymers in the form of sheets, rods, tubes, and plates never melt and have outstanding ablative properties; they are used and constantly improved as heat shields, protective inlets for rocket and jet engines, and all other applications where an extreme combination of infusibility, rigidity, and abrasion resistance is required. Naturally, intensely composite systems in which these materials are used together with high softening metals, metal carbides and nitrides, and with refractories such as BeO and ThO_2 that are difficult to fuse are also studied.

References

1. Kirk-Othmer, *Encyclopedia of Chemical Technology,* Vol. 3, Interscience, New York, 1950, pp. 72–76; also G. Suits, *Suits: Speaking of Research*, Wiley, New York, 1965, p. 77.
2. Kirk-Othmer, *Encyclopedia of Chemical Technology*, Vol. 3, Interscience, New York, 1950, pp. 84–87.
3. Compare for literature references: N. G. Gaylord and H. F. Mark, *Linear and Stereoregular Addition Polymers*, Interscience, New York, 1959, pp. 219, 220.
4. F. H. Winslow, W. O. Baker, N. R. Pape, and W. Matreyek, *J. Polymer Sci.*, **16**, 101 (1955).
5. For additional references compare best the recent pioneering papers of P. Kovacic; e.g., *J. Polymer Sci. A*, **2**, 1193 (1964); also a very informative but difficultly accessible article of F. Long et al., *Sampe*, **10**, A-1 (1966).
6. For additional references compare A. A. Berlin and H. F. Mark lectures presented at the IUPAC Meeting in Tokyo in Sept. 1965.
7. For a very comprehensive survey and for additional literature references compare: A. H. Frazer, *High Temperature Resistant Fibers* (*J. Polymer Sci. C*, Vol. 19), Interscience, New York, 1967.
8. The original articles of I. Langmuir were mainly concerned with monolayers; e.g., I. Langmuir, *J. Am. Chem. Soc.*, **38**, 2221 (1916); the article of Brunauer, Emmett, and Teller appeared in *J. Am. Chem. Soc.*, **60**, 309 (1938).

Ultra-High Neutron Fluxes: Their Production and Use

FREDERIC DE HOFFMANN and WILLIAM L. WHITTEMORE

Gulf General Atomic, Inc., San Diego, California

1. Introduction

In trying to think of the subject of a survey article in fields where matter is under unusual conditions, the field of ultra-high neutron fluxes seems particularly appropriate in connection with Edward Teller's interests. To begin with, high neutron fluxes give rise to interesting effects not only in one field but in many fields: in neutron physics itself, in nuclear chemistry, and even, as we shall show, in fundamental particle physics. Remembering Edward Teller's great contributions in so many fields of science and technology, this therefore seems an appropriate topic. Furthermore, these may be particularly appropriate because high neutron fluxes are achieved in two fields: namely, the field of nuclear explosives with which Teller has been so intimately connected, and interestingly enough, the field of reactor safety, an outgrowth of which has permitted safe high-flux reactor pulses—and again, Edward Teller was the first chairman of the U.S. Reactor Safeguards Committee. Our article then will deal with the second of these fields, namely, with the production of high neutron fluxes from nuclear reactors whose inherent safety permits the attainment of such high fluxes.

In the summer of 1956 Edward Teller participated at General Atomic[1] in a session that, to begin with, did not concern itself with the production of ultra-high neutron fluxes but merely with the search for a research reactor of a low level, say 10 kW, but with inherent safety. What resulted from this session was the TRIGA reactor, a novel idea originated by T. B. Taylor, A. W. McReynolds, and Freeman Dyson (see, for instance, Freeman Dyson's paper)[2], using uranium zirconium hydride as its special new fuel to achieve a large prompt negative temperature coefficient. Before long it was recognized that the special safety feature of the TRIGA could in fact lead to the attainment of ultra-high pulses in a safe way, and as of this writing, a special version of this uranium zirconium hydride reactor has been built that is shortly expected to attain 20 million kilowatts at peak power with peak neutron fluxes of the order of 10^{17} cm^{-2} sec^{-1}. The safe attainment of large neutron pulses can be exploited in a number of ways, including, for example, the possibility of pulsing a uranium zirconium hydride reactor many times per second ($\sim$50) to produce from a small-fuel-inventory core pulsed thermal neutron fluxes as high as 10^{15} cm^{-2} sec^{-1} for research purposes.[3]

We shall divide this paper into the following sections:
Description of U-ZrH pulsing reactors
Transient behavior of U-ZrH reactors
Physical and chemical experiments with large neutron pulses

2. Description of U-ZrH Pulsing Reactors

2.1 General

The pulse characteristics of U-ZrH reactors have been gradually improved during the past ten years, progressing from the TRIGA Mark I,[4] through the TRIGA Mark II, III, and F cores[5] to the present TRIGA-ACPR core (Annular Core Pulsed Reactor). The TRIGA Mark I reactors generally were not at first designed for pulsing and operated on UZrH$_{1.0}$ fuel clad with aluminum. Subsequently, the stainless steel clad U-ZrH$_{1.65}$ fuel elements were developed for higher temperature operation. The principal operating temperature limit of the fuel material itself is set by the hydrogen dissociation pressure. The specification of the ZrH$_{1.65}$ fuel for a typical TRIGA Mark F is given in Table I. The ZrH$_{1.65}$ is essentially a

TABLE I. Triga Mark F Fuel-Moderator Element Specifications

Overall length	28.37 in.
Outside diameter	1.47 in.
Fuel outside diameter	1.43 in.
Fuel length	15 in.
Fuel composition	U-ZrH$_{1.65}$
Weight of U^{235}	36.6 g (avg)
Uranium-235 content	8.5 wt %
Uranium-235 enrichment	20%
Cladding material	304 stainless steel
Cladding thickness	20 mils
Graphite end reflector length[a]	3.4 in.
Maximum fuel temperature	1000°C

[a] Graphite slugs at each end of the fuel moderator material act as top and bottom reflectors.

single delta-phase material that has been operated at calculated temperatures exceeding 1000°C in tests with reactivity insertions of $5.00. Peak power levels as high as 8400 MW have been achieved in the Mark F reactor.

2.2. The ACPR Reactor

The ACPR uses slightly different TRIGA fuel elements than those in the previous TRIGA reactors. Tables II and III show the contrasting physical characteristics of the ACPR core and the expected performance. A physical view of such a facility is shown in Fig. 1. In its simplest version the ACPR reactor is located in an open, below-ground, water-filled tank. The core is arranged in the form of an annulus, as its name implies, so that there is a 9-in.-diameter cylindrical void for experiments. Thus, experiments can be placed to take advantage of the maximum fluxes.

TABLE II. ACPR Physical Data and Estimated Parameters

Fuel compositions	U-$ZrH_{1.625}$
Uranium-235 content	12 wt %
Uranium-235 enrichment	20%
Operational loading ($k = 1.08$)	169 fuel elements, 6 fueled followers (9.36 kg U^{235})
Active core diameter	28 in.
Diameter of dry, central void	9 in.
Effective delayed neutron fraction β_{eff}	0.0073
Prompt neutron lifetime l	30 μsec
Prompt temperature coefficient α (average between 23° and 700°C)	$\sim -9 \times 10^{-5}\, \delta k/\delta t$
Maximum fuel temperature	1000°C

TABLE III. ACPR Performance Data

Pulsing Characteristics	
Total reactivity insertion	$4.60 (3.36% $\Delta k/k$)
Peak power	20,000 MW
Energy release	100 MW-sec
Minimum reactor period	1.14 msec
Pulse width	4.00 msec
Performance	
Maximum fast flux ($E > 10$ keV)	$2.0 \times 10^{17}\, n\, cm^{-2}\, sec^{-1}$
Maximum integrated fast flux ($E > 10$ keV)	$1.1 \times 10^{15}\, n\, cm^{-2}\, sec^{-1}$

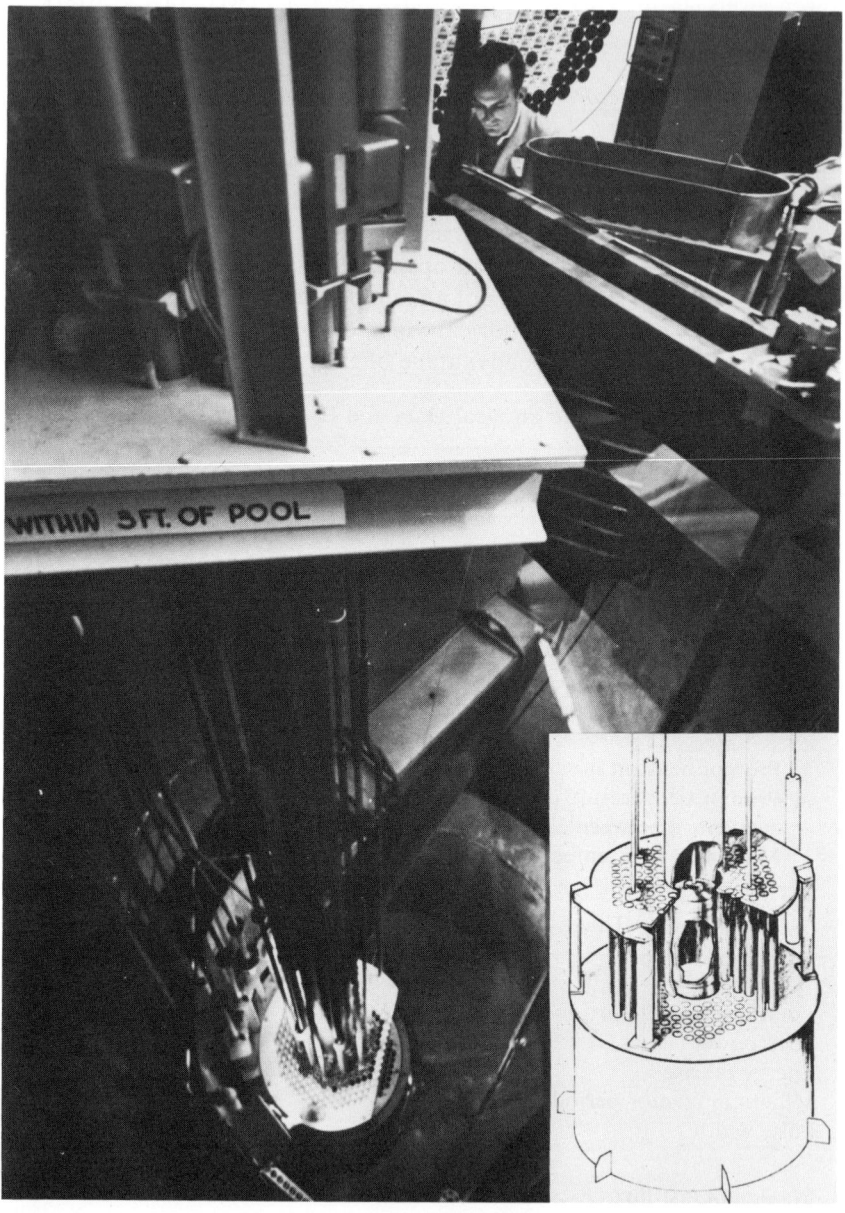

Fig. 1. TRIGA-ACPR, annular core pulse reactor.

3. Transient Behavior of U-ZrH Reactors

3.1. Fundamentals

We shall first describe the transient behavior in a TRIGA reactor by making certain simplifying assumptions in order to bring out the essential physics of the situation; later on in this section we describe the more refined calculations and checks made with experimental data.

We first turn quite generally to the physical behavior of a 20% enriched U-ZrH* TRIGA consisting of rods of approximately 3.75-cm diameter, spaced with their centers 4.25 cm apart and cooled as well as partially moderated by light water. The neutrons are created in the U-ZrH, are slowed down by both the hydrogen in the fuel and the hydrogen in the water, and are lost by a number of processes as follows:

 (a) Fast neutrons above 100 keV tend to leak out of the core
 (b) Intermediate energy neutrons in the 1 eV to 1 keV region are lost by resonance capture in U^{238} and in zirconium
 (c) Thermal neutrons are lost by one of two processes, absorption or leakage. In particular,

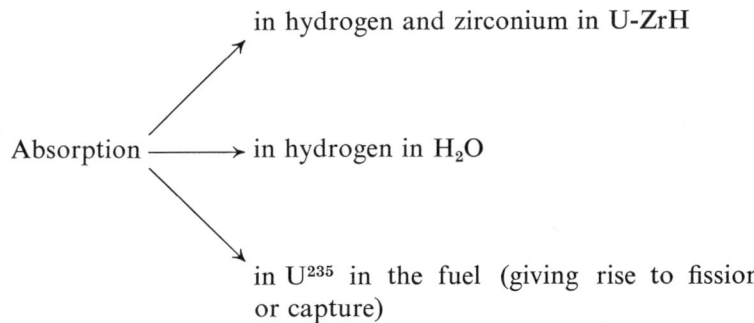

Actually, with respect to thermal leakage it is worth noting that there is a net inleakage into the core because there are more neutrons that are thermalized in the reflector surrounding the core and returned to the core than the thermal neutrons that leak out of the core into the reflector.

All these effects combine to give the reactor an integral temperature coefficient α_t (i.e., the rate of change of reactivity with a change in the fuel temperature), where

$$\alpha_t = \frac{\Delta k_{\text{eff}}/k_{\text{eff}}}{\Delta T} \tag{1}$$

---

* For ease of notation, when "U-ZrH" is used it denotes a typical TRIGA fuel element hydride like $U\text{-}ZrH_{1.65}$ used in a TRIGA Mark F reactor.

and k_{eff}, the effective multiplication of a reactor, is defined as

$$k_{\text{eff}} = \frac{\text{average number of neutrons produced}}{(\text{avg. number of neutrons absorbed}) + (\text{avg. number of neutrons escaping})} \quad (2)$$

In turn, α_t consists of two parts,

$$\alpha_t = \alpha_p + \alpha_s \quad (3)$$

where α_p denotes the prompt temperature coefficient and α_s the slow one. Since the slow temperature coefficient in the TRIGA reactor is due to heating of the water, and since heat transfer between the metal rods and water is characterized by times of the order of, say, seconds, whereas our TRIGA pulses have half-width of the order of milliseconds, we are concerned only with the *prompt* temperature coefficient. For simplicity of notation, hereafter we shall denote the prompt temperature coefficient simply by α.

The important characteristic of our U-ZrH reactor which enables it to pulse safely is its very large prompt negative temperature coefficient. In typical TRIGA reactors it is of the order of $10^{-4}/°C$ in the operating temperature range compared with typical UO_2 lattices which would have an α in the range 5 to 10 times smaller.

3.2. The Components of the Prompt Temperature Coefficient

There are three components of α, typically of about the following magnitude:

Regular Doppler resonance effect	$\sim 2 \times 10^{-5}/°C$
Changes in thermalization in ZrH	$\sim 3 \times 10^{-5}/°C$
The Dyson effect	$\sim 5 \times 10^{-5}/°C$

It is the latter two effects which need discussion.

Thermalization in ZrH

ZrH is a substance in which the zirconium is a tetrahedral crystal with the hydrogen placed in the centers of these tetrahedrons[6] so that in first approximation the hydrogen acts like a particle in a simple potential well, i.e., an isotropic harmonic oscillator. The lowest excited state is in fact at an energy $h\nu = 0.14$ eV, as can clearly be seen from the total cross section of ZrH as a function of neutron energy given in Fig. 2. Higher excited states also can be seen in this figure. Double differential scattering measurements[7] on zirconium hydride confirm the existence and uniform energy spacing of the excited states. The neutrons are scattered to energies *below* 0.14 eV due to the fact that they interact with hydrogen in the core water. The neutrons, however, can also be scattered upward to a higher level by the ZrH, and the likelihood of this occurring is increased as the temperature of the ZrH increases since the population of the levels goes as $e^{-h\nu/kT}$. This part of the temperature coefficient thus comes about not because there is *less* slowing down as the temperature rises but, rather

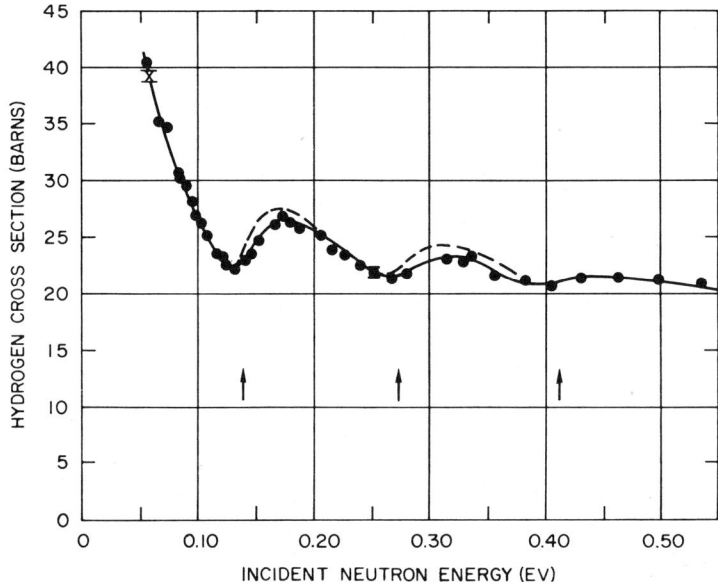

Fig. 2. The solid curve gives the total neutron cross section of hydrogen in $ZrH_{1.5}$ obtained from total cross section by subtracting the constant cross section of zirconium. The Fermi theoretical treatment corrected for Doppler broadening has been fitted to the above data and agrees with the experimental curve except as shown by the dashed curve.

more precisely, because there is increased scattering upward to higher levels. Because the fission cross section decreases somewhat more rapidly with increasing neutron energy than that of the $1/v$ absorbers, an increase in neutron energy due to higher moderator temperature results in a negative contribution from the moderator.

The Dyson Effect

As the fuel temperature rises we have seen that the neutrons *gain* energy from the ZrH and thus more neutrons escape from the fuel into the water. It turns out that each of the neutrons, in first approximation, has nearly the same subsequent history after entering the water. Because the spacing between fuel rods on the average is of the order of the scattering mean free path in water, it is very likely that neutrons entering the water make at least one collision. Since this first collision in H_2O leads to a very large energy loss, neutrons entering the water are rapidly thermalized.

The effect then is that more neutrons are thermalized when the fuel temperature rises, and of these thermal neutrons a fixed fraction is lost by absorption. Hence, more neutrons are ultimately lost by absorption in H_2O at higher ZrH temperature—this large contribution to the negative temperature was first suggested by Dyson.[2] Its calculation is rather complicated because the scattering mean free path in water is energy-dependent and because we have quite the opposite situation from an infinite medium, i.e., geometric changes within the distance of a mean free path. However, modern calculating machines and neutron transport theory are able to calculate the Dyson effect quite well quantitatively.

3.3. The Calculation of the Pulse Characteristics

The Simple Fuchs-Nordheim Model

A good description of the Fuchs-Nordheim model is given by Scalettar.[8] This model assumes (a) that the heat capacity of the U-ZrH does not vary with temperature, (b) that the temperature coefficient α is constant over the temperature range considered, (c) that the effect of the delayed neutrons can be neglected (which is a good approximation if the pulse is fast rising, say for reactivity insertion greater than $2.00), and (d) that energy and space dependence of the neutrons is neglected in this first approximation. Furthermore, it is of course reasonable to neglect the heat transfer to the water during the duration of the pulse, inasmuch as in typical pulses this heat transfer represents only a very small part of the energy generated in the fuel elements during the pulse.

Under these assumptions we shall now follow Scalettar[8] in deriving the peak power and width of the pulse. The prompt excess reactivity is defined by

$$\Delta k_p = k_{\text{eff}} - (1 + \beta_{\text{eff}}) \qquad (4)$$

where β_{eff} is the effective fraction of delayed neutrons.[9,10] If for simplicity's sake we write the prompt excess reactivity from now on merely as Δk and denote the power by P, then

$$\frac{dP}{dt} = \left(\frac{\Delta k - \alpha T}{l}\right) P \qquad (5)$$

where l is the appropriate mean prompt neutron lifetime.* On the other hand, by definition and since the heat capacity S is assumed constant,

$$P = \frac{d(ST)}{dt} = \frac{S\,dT}{dt} \tag{6}$$

or

$$\frac{dT}{dt} = \frac{P}{S} \tag{7}$$

Note that (5) and (7) are a nonlinear set of equations that are best solved by introducing the integral of P; i.e., the energy E

$$dE = P\,dt \tag{8}$$

With this substitution, (5) becomes

$$\frac{dP}{dE} = \frac{\Delta k - \alpha T}{l} \tag{9}$$

Now, with S constant at any given time, use of (6) gives

$$T = \frac{1}{S}\int P\,dt = \frac{E}{S} \tag{10}$$

therefore,

$$\frac{dP}{dE} = \frac{\Delta k}{l} - \frac{\alpha E}{lS} \tag{11}$$

or

$$P - P_0 = \left(\frac{\Delta k}{l}\right)E - \frac{\alpha}{2lS}E^2 \tag{12}$$

Now we can proceed to calculate the following properties of the pulse: $P_{\max}$, the peak power; T_∞, the increase in average core temperature during the pulse; E_∞, the total energy released in the pulse; and, finally, the width of the pulse at $\tfrac{1}{2}P_{\max}$, termed the pulse width.

Peak Power

Maximum power conditions are reached when $dP/dt = 0$, at which point, from Eq. (5), $\Delta k/l = \alpha T/l$, so that

$$T(P_{\max}) = \Delta k/\alpha \tag{13}$$

* The meaning of l can easily be seen if we merely recall the definition $l =$ (total neutrons produced promptly per second)/(number lost per second). It follows that, if we call the number of prompt neutrons produced A, and the number lost per unit time B, then

$$\frac{dN}{dt} = \left(\frac{A-B}{N}\right)N = \frac{(A/B - 1)N}{N/B} = \left(\frac{k_p - 1}{l}\right)N$$

and from Eq. (10)
$$E(P_{max}) = ST(P_{max}) = S\,\Delta k/\alpha \tag{14}$$

From Eq. (12) the equivalent P_{max} is

$$P_{max} - P_0 = \frac{S}{2\alpha l}(\Delta k)^2 \tag{15}$$

For further discussion, it is useful to define the quantity

$$(\Delta k')^2 = \Delta k^2 + \frac{2\alpha l}{S}P_0 \tag{16}$$

In practice $\Delta k' \approx \Delta k$ since the second term of (16) is small, or if $P_0 = 0$ actually is zero. Using Eq. (16) we may eliminate P_0 in (15) and obtain the general result that

$$P_{max} = \frac{S(\Delta k')^2}{2\alpha l} \tag{17}$$

The Values E_∞ and T_∞

The pulse will have run its course when the power has decreased again, i.e., when $P \simeq 0$. Under this condition, and using (16), Eq. (12) becomes a simple quadratic equation in E_∞ that yields

$$E_\infty = \frac{S}{\alpha}(\Delta k + \Delta k') \tag{18}$$

and T_∞ is determined by

$$T_\infty = E_\infty/S \tag{19}$$

Pulse Width

To determine the pulse width, we need to solve Eqs. (5) and (7) for P; it has been shown by Scalettar[8] that in fact

$$P(t) = P_{max}\,\text{sech}^2\left(\frac{t - t_{max}}{2\tau}\right) \tag{20}$$

where t_{max} is the time at which the reactor power is at a maximum and τ represents the asymptotic reactor period

$$\tau = l/\Delta k' \tag{21}$$

To obtain the full pulse width at half maximum power, we set $P(t) = \tfrac{1}{2}P_{max}$ and solve for $(t - t_{max})$. This gives

$$\text{Pulse width} = 3.52\tau \tag{22}$$

Comparison of Simple Model with Experiment

To illustrate that even the simple model gives remarkably good agreement with experiment, we give some calculated and experimental values for a TRIGA Mark F core consisting of stainless steel clad U-ZrH$_{1.65}$ fuel elements. For this case, the heat capacity of the total core is

$$S = 112 \text{ kW-sec/}°\text{C} \tag{23}$$

The computed temperature coefficient is

$$\alpha = 1.1 \times 10^{-4}/°\text{C} \tag{24}$$

A good fit to the various experimental power curves is obtained with a value

$$l = 39 \text{ μsec} \tag{25}$$

as determined experimentally from the shape of the early part of the power curves that follow $e^{-(\Delta k/l)t}$ before the rising temperature affects the results.

TABLE IV. Comparisons between Simple Theory, Machine Calculations, and Experiments, for Varying Reactivity Insertions

Reactivity Insertion		P_{max} (MW)			Pulse width (msec)		
		Experiment	Theory		Experiment	Theory	
Dollars	Δk_p		Simple calc.	Machine calc.		Simple calc.	Machine calc.
3	0.014	2,000	2,500	2,100	11.1	9.9	10.3
4	0.021	4,800	5,700	4,900	7.4	6.7	7.0
5	0.028	8,400	10,000	9,000	5.5	4.9	5.2

		E_∞ (MW-sec)			T_∞ (°C)		
		Experiment	Theory		Experiment	Theory	
			Simple calc.	Machine calc.		Simple calc.	Machine calc.
3	0.014	24	28.5	25	225	255	236
4	0.021	39	42.8	40	340	382	349
5	0.028	54	57.0	57	460	510	462

Table IV shows the comparison of theory and experiment for different reactivity insertions from a very low starting power in each case. It can be seen from the comparison with simple theory that agreement is better for the total energy release and average temperature increase than for the peak power. This is to be expected since the total energy release is an overall integral quantity related to the pulse, and therefore the fine structure of the theory is not likely to affect this quantity as greatly as the exact magnitude of the pulse. Corrections for the variation in heat capacity with change in temperature tend to bring larger changes to the peak power than to the temperature change and energy release. One must turn to more exact calculations[11] to bring theory and experiment into agreement. This agreement is also illustrated in Table IV.

3.4. Machine Calculations of Pulse

The machine calculations that have been performed in the last few years at Gulf General Atomic take a great number of effects into account and are much more complete than the simple Fuchs-Nordheim description discussed above. Among other things these calculations treat the variation of the fuel heat capacity with temperature, as well as a variable temperature coefficient, rather than a constant temperature coefficient during the pulse; they also include delayed-neutron effects (which make about a 1% difference in the results) and an allowance for heat transfer. Also, the results of various sophisticated methods used to measure the thermalization properties of both the ZrH and the H_2O are used in a set of computing codes—FLANGE[12] and GASKET[13]—to generate the scattering kernels that properly describe the interactions of the neutrons with the chemically bound hydrogen. Gulf General Atomic has developed cross-section averaging and space-dependent reactor codes that have been used in recent calculational work on the temperature coefficient; they include GAM-3,[14] GATHER-3,[14] GAZE-2,[15] and GAMBLE-5,[16] as well as DTF-IV,[17] an S_n multigroup transport code written at Los Alamos.

Use of the data from the above codes in machine calculations does indeed give excellent checks with experimental results. In Fig. 3 we show a typical observed pulse for a TRIGA Mark F whose core consists of the elements described in Table I (Section 2). A discussion of the theoretical and experimental results for the TRIGA Mark I cores[4] and the stainless steel clad, higher hydride cores has been published.[11]

4. Physical and Chemical Experiments with Large Neutron Pulses

4.1. Nuclear Physics and Chemistry Experiments

Large pulses of neutrons with high fluxes such as produced in a TRIGA reactor are especially useful in studying those effects in radiation chemistry where the end product is the result of double processes. These are of course proportional to the square of the neutron flux. Maier-Leibnitz[18] has pointed out that the large neutron fluxes from a TRIGA

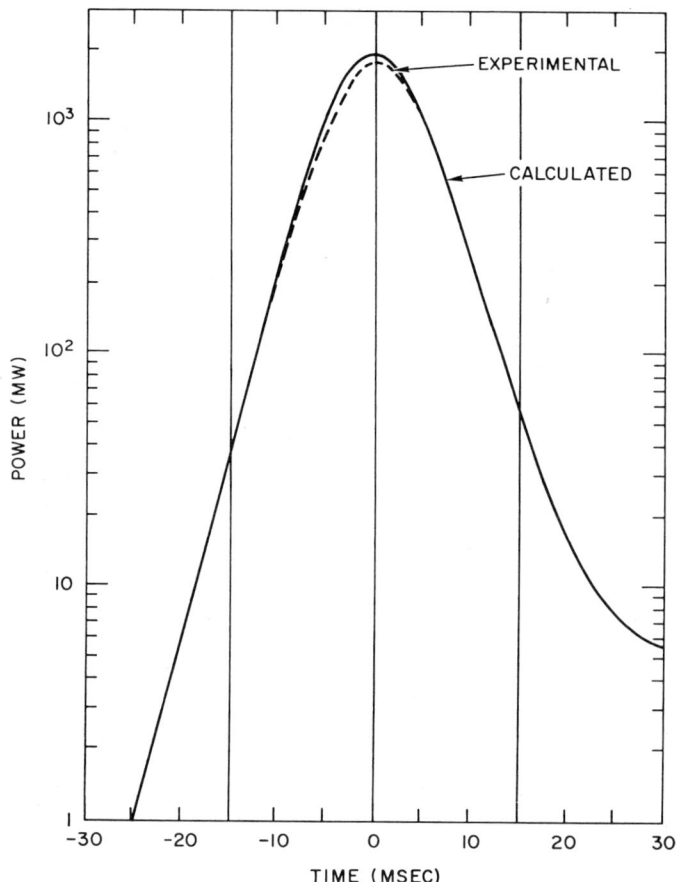

Fig. 3. Power versus time for TRIGA Mark F, $3.00 pulse.

reactor can offer significant improvement compared with a steady-state reactor for many cases involving double events. In his development, the ratio of decay rates for the second isotope $A_{\text{pulse}}/A_{\text{cont}}$ is given as

$$\frac{A_{\text{pulse}}}{A_{\text{cont}}} = \frac{\varphi^2 \Delta^2}{\phi \tau_1 \tau_2} \tag{26}$$

where φ and Δ are the peak flux in the pulse and the pulse width, respectively, ϕ is the steady-state flux, and τ_1, τ_2 are the mean life of the first and second isotopes, respectively, providing that τ_1 is smaller than the pulse width Δ. For a typical in-core irradiation facility using a TRIGA Mark F reactor[5] capable of producing a pulsed peak flux of $\sim 10^{17}\ nv$, a pulse width Δ of 7×10^{-3} sec, and a steady-state flux ϕ of $\sim 2 \times 10^{13}\ nv$, the pulsed experiment is superior whenever $(\tau_1 \tau_2)^{\frac{1}{2}} < 35$ sec, and if $\tau_1 < 7 \times 10^{-3}$ sec. As higher fluxes are produced from the pulsed reactor, the advantage of the pulsed experiment will increase so that the half-lives investigated can be even longer, that is, $(\tau_1 \tau_2)^{\frac{1}{2}}$ can be greater than indicated above.

Maier-Leibnitz and Springer[19] have indicated a large number of beam experiments where neutron optical devices provide a decided experimental advantage. Some of these experiments, such as investigation of (n, γ) radiation and total cross section measurements for low energy to allow determination of low absorption cross sections, may be further improved by the use of large pulses to produce intense neutron bursts. Although the TRIGA Mark F pulse is relatively long for the usual time-of-flight experiments on thermal neutrons, the use of reflecting beam pipes (tube) makes possible the use of long flight paths (≥ 100 m) without significant loss of intensity in performing time-of-flight analyses. For such a case, Maier-Leibnitz has shown that a single pulse will produce more than 10^8 neutrons with $\lambda \geq 4$ Å using a tube with 10 cm^2 cross section. The time resolution of the neutrons drifting along the flight path would be better than 10%.

4.2. Particle Physics Experiments

The large void internal to the pulsed uranium-zirconium hydride reactor described in Section 2 provides a unique possibility for measuring in a direct manner the neutron–neutron scattering cross section, one of the few remaining cross sections fundamental to particle physics yet to be measured directly. A group* is already developing a detailed plan to perform this measurement using a variation of the above-described uranium-zirconium hydride reactor. A measurement of this cross section at thermal energies using the pulsed reactor appears feasible and is expected to yield an accuracy of about 4%, which would correspond to an accuracy of about 2% in the scattering amplitude. The latter quantity is of fundamental importance for comparison with p–p and n–p scattering.

* This group consists of: C. D. Bowman and W. C. Dickinson, Lawrence Radiation Laboratory; C. O. Muehlhause, National Bureau of Standards; V. L. Sailor, Brookhaven National Laboratory; and W. L. Whittemore, Gulf General Atomic. Several informal and as yet unpublished documents generated by this group treat in greater detail the material presented here. The theoretical treatment of the experiment here presented closely parallels some of that in these documents.

Of the three possible nucleon–nucleon interactions at low energies, n–n is the simplest since the interaction takes place wholly in the singlet state. Furthermore, in n–n interactions no long-range Coulomb scattering exists to complicate measurements, although additional electromagnetic effects need to be considered. The direct observation of n–n scattering, however, has been made difficult heretofore by the availability of only low-density neutron gases. The use of pulsed reactors for this experiment should significantly improve the experimental situation.

In the low-energy limit, the potential between two nucleons is customarily characterized by its Fermi scattering length a, so that the total cross section at zero energy is given by

$$\sigma = 4\pi a^2 \qquad (27)$$

We first turn our attention to the neutron–neutron interaction a_{nn}.

One is of course interested both in the value of a_{nn} and also, equally significantly, in the question of whether there is in fact charge symmetry. By charge symmetry we mean that the nuclear part of the neutron–neutron interaction is the same as the nuclear part of the proton–proton interaction.

The best experimental data so far on the n–n amplitude[20] come from the $D(\pi^-, \gamma)2n$ reaction, which yields

$$a_{nn} = -16.47 \pm 1.27 \text{ F} \qquad (28)$$

$$r_0^{(nn)} = 2.65 \text{ F} \qquad (29)$$

Until a direct measurement is made, this is the value for a_{nn} that one has to adopt for the neutron–neutron interaction, and we shall do so for the purpose of the comparisons that follow. Šlaus[20] notes that to obtain the purely nuclear portion of the n–n interaction the value from Eq. (28) must be corrected for two effects, namely, (*1*) the interaction energy due to the two magnetic moments of the two point nucleons and (*2*) electromagnetic effects due to spreading out of the magnetic moment of the two nucleons.* Calculations by Riazuddin[21] and by Schneider and Thaler,[22] who use a hard-core model, indicate that corrections due to (*1*) and (*2*) are about -0.8 F, so that

$$a_{nn} \text{ (nuclear portion only)} = -17.2 \text{ F} \qquad (30)$$

We now turn to the data that can be obtained about the purely nuclear part of the p–p force from the measurement of proton–proton scattering. Rather accurate data exist for p–p scattering below 10 MeV; a recent analysis of these data is given by Heller.[23] He defines his phase

---

*In the case of p–p scattering one must also consider charge spreadout and vacuum polarization.

shifts $\delta_L{}^E$ to be phase shifts with respect to the entire electromagnetic interaction (i.e., coulomb plus other electromagnetic effects). His total phase shifts with respect to the coulomb functions are then $K_L = \delta_L{}^E + \tau_L$, where τ_L is the phase shift produced by the electromagnetic potential with respect to coulomb functions. His calculation is simplified further by neglecting any contribution to τ besides vacuum polarization. Heller thus obtains a set of $\delta_L{}^E$ vs. energy and in particular singlet S phase shifts $\delta_0{}^E$.

Heller then makes an effective range function fit through the relation

$$X(k^2) = \left(-\frac{1}{a_{pp}}\right) + \frac{1}{2} r_0^2 k^2 - P r_0^3 k^4 \qquad (31)$$

where $X(k^2)$ is the function corresponding to $k^2 \cot \delta$ in the simple effective range treatment, but with coulomb interaction and vacuum polarization included (see Refs. 23 and 24). He finds

$$a_{pp} = 7.817 \pm 0.007 \text{ F} \qquad (32)$$

$$r_0 = 2.820 \pm 0.018 \text{ F} \qquad (33)$$

As Heller points out, the phenomenological models for the nuclear interaction have generally fitted the constants obtained from high-energy measurements without the low-energy data being taken into account in detail except to have the scattering length and effective range approximately correct. Heller took such fit to the high-energy data, namely, the Hamada-Johnston[25] potential (which has a hard core), and from it found an S-phase shift $\delta_0{}^c$ with respect to coulomb functions, neglecting *any* of the other electromagnetic effects, i.e., even without vacuum polarization. He found good agreement between $\delta_0{}^c$ and the measured $\delta_0{}^E$.

Now, in turn, from the Hamada-Johnston potential one can compute the purely nuclear part of the neutron–neutron scattering length and find

$$a_{nn} \text{ (nuclear portion only)} = -17 \text{ F} \qquad (34)$$

Heller, Signell, and Yoder[26] have similarly taken other potential models and found a_{nn} (nuclear portion only) to be about -17 F. Thus, it seems that at least in first approximation charge symmetry is valid.

Using $a_{nn} \approx -17$ F, one finds from Eq. (27) that $\sigma_{nn} = 35$ b, a value that we shall use in our estimates for the n–n scattering experiments under consideration.

The value of a_{nn} is interesting not only for the question of charge symmetry but also for testing charge independence. If charge independence is valid, the purely nuclear portion of the n–p force would be equal to that of the n–n force. Currently, the n–p scattering length[27] is experimentally found to be more negative than the n–n length:

$$a_{np} - a_{pp,nn} \approx -7 \text{ F} \tag{35}$$

An accurate direct measurement of a_{nn} independent of the p–p measurements may be useful to tie down this difference fairly exactly. In turn then, at some time in the future good phenomenological potentials (perhaps from models with π, vector, and scalar mesons) may be developed such that when all electromagnetic effects are properly taken into account one could theoretically account for part or all of this difference. Therefore, one could then decide whether or not charge independence holds.

Several possible experiments have been proposed in the past to measure the neutron–neutron scattering cross section. Muehlhouse[28] has shown that the flux required to perform this measurement in a large internal void in a steady-state reactor is at least $10^{15}\ n\ \text{cm}^{-2}\ \text{sec}^{-1}$.

The use of a pulsed reactor for measuring the neutron–neutron scattering cross section has been proposed.[29,30] A pulsed uranium-zirconium hydride reactor is currently being evaluated in detail.[31] This pulsed system will be discussed below.

A method utilizing a nuclear explosive device to produce simultaneously two crossed beams of fission energy has been proposed[32] wherein it is claimed that a_{nn} would be determined to within 3%. The experiment is quite difficult and, since it has a one-shot nature, all of the experimental uncertainties must be fully anticipated. For example, even though the two crossed beams originate from a single detonation, the timing of their interaction and the evaluation of energies as a function of time for neutrons that are interacting are difficult. The experiment would attempt to measure $\sigma_{nn}(E)$ as a function of energy for larger energies and would utilize effective range theory to extrapolate to zero energy to obtain a value comparable to that attained for thermal neutron energies.

The proposed method for direct measurement of the thermal neutron–neutron scattering cross section using a pulsed reactor is essentially an experiment to determine the collision rate in an ideal neutron gas of known density. A schematic arrangement is shown in Fig. 4. The detector is placed beyond the virtual apex of the cone formed by the tapered tube so that it cannot view the walls of the tube. Thus, neutrons emanating from these walls should not be detected. Moreover, at sufficiently low gas pressures in the pipe, few neutrons will be detected from gas scattering. Finally, few neutrons should be detected from the far end of the tube. However, neutrons will reach the detector as a result of self-scattering in the void, i.e., from n–n scattering.

The neutron flux in the cavity comprises an ideal gas within which the collision rate can be calculated. If the neutron density n has a Maxwellian

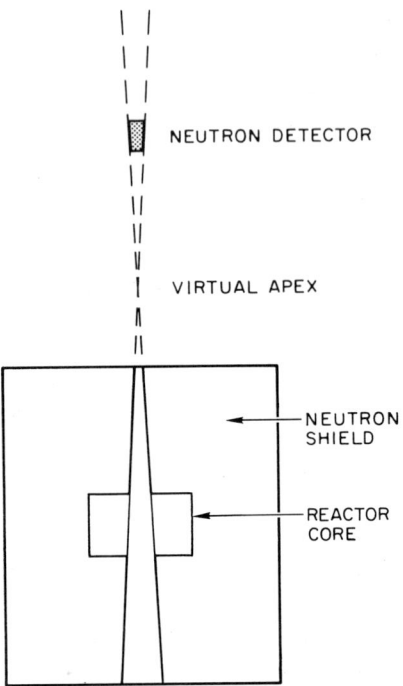

Fig. 4. Vertical cross section of neutron–neutron scattering experiment showing annular reactor core, neutron throughpipe, and detector location.

velocity distribution characterized by a most probable speed v_0, the collision rate per unit volume is given by

$$N = (2/\pi)^{\frac{1}{2}} n v_0 \sigma_{nn} = (\pi/8)^{\frac{1}{2}} (\varphi^2 \sigma_{nn}/v_0) \tag{36}$$

where σ_{nn} is the scattering cross section and $\varphi = n\bar{v}$ is the flux. Since there are two scattered neutrons per collision, the counting rate of the detector is given by

$$C = 2\bar{N} V (\Omega/4\pi) \varepsilon \tag{37}$$

where $\bar{N}$ is the collision rate averaged over the cavity volume V, $\Omega/4\pi$ is the effective solid angle subtended by the detector, and ε is the counting efficiency of the detector.

The actual counting rate at the detector is significantly affected by the effective solid angle of view and the volume of the high flux region within the beam pipe. Each of these last two factors varies as the square of the diameter of the pipe in the region of high flux. The *n–n* signal at the detector thus varies as the fourth power of the through-hole diameter as well as by the square of the flux. It is therefore evident that not only is a high peak flux required but also as large a throughpipe diameter as possible. The fact that the true signal varies as the square of the flux or reactor power should aid in its identification, since most backgrounds are either constant or vary linearly with power.

A study has been performed[33] to adapt the nuclear design of the pulsed reactor described in Section 2 to the requirements of the neutron–neutron scattering experiment. Primarily, this required that the thermal flux be optimized, whereas the earlier reactor required optimization of the flux above 10^4 eV. The results of this study have shown that thermal fluxes of about 10^{17} nv can be obtained with a pulse width (at half maximum power) of about 0.005 sec. The same study has indicated that a further increase by a factor of 2 in the resulting detected neutrons may be obtained by using specially designed U-ZrH fuel to decrease the pulse width.

The proposed uranium-zirconium hydride reactor with a large central hole will provide quite suitable experimental counting rates. The intensity of scattered neutrons arriving at the detector is estimated to be about 100 per pulse with suitably small background. This and other related parameters are listed in Table V. A relatively small number of pulses will thus be required to provide an accuracy of 2% in the desired neutron–neutron scattering amplitude.

TABLE V. Modified ACPR Conditions

φ_{max}	0.89×10^{17} cm^{-2} sec^{-1}
v_0	2.40×10^5 cm sec^{-1}
Half-width	5 msec
Diameter of central void	9 in.
Detected neutrons per pulse	100

One concludes that the development of pulsed uranium-zirconium hydride reactors makes possible a large class of experiments that otherwise could not be performed with steady-state reactors. The proposed neutron–neutron scattering experiment is an excellent example of use of ultra-high neutron fluxes.

Acknowledgment

The authors would like to express gratitude to G. B. West for assistance in making comparisons between theoretical and experimental results for pulsed U-ZrH systems and for helpful comments on this chapter.

References

1. E. Teller, "Safe Reactor Concepts—I," General Atomic Summer Symposium Lecture No. 1, June 18, 1956.
2. F. Dyson, "A Design for a Safe Reactor," General Atomic International Symposium on Reactor Problems, Lecture No. 3, August 27, 1956.
3. W. L. Whittemore and G. B. West, "A Multiple Pulsed TRIGA-Type Reactor for Neutron Beam Research," in *Seminar on Intense Neutron Sources, Sante Fe, New Mexico, September 19–23, 1966*, (CONF 660925), U.S. At. Energy Comm. Div. Tech. Inf. Services, 1967, p. 413.
4. R. S. Stone, H. P. Sleeper, Jr., R. H. Stahl, and G. B. West, "Transient Behavior of TRIGA, a Zirconium Hydride, Water-Moderated Reactor," *Nucl. Sci. Eng.*, **6**, 255 (1959).
5. "TRIGA, Experimental and Irradiation Facilities for Research and Development," General Dynamics, General Atomic Division Rept. GA-1695 (Rev. 5), November 1, 1966.
6. R. E. Rundle, C. G. Shull, and E. O. Wollan, "The Crystal Structure of Thorium and Zirconium Dihydrides by X-ray and Neutron Diffraction," *Acta Cryst.*, **5**, 22 (1952); R. E. Rundle, A. S. Wilson, R. Nottorf, and R. F. Raeuchle, "The Crystal Structures of ThH_2 and ZrH_2," U.S. At. Energy Comm. Rept. AECD-2120, n. d., decl. July 19, 1948.
7. W. L. Whittemore, "The Nature of Hydrogen Motion in ZrH Determined from an Experimental Neutron Study of Large, Bound Energy Levels," in *Inelastic Scattering of Neutrons in Solids and Liquids, Proceedings of a Symposium held in Bombay, 15–19 December 1964*, Vol. II, International Atomic Energy Agency, Vienna, 1965, pp. 305–315.
8. R. Scalettar, "Space Dependent Corrections to the Fuchs-Nordheim Model," in *Symposium on Reactor Kinetics and Control, Proceedings of the Symposium held at Tucson, Arizona, 25–27 March 1963*, U.S. At. Energy Comm. Rept. TID-7662, 1963, pp. 253–264.
9. F. de Hoffmann, "Criticality of the Water Boiler and Effective Number of Delayed Neutrons," U.S. At. Energy Comm. Rept. AECD-3051, Los Alamos Scientific Laboratory, Dec. 8, 1944.
10. A. M. Weinberg and E. P. Wigner, *The Physical Theory of Neutron Chain Reactors*, University of Chicago Press, Chicago, 1958.
11. G. B. West, W. L. Whittemore, J. R. Shoptaugh, Jr., J. B. Dee, and C. O. Coffer, "Kinetic Behavior of TRIGA Reactors," General Dynamics, General Atomic Division Rept. GA-7882, Mar. 31, 1967.
12. J. R. Beyster et al., "Integral Neutron Thermalization, Annual Summary Report, Oct. 1, 1964, through Sept. 30, 1965," U.S. At. Energy Comm. Rept. GA-6824, General Dynamics, General Atomic Division, Nov. 1, 1965.
13. J. U. Koppel, J. R. Triplett, and Y. D. Naliboff, "GASKET, A Unified Code for Thermal Neutron Scattering," U.S. At. Energy Comm. Rept. GA-7417, General Dynamics, General Atomic Division, Sept. 22, 1966; and GA-7417 (Rev.), Mar. 10, 1967.
14. J. Adir, S. S. Clark, R. Froehlich, and L. J. Todt, "Users' and Programmers' Manual for the GGC-3 Multigroup Cross Section Code," Parts 1 and 2, General Dynamics, General Atomic Division Rept. GA-7157, July 25, 1967.
15. S. R. Lenihan, "GAZE-2, A One-Dimensional, Multigroup, Neutron Diffusion Theory Code for the IBM-7090," General Dynamics, General Atomic Division Internal Rept. GA-3152, Aug. 3, 1962.

16. J. P. Dorsey, "GAMBLE-5, A Program for the Solutions of the Multigroup Neutron-Diffusion Equations in Two Dimensions," U.S. At. Energy Comm. Rept. GA-8188, General Dynamics, General Atomic Division, Dec. 4, 1967.
17. K. D. Lathrop, "DTF-IV—A Fortran-IV Program for Solving the Multigroup Transport Equation with Anisotropic Scattering," U.S. At. Energy Comm. Rept. LA-3373, Los Alamos Scientific Laboratory, July 15, 1965.
18. H. Maier-Leibnitz, "Uses of the Pulsed TRIGA Reactor," General Dynamics, General Atomic Division Rept. GA-3075, April 5, 1962.
19. H. Maier-Leibnitz and T. Springer, "The Use of Neutron Optical Devices on Beam-Hole Experiments," *J. Nucl. Energy, Pt. A/B*, **17**, 217 (1963).
20. I. Šlaus, "Neutron-Neutron Interaction," *Rev. Mod. Phys.*, **39**, 575 (1967).
21. Riazuddin, "Charge Dependent Effects on Scattering Lengths of np and pp Systems," and "On the Difference of π^0-Nucleon and $\pi^\pm$-Nucleon Coupling Constants," *Nucl. Phys.*, **7**, 217, 223 (1958)
22. R. E. Schneider and R. M. Thaler, "Effect of Electromagnetic Corrections to Low-Energy Nucleon–Nucleon Scattering on Charge Independence," *Phys. Rev.*, **137**, B874 (1965).
23. L. Heller, "Interaction of Two Nucleons at Low Energies; $I = 1$," *Rev. Mod. Phys.*, **39**, 584 (1967).
24. L. Heller, "Proton–Proton Effective-Range Theory with Vacuum Polarization," *Phys. Rev.*, **120**, 627 (1960).
25. T. Hamada, and I. D. Johnston, "A Potential Model Representation of Two-Nucleon Data Below 315 MeV," *Nucl. Phys.*, **34**, 382 (1962).
26. L. Heller, P. Signell, and N. R. Yoder, "Charge Symmetry, Charge Independence, and the Nucleon–Nucleon Scattering Lengths," *Phys. Rev. Letters*, **13**, 577 (1964).
27. R. Wilson, *The Nucleon–Nucleon Interaction*, Interscience, New York, 1963.
28. C. O. Muehlhause, "Proposed Experiment to Observe n–n Scattering," in *Proceedings of the International Conference on Nuclear Physics with Reactor Neutrons, held at Argonne National Laboratory, October 15–17, 1963*, U.S. At. Energy Comm. Rept. ANL-6797, pp. 21–26.
29. V. Sailor, see "Panel Discussion on Neutron and Fission Physics," in *Seminar on Intense Neutron Sources, Sante Fe, New Mexico, September 19–23, 1966*, (CONF 660925), U.S. At. Energy Comm. Div. Tech. Inf. Services, 1967, p. 810.
30. W. L. Whittemore, "Considerations of a Pulsed Neutron-Neutron Scattering Experiment with a Pulsed TRIGA," General Dynamics, General Atomic Division Internal Rept. GAMD-3485, Oct. 22, 1962.
31. C. D. Bowman, W. C. Dickinson, C. O. Muehlhause, V. L. Sailor, and W. L. Whittemore, unpublished documents.
32. W. C. Dickinson and C. D. Bowman, "Concept for an Energy Dependent Neutron–Neutron Scattering Experiment Using a Nuclear Explosion," U.S. At. Energy Comm. Rept. UCRL-70376, Lawrence Radiation Laboratory, Mar. 17, 1967; abstracted in *Rev. Mod. Phys.*, **39**, 592 (1967).
33. G. B. West, "Study of the Adaptation of the Annular Core Pulsing Reactor for a Neutron–Neutron Scattering Experiment," General Dynamics, General Atomic Division Rept. GA-8310, Oct. 1967.

The Physics of High Temperature Plasma

RICHARD F. POST

*Lawrence Radiation Laboratory, University of California,
Livermore, California*

I. Introduction

Though its roots go far back in the history of physics research, the study of that unique form of matter, high temperature plasma, began not more than about 15 years ago. In terms of rate of growth of knowledge and of evolving potentialities this research field is today one of the most rapidly developing of any in physics. Two factors have played major roles in this rapid evolution. The first of these, the one which sparked the new interest in the physics of high temperature plasma, is the challenging but as-yet unsolved problem of extracting power from controlled thermonuclear reactions. The second source of the reawakened interest in plasma physics comes from space research. Interplanetary space, rather than being empty and devoid of interest, has been found to be filled with high temperature plasma—its origin, the sun itself. In this huge volume of tenuous plasma surrounding the sun, fascinating and complex processes take place, many of which have a direct influence on terrestrial phenomena, such as the aurorae, magnetic storms, and the magnetically trapped plasmas of the earth's radiation belts. Indeed, though the basic laws governing the motion of individual charged particles have long been known, the behavior of collections of charged particles—i.e., plasmas—can be enormously complex. An analogy is the complicated behavior exhibited by biological systems, as contrasted to the relative simplicity of the individual atoms that go to make up such systems.

It is, of course, impossible to predict in detail what the eventual impact on science and technology of our growing understanding of the plasma state will be. But certainly this understanding will influence the field of space science in many ways, both technological and scientific. And on earth, though there may arise new and unsuspected applications of plasma technology, the one application of clearest and greatest significance is that already cited—controlled thermonuclear fusion reactions. The promise of this new source of nuclear energy—safe, potentially inexpensive, and virtually inexhaustible—is so great as to warrant many years of intensive research. Indeed, since its emergence from secrecy restrictions in 1958, fusion research has become a major international field of study, involving a world-wide level of support equivalent to about 100 million dollars per year. No one contends that fusion power will become an economic reality in the next decade, but solution of the *scientific* problems of controlled fusion, now seen to be intimately related to fundamental problems in the physics of high temperature plasmas, is probably only a relatively few years away.

A common thread runs through all aspects of high temperature plasma research, both in space and in the laboratory. This factor is the existence of electromagnetic, particularly magnetic, fields within the plasma. Furthermore, the complexity of plasma behavior arises almost entirely from the interplay between these fields and the plasma itself. In fusion research, the ubiquity of magnetic fields reflects the practical necessity for confining the hot plasma free from contact with material walls. As a matter of fact, achieving adequate plasma confinement is the central problem of fusion research. In space, magnetic fields surround the sun and many of the planets, and therefore all aspects of space plasma research are concerned with some facet of the interaction between plasmas and magnetic fields.

Up to this point, plasma, sometimes called "the fourth state of matter," has been defined merely as "a collection of charged particles." Such a definition is inadequate on at least two counts. First, in almost all circumstances of interest plasma is an electrically neutral entity, or nearly so. This is to say that in a plasma the total amount of positive charge, contributed by the ions, will be closely equal to the total amount of negative charge, contributed by the electrons. In fact, preserving the status quo in this respect is the strongest personality trait of plasma. This trait is, however, about the only stable one the beast possesses! With respect to the question of charge neutrality, we can understand this property from the fact that in other than very tenuous plasmas any tendency toward a local deviation from charge neutrality will result in the creation of electric fields within the plasma opposing the change.

The second point of amplification needed in a definition of plasma is a matter of size and scale lengths. Just as one will not discover biological phenomena in molecular systems whose overall dimensions are not very large compared to interatomic spacings, it will also be found that plasma effects are unimportant in charged particle systems the dimensions of which are too small. Here the unit of "smallness" is not the average spacing between the particles, but a longer length, the "Debye shielding distance." The Debye shielding length, or Debye radius is defined by the equation

$$\lambda_D = [\kappa T_e/4\pi n_e e^2]^{1/2} = 6.90[T_e/n_e]^{1/2} \qquad (1)$$

where T_e is the kinetic temperature of the electrons and n_e is their particle density. The length λ_D is the range beyond which the local coulomb electrostatic field of a given charged particle (a field of infinite range in a vacuum) will no longer be felt by other charged particles in the plasma. This screening effect arises because the local field will be canceled by compensating cooperative motions of plasma particles that lie in the intervening distance. Marking the onset of the most elementary dynamical plasma effect, cooperative shielding, the Debye radius can be taken as defining the smallest significant characteristic dimension in a high temperature plasma. Thus collections of charged particles whose overall dimensions are large compared to λ_D will exhibit plasma-like properties; those of appreciably smaller dimensions will behave merely as independently moving particles as in the usual regime of the particle accelerator or the ordinary electron tube. As encountered in natural plasmas, or in the laboratory, λ_D may vary by orders of magnitude, depending on the circumstances. To illustrate the range encountered, λ_D is plotted in Fig. 1 as a function of T_e and n_e. As seen from the figure, in interplanetary space, for example, λ_D may be as large as 10^5 cm, even so still microscopic compared to the relevant astronomical distances. In laboratory plasmas λ_D may lie

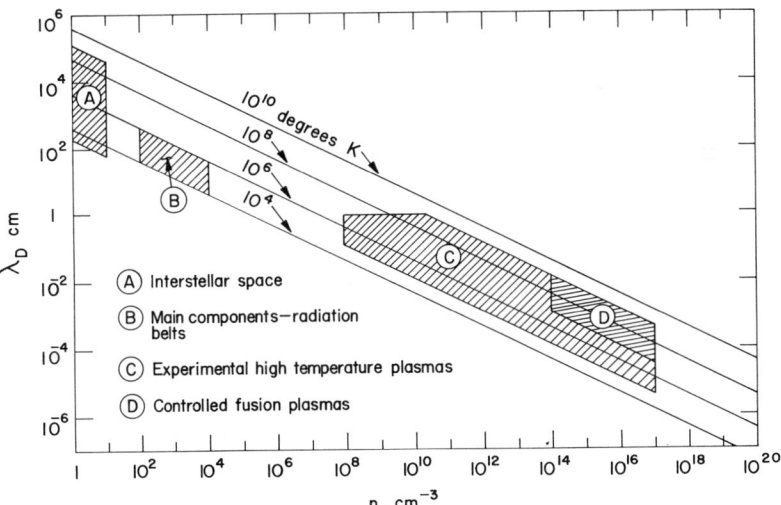

Fig. 1. The Debye length, λ_D, as a function of electron density, for several different values of the electron temperature.

between a centimeter or so for low density plasmas, down to as little as 10^{-5} cm in dense plasmas.

Whatever the circumstance, λ_D defines a lower size limit for the "fine-grainedness" of all cooperative phenomena in a plasma, i.e., for waves, shocks, or fluid-like motions. In the context of space research this circumstance implies that here the detailed structure of the plasma phemomena observed, though on a scale small compared to interplanetary dimensions, will almost always be on a scale large compared to any space probe. Quite the opposite is true for laboratory plasma experiments, except those at low densities.

What should one attempt to cover in writing today concerning "the physics of high temperature plasma"? Only a few years ago it would have been possible with modest effort to write a concise, clear (and in many ways, incorrect) discussion of the topic. Today one cannot get off so easily. The amount of work, especially theoretical, that has been accomplished in this field is really prodigious, and one cannot do justice to the field without attempting to explain the impact of this theory, and its relationship to experimental observation, that defines our picture of the plasma state. In the discussion to follow we will attempt to give the reader a feeling for both the breadth and the coherency of this picture, the result of a decade's intense study. If the reader feels somewhat overwhelmed by the variety of topics that it appears must be discussed, he is not alone—the author has had the same problem!

Our approach to a rational description of modern plasma physics will first lie through the development of some of the fundamental equations describing the plasma state, followed by their application to the areas of greatest importance to the mastery of the subject. These areas are:

1. The interaction of plasma and magnetic field—and especially the problem of magnetic confinement of plasma.

2. The effects of collisional relaxation in a high temperature plasma.

3. The dynamics of particle motions in plasmas immersed in magnetic fields.

4. Collective wave motions in a high temperature plasma.

5. The variegated phenomena of plasma instabilities.

After an apparently unavoidably lengthy discussion of the above topics, we shall attempt to put the entire field in perspective by giving a brief resumé of the present status of experimental knowledge in plasma physics. Though its brevity will not properly credit the great volume of experimental work that stimulated and confirmed (or sometimes confounded) the theoretical work, our discussion will attempt to bring into focus the entire subject as it relates to the twin motivations for research into high temperature plasma phenomena—fusion power and space research.

Before launching into our discussion, one item of business needs to be settled: what do we mean by a *high temperature* plasma? We mean simply a plasma in which the kinetic temperature is high enough (or the density is low enough) to insure that randomizing collisions between the individual particles occur on a time scale long compared to all plasma-like effects. The importance of this restriction will become more apparent as we proceed.

II. The Two Faces of Plasma—Particle vs. Fluid

In seeking an understanding of an unfamiliar form of matter, two needs should be satisfied: first, one should have a descriptive "model" to guide the thinking, and second, one should have a set of applicable physical laws stated in concise mathematical form. In the case of high temperature plasma the second need was in principle satisfied before the first. That is to say, the basic equations describing the behavior of a high temperature plasma were known many years before attempts were made in the laboratory to study this behavior. Yet despite the prior availability of the equations, it was not until experimentation began that a meaningful picture of the plasma state started to emerge. Part of the reason for this circumstance is again the matter of complexity. Though concise and rigorous, the plasma equations in their pristine form are virtually useless. Even when simplified to the point of tractability, the resulting equations contain such a wealth of possible solutions that without experience to guide the intuition, one "could not see the forest for the trees."

The exact (and intractable) equations for plasma behavior can be written down directly from electromagnetic theory, combined with results from special relativity. Though we shall not attempt here to do this in detail, we shall sketch the procedure, since it illustrates a rigorous model of a plasma—namely, the "exact particle model." According to this model we are to look on plasma as a gas of individual charged particles in which the motion of each particle is the self-consistent resultant of the totality of all forces acting on it—i.e., electromagnetic or gravitational forces. Thus, given an initial state of the plasma, i.e., given the location and the instantaneous velocity of all of its particles at time zero, one could in principle determine the subsequent behavior of the plasma down to the minutest detail by solving the equations of motion for each of its particles simultaneously, using the self-consistent electromagnetic fields in which each particle moves. The fly in the ointment is of course the fact that the electromagnetic field at each particle is influenced by the position and the velocity of *all* the other particles—the "N-body problem" in its most virulent form!

According to the stated model we adopt a set of coupled equations of motion for the plasma particles, supposed N in number, so that

$$\left\{\begin{array}{l} \dfrac{d\mathbf{p}_1}{dt} = \mathbf{F}_1 \\ \dfrac{d\mathbf{p}_2}{dt} = \mathbf{F}_2 \\ \hrulefill \\ \dfrac{d\mathbf{p}_N}{dt} = \mathbf{F}_N \end{array}\right. \quad (2)$$

where

$$\mathbf{p}_j = \frac{m\dot{\mathbf{r}}_j}{[1 - \dot{\mathbf{r}}_j{}^2/c^2]^{1/2}}$$

and

$$\mathbf{F}_j = e_j\left[\mathbf{E}(\mathbf{r}_j) + \frac{\dot{\mathbf{r}}_j}{c} \times \mathbf{B}(\mathbf{r}_j)\right] + \mathbf{G}(\mathbf{r}_j) \quad (2a)$$

$\mathbf{E}_j$ and $\mathbf{B}_j$ are the electric and magnetic fields at the jth particle, consisting of the superposition of any externally applied fields plus the fields produced by all other particles of the plasma. The term $\mathbf{G}$ represents any additional force field, such as gravity (usually of negligible importance compared to the electromagnetic forces).

At this point we could complete the set of Eq. (2) by writing down expressions for the electromagnetic fields felt by each of the N particles, utilizing retarded potentials appropriate to the fields produced by the $N - 1$ remaining particles, and then adding to these all externally generated fields. We would in this way obtain a closed, self-consistent set of relativistically invariant equations that would completely describe the behavior of fully ionized high temperature plasma. In this sense the physics of high temperature plasmas contains no fundamental questions that cannot be answered by solution of these equations; in principle it is a determined subject.

Though the existence of a rigorous theoretical description settles a philosophical question in our search for understanding of the plasma state, it settles little else. In fact the task of modern plasma physics is to build a physical intuition and an operational understanding of the plasma state despite the refractory nature of the rigorous plasma equations. For the theoretists this task has meant finding valid approximations to the fundamental equations that are tractable analytically. For the experimentalists it has meant the twin task of discovering the important modes of behavior of plasma and then attempting to reconcile these with the best available theoretical model—a process in turn influencing the course of theory.

Though in general the use of approximate equations has been the lot of plasma researchers, there is one area where something approaching a direct solution to the exact plasma equations has been successfully attempted. In recent years digital computers have been developed to the point where their speed and memory capacity permit the simultaneous self-consistent calculation of the trajectories of some hundreds or even thousands of particles in interaction with each other. Though thus far only one- and two-dimensional problems have been tackled, it has nevertheless been found possible to reproduce many important features of plasma behavior, including unstable modes. Computer simulation of plasma is especially valuable in the investigation of nonlinear phenomena, an area where the analytical approach finds its most difficult tasks.

We turn now to a discussion of some of the approximate theoretical models that have been introduced to analyze and describe plasma behavior. As our discussion continues, the reader may have the feeling that he is being led deeper and deeper into a dark forest of complexity, only to be abandoned. Take heart—plasma behavior is only 90% as complicated as it seems at first!

A. Plasma as a Fluid—Hydromagnetics

Among the first attempts to simplify the plasma equations was the introduction of the hydromagnetic equations, associated with the "fluid model" of a plasma. According to this somewhat myopic view of a plasma, one pictures it as a structureless electrically conducting fluid, flowing, expanding, or compressing in response to electromotive or inertial forces acting on it. In plasmas where the collision mean free path is very short (i.e., at low temperatures or at very high densities), such a model would be a natural one, much as it is in a dense gas or an ordinary fluid, described by the equations of hydrodynamics. Here we must add the element of electrical conductivity, of course. In fact, some of the first experimental checks on the validity of the hydromagnetic equations were performed using liquid metals, such as mercury.

The basic equations of hydromagnetics were developed some years before intensive study of high temperature plasmas was undertaken, and it was therefore natural to attempt to apply these equations to the high temperature regime, despite the fact that in this regime the collision mean free path is typically very large compared to the plasma dimensions—a circumstance completely outside the assumptions made in the first postulation of the equations. Nevertheless, the equations of hydromagnetics were found to be remarkably useful and successful in treating such problems.

This success was later explained when subsequent theoretical developments showed that their applicability does indeed extend to the collisionless regime.

To introduce the hydromagnetic equations we shall here utilize a simplified derivation, based on elementary continuity and force balance conditions. Consider the plasma to be made up of two coexisting fluids, electronic and ionic, and then write down force balance equations for each fluid in a typical volume element of the plasma.

For the ions the equation is

$$-n_i M \dot{\mathbf{v}}_i + n_i e Z \mathbf{E} + n_i(eZ/c)\mathbf{v}_i \times \mathbf{B} - \nabla \cdot \mathbb{P}_i - n_i M \nabla \psi + \alpha(\mathbf{v}_e - \mathbf{v}_i) = 0 \tag{3}$$

Similarly, for the electrons one has

$$-n_e m \dot{\mathbf{v}}_e - n_e e \mathbf{E} - n_e(e/c)\mathbf{v}_e \times \mathbf{B} - \nabla \cdot \mathbb{P}_e - n_e m \nabla \psi - \alpha(\mathbf{v}_e - \mathbf{v}_i) = 0 \tag{4}$$

Here the first term represents the inertial force; the second the force exerted by the electric field; the third term that by the magnetic field; the fourth the divergence of the pressure tensor (giving momentum transfer by particle flux); the fifth the gravitational force; and the sixth, vanishing in the limit of infinite collisional mean free path, the dynamical friction resulting from net relative motion between electrons and ions (such as caused by directed currents in the plasma).

Note now that, except at very low densities, $(n_e - Zn_i) \ll n_e$ or n_i, the charge neutrality condition for the plasma. This condition does not imply that $n_e = Zn_i$ identically, but merely that in a well-developed plasma a vanishingly small fractional deviation from charge neutrality is capable of producing finite electric fields.

Using the charge neutrality assumption, we may now greatly simplify the equations. Adding Eqs. (3) and (4) we are led to a simple-appearing equation applying to the plasma as a whole;

$$-\rho \dot{\mathbf{v}} + \frac{1}{c}\mathbf{j} \times \mathbf{B} - \nabla \cdot \mathbb{P} - \rho \nabla \psi = 0 \tag{5}$$

where

$\rho = n_e m + n_i M = n_i(mZ + M)$ the mean mass density

$\mathbb{P} = \mathbb{P}_e + \mathbb{P}_i$ the total plasma pressure tensor

$\mathbf{v} = \dfrac{mZ\mathbf{v}_e + M\mathbf{v}_i}{mZ + M}$ the net mass velocity

$\phantom{\mathbf{v}} = n_i e Z(\mathbf{v}_i - \mathbf{v}_e)$ the electric current density

In Eq. (5) the gravitational term, $\rho \nabla \psi$, is usually small compared to the other terms, and may be neglected, leading to the useful equation (Newton's first law!);

$$\rho \dot{\mathbf{v}} = (1/c)\mathbf{j} \times \mathbf{B} - \nabla \cdot \mathbb{P} \qquad (6)$$

Inspection of this equation shows immediately why magnetic field is a ubiquitous ingredient of high temperature plasma research. In the absence of a magnetic field Eq. 6 implies that a finite plasma ($\nabla \cdot \mathbb{P} \neq 0$) will simply expand at a rapidly accelerating rate, as does any unrestrained gas. This expansion will continue until the plasma either encounters a restraining magnetic field or is dissipated by striking a material surface. At the particle speeds implied by "high temperature," the result will be that the plasma would not exist for more than a few microseconds, even in a large chamber.

On the other hand, a second look at Eq. (6) shows that if a sufficiently strong magnetic field is present the plasma may be confined in magnetostatic equilibrium ($\dot{\mathbf{v}} = 0$), provided the pressure balance condition

$$\nabla \cdot \mathbb{P} = (1/c)\mathbf{j} \times \mathbf{B} \qquad (7)$$

is satisfied throughout the plasma. Since electrical currents through the plasma imply energy dissipation (however slow the time scale), magnetic confinement is seen to be a nonequilibrium process. That is to say, the currents within the plasma required to permit pressure balance and confinement necessarily involve a flow of energy which must be supplied either by diffusion of magnetic field into the plasma or by a continuous influx of new particles.

Equation (6) contains five variables, ρ, $\mathbf{v}$, $\mathbf{j}$, $\mathbf{B}$, and $\mathbb{P}$. To complete the set of hydromagnetic equations it is necessary to correlate these variables through auxiliary equations. These equations are: (*1*) an equation of continuity for the fluid and an assumed equation of state and (*2*) Maxwell's equations to interrelate the fields and currents. However, since the full set of hydromagnetic equations have been extensively discussed in the literature[1,2] and since they have been largely supplanted by more sophisticated treatments in recent years, we shall not list the complete set here, but will instead treat some special problems for which a contracted set will be adequate.

Returning to Eq. (6) we will rewrite it in terms of a "natural" coordinate system, in which the direction of the coordinate axes is related to the local direction and curvature of the field lines, in the following way: The unit vector **b** lies in the direction of **B**; **n**, normal to the local flux surface, is directed toward the local center of curvature of the field line; and **t**, tangent to the local flux surface, is perpendicular to both **n** and **b**. In this coordinate system the pressure tensor takes on a diagonal form and may be written as

$$\mathbb{P} = p_\| \mathbf{bb} + p_\perp \mathbf{nn} + p_\perp \mathbf{tt} \tag{7}$$

corresponding to two equal perpendicular pressure components of magnitude, $p_\perp$ normal to the field lines and a parallel pressure component $p_\|$ along the field lines. By use of the unit dyadic $\mathbf{II}$, Eq. (7) may also be written as

$$\mathbb{P} = p_+ \mathbf{II} + p_- \mathbf{bb} \tag{8}$$

where

$$p_+ = p_\perp \quad \text{and} \quad p_- = p_\| - p_\perp$$

This representation shows that $\mathbb{P}$ can be thought of as the sum of an ordinary scalar pressure of magnitude $p_\perp$ plus a single nonscalar component of magnitude $p_- = (p_\| - p_\perp)$ parallel to the field lines.[2]

Noting that $\mathbf{b} = \mathbf{B}/|\mathbf{B}|$, $\nabla \cdot B = 0$, and that $\mathbf{j} \times \mathbf{B}$ is perpendicular to $\mathbf{B}$, we may write for the $\mathbf{b}$ (parallel) component of Eq. (6),

$$[\rho \dot{\mathbf{v}} + \nabla p_\| + (p_\perp - p_\|)\nabla B/B]_\| = 0 \tag{9}$$

We may similarly evaluate the $\mathbf{n}$ and $\mathbf{t}$ components, finding for the nonvanishing components

$$\{\rho \dot{\mathbf{v}} - (1/c)\mathbf{j} \times \mathbf{B} + \nabla p_+ + p_-[(\mathbf{b} \cdot \nabla)\mathbf{b}]\} = 0 \tag{10}$$

Now from Maxwell's equation (ignoring displacement currents, small at hydromagnetic velocities) we have

$$\nabla \times \mathbf{B} = (4\pi/c)\mathbf{j} \tag{11}$$

so that one obtains

$$\{\rho \dot{\mathbf{v}} + \nabla[p_\perp + B^2/8\pi] - [B^2/4\pi + (p_\perp - p_\|)](\mathbf{b} \cdot \nabla)\mathbf{b}\} = 0 \tag{12}$$

The vector $\mathbf{c} = (\mathbf{b} \cdot \nabla)\mathbf{b}$ lies in the $\mathbf{n}$ direction and is numerically equal to the local reciprocal radius of curvature of the lines. Thus $\mathbf{c} \cdot \mathbf{n} = 1/R$ and $\mathbf{c} \cdot \mathbf{t} = 0$.

The three equations for the parallel, perpendicular, and tangential components of (6) are therefore

$$\rho \dot{\mathbf{v}}_\| = -\{\nabla p_\| + (p_\perp - p_\|)\nabla B/B\}_\| \tag{13}$$

$$\rho \dot{\mathbf{v}}_\perp = -\{\nabla(p_\perp + B^2/8\pi) - [B^2/4\pi + (p_\perp - p_\|)](1/R)\}_\perp \tag{14}$$

$$\rho \dot{\mathbf{v}}_t = -\{\nabla p_\perp + B^2/8\pi\}_{\tan} \tag{15}$$

Though these equations do not form a complete set, they are nevertheless useful and informative. It can be seen that the presence of a magnetic field introduces an essential anisotropy into the situation. For example, consider an axially symmetric dipole magnetic field, such as the earth's magnetic field above the stratosphere. Suppose that a localized, isotropic ($p_\perp = p_\parallel = p > B^2/8\pi$) burst of plasma were created within this field. Equation (13) shows that expansion would immediately begin to occur along the field lines, toward the poles, unimpeded by the positive magnetic gradient (since $p_\perp = p_\parallel$ initially). Similarly, because of the symmetry of the field it can be seen from Eq. (15) that expansion will proceed freely tangential to the flux surfaces, the density and pressure tending toward a state of azimuthal symmetry. In the radial direction, however, Eq. (14) suggests that expansion will only occur to the point where the kinetic pressure, decreasing with the expansion, becomes comparable to the magnetic pressure terms. A prime example of the process just described was the audacious ARGUS experiment,[3] in which a plasma was injected into the earth's magnetosphere by exploding a small atomic bomb at high altitude. The end result of this experiment was the creation of a long-lived shell of confined plasma in magnetostatic equilibrium with the earth's field. But here we note, from Eq. (13) that in the ARGUS experiment the final state reached by the plasma, during its long decay, could not be one of isotropic pressure. The negative pressure gradient along the field lines, in either direction from the equator, necessary for the vanishing of $p = p_\perp + p_\parallel$ at the poles, can only be sustained if the pressure is anisotropic, with $p_\perp > p_\parallel$. Such a circumstance would be expected to arise naturally in this case, however, since it is the components of the plasma moving most nearly along the field lines that tend to escape most readily. Thus the pressure tensor of the remaining plasma cannot be isotropic.

B. Magnetostatic Equilibrium—Plasma Confinement

The example just given illustrates one facet of that most important subject in high temperature plasma physics, magnetic confinement. The condition for a confined state, in magnetostatic equilibrium, to exist is given simply by setting $\dot{v} = 0$ in Eqs. (13)–(15).

We rewrite these equations in their equilibrium form ($\dot{v} = 0$) as

$$\nabla p_\parallel = -(p_\perp - p_\parallel)\frac{\nabla B}{B} \qquad \text{parallel to } B \qquad (16)$$

$$\nabla\left(p_\perp + \frac{B^2}{8\pi}\right) = \left[\frac{B^2}{4\pi} + (p_\perp - p_\parallel)\right]\frac{1}{R} \qquad \begin{array}{l}\text{perpendicular to } \mathbf{B} \text{ and toward} \\ \text{the center of curvature}\end{array} \qquad (17)$$

$$\nabla\left(p_\perp + \frac{B^2}{8\pi}\right) = 0 \qquad \text{tangent to the flux surface} \qquad (18)$$

Inspection of these equations shows that it is possible to make an important separation between two topologically distinct classes of confining field, on the basis of the type of plasma that may be confined in each.

To be less cryptic and more specific, we can deduce from Eq. (16) that the topology of any magnetic field configuration capable of confining plasmas characterized by an isotropic, i.e., a scalar pressure (the state toward which collisions will drive any initial plasma state) must be toroidal, i.e., the topology of a doughnut. All field lines on which plasma is to be contained must remain within the plasma, as shown schematically

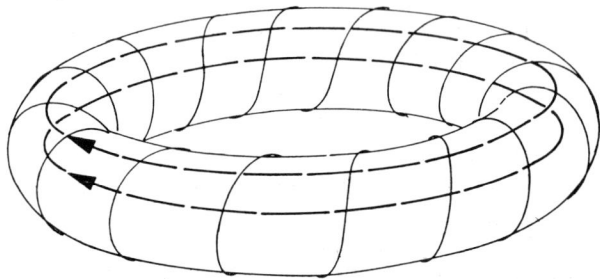

Fig. 2. Geometry of a simple axially symmetric toroidal field. Note that the field lines are circles closing on themselves inside the torus.

in Fig. 2. This result, of course, follows from the fact that when $p_\perp = p_\parallel = p$, ∇p must be zero along a field line. Among the class of experimental devices employing toroidal fields are devices such as the stellarator, and Astron, and the multipole, to all of which we shall later refer.

As is now well known, plasma can also be held confined by the "magnetic mirror" effect,[1,4] i.e., it may be held between two regions of stronger-than-average magnetic field which prevent it from escaping freely along the lines of force. Such systems are of "open-ended" topology—i.e., the topology of a cylinder, since field lines on which plasma is confined may leave the system. Four examples of open-ended fields are shown in Fig. 3. They include: (a) one we have already mentioned, a dipole field such as the earth's field above the atmosphere, (b) a simple axially symmetric mirror field, (c) a "quadrupole" mirror field, and (d) a cusp magnetic field. Fields (a), (b), and (c) all have nonzero minima along the lines between the mirrors, while in (d) the field is zero at the midplane for the central line. Field (c) is an example of a very important class of open-ended mirror configurations—"magnetic well" field configurations—that are highly advantageous with respect to a whole class of plasma instabilities.

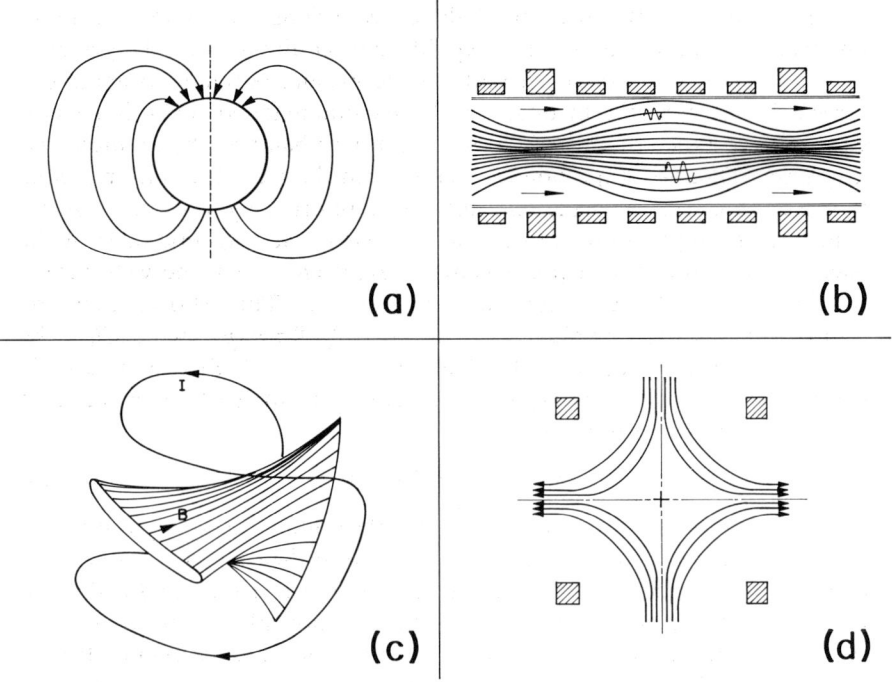

Fig. 3. Examples of "open-ended" confinement fields. (*a*) Dipole mirror field similar to Earth's magnetic field. (*b*) Axially symmetric magnetic mirror fields produced by circular coils. (*c*) Quadrupole magnetic well mirror fields, here shown as produced by a coil shaped like the seam on a baseball. (*d*) Axially symmetric cusp field. Produced by two circular coils carrying opposing currents.

It can be seen from Eq. (16) that open-ended mirror systems, as noted earlier, can only contain plasmas characterized by a nonscalar pressure, in particular ones in which $p_\perp > p_\parallel$. It follows from this fact that open-ended systems are necessarily more susceptible to losses caused by the randomizing effects of collisions and other stochastic processes than are the toroidal systems. This is because such randomizing processes tend to create an isotropic pressure distribution, and thus lead directly to losses through the mirrors. In a torus the pressure may become isotropic without implying losses; these require actual transport across the field. As we shall see, however, the advantage of mirror systems relative to toroidal ones lies in the fact that magnetostatic equilibria are much more readily achieved, as well as in the fact that the magnetic lines are not constrained to remain within the plasma. This latter circumstance permits much greater flexibility, for example in the choice of field configurations, such as the magnetic well, that enhance plasma stability.

Returning to the pressure balance equations, we note one more important limitation imposed by equilibrium requirements. The presence of the transverse electrical currents in the plasma required to achieve a force balance implies a distortion of the vacuum magnetic field, in general weakening it (since the kinetic pressure must be balanced by an increased magnetic gradient). Equation (17) shows that there exists a limiting value to the total kinetic pressure that may be supported by a given magnetic field. For example, in regions of the field where the magnetic lines of the vacuum field (no plasma present) are straight ($R \to \infty$), the well-known relation $(p_\perp + B^2/8\pi) = $ constant is satisfied. This shows that the limiting pressure confinable is that for which $B_0^2/8\pi = p_\perp$, where B_0 is the strength of the vacuum field. The ratio $p_\perp/(p_\perp + B^2/8\pi)$ is usually designated as β in discussing magnetic confinement, so that in the limit $R \to \infty$, (17) can be written as

$$\beta = (1 - B^2/B_0^2) \leq 1 \tag{19}$$

Plasma stability requirements often place more stringent limitations on β, sometimes requiring that $\beta \ll 1$. Nevertheless, the existence of equilibrium solutions with $\beta \to 1$ has been demonstrated theoretically, and experimental situations where this is the case have also been observed in the laboratory. To see the orders of magnitude of the numbers involved in such "high β" experiments we note that in some of these the vacuum magnetic field was in excess of 50,000 G. Thus $B_0^2/8\pi \approx 10^8$ dyn/cm$^2 \approx$ 100 atm—a very substantial pressure to have been supported by anything as insubstantial as a magnetic field!

1. Equilibria in Toroidal Systems

We have thus far discussed only the necessary conditions for magnetostatic equilibrium. The question of finding the sufficient conditions for the existence of such equilibria is surprisingly subtle, one that has not yet been satisfactorily settled in many instances. As in many other problems in plasma physics, in considering magnetostatic equilibria one's physical intuition is rarely a reliable guide, often leading one astray. We may illustrate this point by considering the problem of magnetostatic equilibrium in the most elementary toroidal confining field, i.e., one formed by wrapping a solenoidal coil of arbitrary cross section around in a circle to form a figure of revolution. From its simplicity and symmetry you might expect that such fields would be nearly ideal from the standpoint of confinement. Yet the hydromagnetic equations show that these fields cannot even provide magnetostatic equilibrium! This now classic problem in plasma physics illustrates the care with which the problem of magnetic confinement must be approached.

The proof proceeds as follows: write the pressure balance equation in the form

$$\nabla p = (1/c)\mathbf{j} \times \mathbf{B} \tag{20}$$

(Since $\nabla B = 0$ along the field lines, the use of a nonscalar pressure would not alter the end result.) Taking the z component of this equation (z is the axis of revolution of the torus, θ the azimuthal position, and r the distance from the z axis), we find

$$\frac{1}{c} j_r B_\theta - \frac{\partial p}{\partial z} = 0 \tag{21}$$

From Maxwell's equation $\nabla \times \mathbf{B} = (4\pi/c)\mathbf{j}$, taking the r component

$$\frac{4\pi}{c} j_r = -\frac{\partial B_\theta}{\partial z} \tag{22}$$

since $\partial B_z/\partial \theta = 0$. It follows that

$$\left\{ \frac{B_\theta}{4\pi} \frac{\partial B_\theta}{\partial z} + \frac{\partial p}{\partial z} \right\} = 0 \tag{23}$$

i.e.,

$$\frac{\partial}{\partial z}\left[p + \frac{B_\theta^2}{8\pi} \right] = 0 \tag{24}$$

so that the quantity $[p + B_\theta^2/8\pi]$ is a function of r only.

Similarly, from the r component of (20) and from the z component of $\nabla \times \mathbf{B}$, we find

$$\frac{\partial}{\partial r}\left[p + \frac{B_\theta^2}{8\pi} \right] + \frac{1}{4\pi} \frac{B_\theta^2}{r} = 0 \tag{25}$$

so that B_θ^2/r is a function of r only, since $[p + B_\theta^2/8\pi]$ has been shown to be a function of r only. (Remember that the quantity B_θ is not simply the vacuum field, but the magnetic field as it may be modified by the presence of the plasma.)

Combining the results from (24) and (25) it follows that p is itself a function of r only, and thus is independent of z. Thus we have show that no equilibrium solution bounded in z is possible, independent of the relative values of p and $B^2/8\pi$.

Thus toroidal equilibria are impossible unless the symmetry of the simple toroidal field is somehow destroyed. One way to accomplish this transiently is to induce a circulating electric current within the plasma. This current could be carried by the plasma itself or by a metallic conductor imbedded within the plasma. Both of these ideas have their difficulties, but both have been successfully employed in experiments.

Another method, the one employed in the so-called stellarator devices,[5] uses helical windings of a multipole nature (adjacent conductors carry oppositely directed currents). The basic idea here, one of the fundamental ones in magnetic confinement, is to use these extra windings to twist the field lines within the torus into nested helices, so that the resultant field possesses what is known as a "rotational transform." This simply means that, starting at any given reference cross section of the torus (see Fig. 4), as one fastens his attention on any given field line passing through

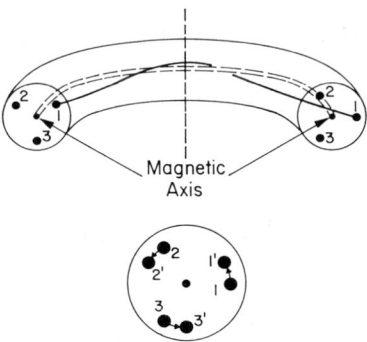

Fig. 4. Topology of the stellarator field. Cross section of torus showing the effect of rotational transform, which causes field lines to twist around magnetic axis. Lower section shows that field line bundles do not return to starting point after circling through torus.

this cross section and then follows this line as it moves around the torus he will find the line twisting helically within the torus so that in general it will not come back to its starting point on the reference surface. Instead, the lines will trace out a set of nested surfaces as they pass repeatedly around the torus. Only one line, the "magnetic axis," lying somewhere near the middle of the reference surface, will return on itself. In the first stellarators (ca. 1953) a rotational transform was achieved by actually twisting the torus into a figure 8 (a situation resembling a three-dimensional Möbius strip, as far as the field lines are concerned). It was not until somewhat later that it was shown that an equivalent result could be achieved by the use of the helical windings we have mentioned, wound on the outside of a simple toroid or on a racetrack-shaped chamber.

One way to understand the physical necessity for rotational transform in a stellarator is to recognize that in a simple torus symmetry requires that $\mathbf{j} \cdot \mathbf{B} = 0$ within the plasma. In this case the condition $\nabla \cdot \mathbf{j} = 0$ cannot be satisfied, and therefore the equation of continuity, $\nabla \cdot \mathbf{j} + \dot{\rho} = 0$, implies that charge accumulations will grow on the magnetic lines with no mechanism for their relief, thus preventing the attainment of a state of pressure equilibrium. Introduction of a rotational transform destroys the symmetry conditions, in effect providing a means for the charge accumulations to be relieved on a given magnetic surface by flow along the lines. It is in fact a necessary property of stellarator equilibria that they exhibit current flow along the field lines.

In the special case of axially symmetric tori with rotational transform (i.e., ones with axially symmetric internal floating conductors), it is possible to demonstrate theoretically the existence of equilibria. However, in the author's opinion, no satisfactorily rigorous proof of the existence of magnetostatic equilibria has as yet been given for the stellarator. In stellarators, the field topology is much more complicated than it is for axially symmetric systems, especially when one takes into account the perturbing effects of discrete, and never error-free, magnet coils and conductors used to create these fields. Although there is a general belief among stellarator researchers that adequate and accessible equilibrium states do exist, it seems fair to say that neither experiment nor theory have as yet completely dispelled the worries on this subject. Certainly in the context of cosmological or interplanetary plasmas it seems highly unlikely that cases of stellarator-like toroidal equilibria exist.

2. Equilibria in Mirror Systems

The situation is very different with respect to equilibria in open-ended configurations. Somewhat surprisingly, the seemingly innocent change from closed to open topology makes possible a wide class of plasma equilibria. The prime examples from nature are, of course, the radiation belts surrounding the earth and other celestial bodies. In the laboratory, not only were robust equilibria observed in the first mirror machine experiments, but their existence has been put on a firm theoretical basis. We may illustrate an important class of these equilibria with a simple example, applied to magnetic well geometry (Fig. 3). For such fields, and in the low β limit ($p_\perp \ll B^2/8\pi$), we need only show that the pressures $p_\perp$ and $p_\parallel$ may be expressed as functions of B alone. To show this we may rewrite the pressure balance equation in the form

$$B \frac{dp_\parallel}{dB} = -(p_\perp - p_\parallel) \tag{26}$$

Solutions satisfying this equation and the boundary conditions will represent bounded equilibrium solutions for plasma confined within closed constant-B contours of the magnetic well field. As a simple example, consider the solution below, verifiable directly by substitution in (26):[6]

$$p_\| = p_0 B(B_m - B)^n$$
$$p_\perp = np_0 B^2(B_m - B)^{n-1} \qquad B < B_m \qquad (27)$$

$$p_\| = p_\perp = 0 \qquad B > B_m \qquad (28)$$

Here B_m is the magnetic field at the plasma boundary. The exponent n may be chosen arbitrarily, and an arbitrary sum of such solutions may be taken.

Solutions of the type given above have been generalized in terms of invariants of the particle motions (discussed later) and have been extended to include a wide class of finite-β equilibria.[7] The present theoretical picture is therefore that it is possible to demonstrate rigorously the existence of equilibria for an extremely wide variety of mirror fields including both simple mirror and magnetic well geometry, and extending to ones for relatively large β values. Further, by contrast with toroidal equilibrium solutions, magnetostatic equilibria in mirror systems do not require the presence of currents along the lines. In fact, from the condition $\nabla \cdot \mathbf{j} = 0$ it can be shown that $\mathbf{j} \cdot \mathbf{B}$ must vanish if the plasma is bounded at the ends by vacuum.

C. Electric Fields; Diffusion and Ambipolar Potentials

We have thus far utilized the hydromagnetic equations in a form not containing the intrinsic electric fields associated with the confinement process itself. But nonvanishing plasma potentials and finite internal static electric fields are necessary for the existence of magnetostatic equilibria. The presence of such fields and potentials arises from the disparity between the mass, mobility, and magnetic rigidity of the electrons and the ions. While a detailed treatment of these effects cannot be accomplished with the fluid equations, we may estimate the magnitude of such fields simply by returning to the fundamental volume force equations, Eqs. (3) and (4). Multiplying the first by M and the second by $-mZ$, we obtain

$$\rho \left[\frac{M^2 \ddot{\mathbf{r}}_i + m^2 Z^2 \ddot{\mathbf{r}}_e}{(M + mZ)^2} \right] = n_e e \left[\mathbf{E} + \frac{1}{c}(\dot{\mathbf{r}} \times \mathbf{B}) \right] - \left[\frac{M \nabla \cdot \mathbb{P}_i - mz \nabla \cdot \mathbb{P}_e}{M + mZ} \right]$$
$$+ \rho \left[\frac{M^2 + m^2 Z^2}{(M + mz)^2} \right] \nabla \psi + \frac{\alpha}{n_e e} \mathbf{j} \qquad (29)$$

Since $M \gg m$ we may drop quantities of order m/M relative to the other terms. Let us also consider the case of magnetostatic equilibrium

($v_i = v_e = 0$) and drop the gravity term $\rho\nabla\psi$, finding

$$\left[\mathbf{E} + \frac{1}{c}(\mathbf{v}\times\mathbf{B})\right] - \frac{1}{n_e e}\nabla\cdot\mathbb{P}_i + \frac{\alpha}{n_e^2 e^2}\mathbf{j} = 0 \tag{30}$$

Let us now consider the various terms, starting from the last. This term represents the "Ohm's law" term arising because of the finite resistivity of the plasma, proportional to the dynamical resistance, α, between electrons and ions. Since α is a measure of the collisional interactions between the electrons and the ions, in a high temperature plasma where collision mean free paths are long, α becomes small so that resistivity may be neglected in lowest order.

Considering now the remaining terms, we may not drop any of them *a priori*. However, in the case where the plasma exists in a static state of equilibrium with no net mass motion ($\mathbf{v} = 0$) the equation reduces to

$$\mathbf{E} = \frac{1}{n_e e}\nabla\cdot\mathbb{P}_i \tag{31}$$

Thus in this particular (and important) case we see that gradients of the ion pressure must be supported entirely by electric fields within the plasma. The reason for this curious result is that here the electric current $\mathbf{j}$ is being carried essentially entirely by the electrons and it is the $\mathbf{j}\times\mathbf{B}$ body force that is balancing the total plasma pressure in equilibrium. Therefore, the role of the electric field in (31) is to provide a means for the electrons to hold back the ions, i.e., to take up the pressure forces arising from the ion density gradients. This electric field is a so-called "ambipolar" field, one arising because of the tendency for one species (here the ions) to separate from the other because of intrinsic differences in the forces exerted on the two species. Internal electric fields of ambipolar nature arise in virtually every case of plasma equilibrium. Their existence in steady state of course implies that there must also exist departures from exact charge neutrality within the plasma. These departures will, however, be small relative to the total charge of either species in all but very low density plasmas.

The first term in Eq. (30) deserves special mention. In those cases where pressure gradients and resistivity effects are small, we obtain the result

$$\mathbf{E} + (1/c)(\mathbf{v}\times\mathbf{B}) = 0 \tag{32}$$

This result allows us to find the polarization electric field within a plasma arising in the presence of a fluid velocity, $\mathbf{v}$, and a magnetic field. But the magnitude of this field, $\mathbf{E} = -(1/c)(\mathbf{v}\times\mathbf{B})$, is just that which must apply within a perfect conductor moving across a magnetic field, since in the frame of the conductor $\mathbf{E} = 0$. We see, therefore, that net mass motions or rotations of high temperature plasma in a magnetic field will always be accompanied by electric fields of the order of those given by Eq. (32), except as modified by strong pressure gradients.

Returning to the full equation (30) we may calculate a very important effect—the "classical" diffusion of plasma across a confining magnetic field. Physically, this diffusion arises from the dissipative effect of collisional drag between the electrons and the ions. The energy loss from this resistivity process must be provided by a compensating energy derived from the slow expansion of the plasma across the field. This means, as we have noted earlier, that magnetic confinement is necessarily a non-equilibrium process, here tending toward the thermodynamic equilibrium state of being uniformly distributed from wall to wall within the chamber. But what makes magnetic confinement attractive is that in high temperature plasmas the *rate* of approach to this final, unconfined, equilibrium state should in theory be exceedingly slow, as we shall see.

To attack the problem let us consider a column of confined plasma, and let us restrict ourselves to the case of scalar pressure, $p = p_\| + p_\perp$, $p_\| = p_\perp$. Consider first the force balance equation in terms of the electric current (see Eq. (7)).

$$\nabla p = (1/c)(\mathbf{j} \times \mathbf{B}) \tag{33}$$

We may find the component of **j** perpendicular to **E** by taking the cross product with **B**, finding

$$j_\perp = c(\mathbf{B} \times \nabla p)/B^2 \tag{34}$$

Also we find from (30), again taking the cross product with **B**:

$$\mathbf{v}_\perp = c\left[\frac{\mathbf{E} \times \mathbf{B}}{B^2}\right] + \mathbf{B} \times \left[\frac{1}{n_e e}\nabla p_i + \frac{\alpha}{n_e e^2}j\right]\frac{c}{B^2} \tag{35}$$

Substituting for **j** from (34) we find

$$\mathbf{v}_\perp = c\left[\frac{\mathbf{E} \times \mathbf{B}}{B^2}\right] + \frac{c}{n_e e}\left[\frac{\mathbf{B} \times \nabla p_i}{B^2}\right] + \frac{\alpha c}{n_e e^2}\left[\frac{\mathbf{B} \times (\mathbf{B} \times \nabla p)}{B^4}\right] \tag{36}$$

The first two terms of this equation represent circumferential drifts that arise in the plasma in order to satisfy the conditions of magnetostatic equilibrium. The first of the two, $c(\mathbf{E} \times \mathbf{B})/B^2$, is of special significance and will be discussed later. In steady state neither of the first two terms will give rise to transport across the field, as both are purely azimuthal terms. The last term, however, is directed radially, i.e., "downhill" on the density gradient, and corresponds to a slow diffusion across the magnetic field. This term depends only on the collisional drag between electrons and ions, since to the order considered here ion–ion collisions do not give rise to net transport across the field. The magnitude of this collisional diffusion velocity is, from Eq. (36),[1]

$$v_c = -(\alpha c/n_e e^2)(\nabla p/B^2) \tag{37}$$

To evaluate this diffusion drift in terms of plasma parameters requires an explicit determination of the dynamical friction term, α. This term is proportional to the transverse electrical resistivity of the plasma, $\eta_\perp$, essentially Ohm's law for the plasma as deduced from (30):

$$\eta_\perp = \alpha c/n_e e^2 \tag{38}$$

Spitzer[1] has evaluated $\eta_\perp$ for a high temperature plasma by performing the appropriate collision integrals. His result gives

$$\eta_\perp = 1.29 \times 10^{13} \frac{Z \ln \Lambda}{T_e^{3/2}} \text{ emu}$$

$$= 1.29 \times 10^4 \frac{Z \ln \Lambda}{T_e^{3/2}} \text{ ohm-cm} \tag{39}$$

Here T_e is the kinetic temperature of the electrons in degrees Kelvin. The term $\ln \Lambda$ is of order 20 for typical plasma regimes. It expresses the "multiple scattering" collisional effects arising when one takes into account the effects of the "distant" collisions in which a colliding electron is infinitesimally deflected as it passes within a distance equal to the Debye length of an ion. Thus the quantity Λ itself, a very large number, is simply equal to the ratio of λ_D to the minimum impact parameter.

Inserting the above value for $\eta_\perp$ in the expression for the diffusion velocity, one finds

$$v_c = -[1.29 \times 10^{13} Z \ln \Lambda](\nabla p/B^2 T_e^{3/2}) \tag{40}$$

Under the conditions usually encountered in high temperature plasma research this velocity is extremely slow. This result corresponds to the physical circumstance that, as shown by Eq. (39) the electrical resistivity of a hot plasma is very small, so that the time required for magnetic field to interpenetrate it is correspondingly long, as it would be with a very good metallic conductor upon which a magnetic field is impressed (the "skin effect").

As an example of the time scale for the plasma diffusion consider an electron-proton plasma in which $T_e = T_i = 10^{8}°K$ (about 10 KeV kinetic temperature*; i.e., about 15 keV mean particle energies). Take

* 1 keV kinetic temperature = $1.17 \times 10^{7}°K$.

the total particle density to be 10^{14} cm^{-3} and the magnetic field to be 10^4 G. For this case we find (with $\ln \Lambda = 20$)

$$v_c \approx 3.6 \, \nabla n/n \text{ cm/sec} \qquad (41)$$

Thus if the characteristic distance for the density gradient is 10 cm, $v_c \approx 0.36$ cm/sec, or less than 10^{-8} of the mean proton velocity! Furthermore, since v_c scales as $B^{-2}T_e^{-3/2}$ it would not be difficult to reduce it by another order of magnitude or two through modest increases in either B or T_e. Thus in theory the confining effect of a magnetic field, as far as cross-field diffusion is concerned, should be extremely good. Alas, in practice no documented case of a high temperature plasma behaving in accord with Eq. (40) has yet been established. In stellarators, for example, where it might be expected to apply, the observed rates of diffusion are some five orders of magnitude larger than the "classical" value given by Eq. (40). The exact cause or causes for this gross discrepancy have not yet been traced, but the chances are that the explanation lies either in difficulties with the basic plasma equilibrium state in the fields that have thus far been employed or in the effects attendant to low frequency plasma instabilities, or both. We will discuss these matters at greater length in connection with the dark problem of plasma instabilities.

Before leaving the subject of electric fields and diffusion effects within a plasma, we must discuss briefly another situation where electric potentials arise in plasma confinement. In the confinement of high temperature plasma in mirror systems, a net positive potential is usually found (as opposed to the net negative potential associated with classical diffusion in closed systems). The origin of this potential lies in the different rates of the collision processes as between electrons and ions—a subject that we will discuss at greater length in the next section. In the present context it is only necessary to recognize that the electrons, because of their greater mobility and their greater collision frequencies are *a priori* more likely to escape through the mirrors than the ions. But as we have seen, wherever disparate loss rates exist, a net charge (the ambipolar potential) will automatically result—of such a sign as to bring the rates to equality. Thus in mirror systems the ions will tend to hold in the electrons from loss along the field lines by virtue of the coupling through the ambipolar electrostatic fields.[8,10] Except in unusual circumstances, we can therefore assume that in mirror-confined plasmas it is the ions that are actually held in by the mirror effect; the electrons are held in by electrostatic forces, arising from the differences in the rates of collision processes between electrons and ions. Thus even in our "collisionless" high temperature plasma, one where all collision processes are assumed to occur on a slow time scale compared to other plasma processes, we cannot ignore the role played by collisions.

D. The Particle Model

1. The Boltzmann Equation and Collisions

Useful though the hydromagnetic equations are, they represent an unrealistic view of plasma and therefore give no information on many important effects. Indeed a plasma is not a fluid; in actuality our first model, N mutually interacting particles, is the only rigorous model. Caught between the intractability of the exact model and the inadequacies of the fluid model, plasma theorists have chosen to adapt the methods of kinetic theory to the plasma problem, with considerable success. In such models the particulate nature of the plasma is retained, but the particles are treated "on the average" by introducing the classical idea of the distribution function in phase space, $f(\mathbf{r}, \mathbf{v})$, where the quantity $n_0 f(\mathbf{r}, \mathbf{v}) d\mathbf{r}\, d\mathbf{v}$ is the average number of particles in the infinitesimal element of phase volume $d\mathbf{r}\, d\mathbf{v}$. Given the distribution functions f_j for the j different species of particles in the plasma (electrons and whatever types of ions are present), one can in principle calculate all the macroscopic quantities of interest—for example, the particle density, the mean square velocity, or the net directed velocity of any volume element of the plasma, given respectively by the integrals[11]

$$n_j = n_{0j} \int f_j\, d\mathbf{v} \qquad (42)$$

$$\bar{v}_j^{\,2} = \frac{n_{0j}}{n_j} \int f_j v^2\, d\mathbf{v} \qquad (43)$$

$$\mathbf{u} = \frac{n_{0j}}{n_j} \int f_j \mathbf{v}\, d\mathbf{v} \qquad (44)$$

Here n_{0j} is the average particle density of the jth species in the plasma.

Similarly, for the pressure tensor (measuring the momentum transfer)

$$\mathbb{P}_j = M_j n_{0j} \int (\mathbf{v} - \mathbf{u})(\mathbf{v} - \mathbf{u}) f_j\, d\mathbf{v} \qquad (45)$$

where $(\mathbf{v} - \mathbf{u})$ represents the "random" component of the particle velocities. One can therefore recover the hydromagnetic equations, in the appropriate limit, by simply taking moments of the Boltzmann equation, and relating these, through Maxwell's equations, to the field quantities.

To determine the values of f for any given situation one needs an equation relating the time variation of f to the dynamical forces operating within the plasma. Such an equation is the Boltzmann equation of kinetic theory.[12] This equation, simply an equation of continuity in phase space, can be represented symbolically for each species as

$$\frac{df_j}{dt} = \left(\frac{\delta f_j}{\delta t}\right)_{coll} \tag{46}$$

The left-hand side of the equation simply represents the total time derivative of f, i.e., its variation as computed along the trajectory in phase space traced out by the infinitesimal group of particles represented by f. In terms of phase space coordinates

$$\frac{df}{dt} = \frac{\partial f}{\partial t} + \mathbf{v} \cdot \nabla f + \left(\frac{\mathbf{F}}{M}\right) \cdot \nabla_v \tag{47}$$

where ∇_v denotes the gradient operator in velocity space coordinates and F is any force field in which the plasma moves.

The right-hand side of Eq. (46), representing the effects of interparticle collisions, must be evaluated by a detailed consideration of the particle kinetics involved. In a high temperature plasma, however, the time scale for collision effects is by definition much longer than that for dynamical processes such as waves, instabilities, or other high frequency phenomena. These are described by the left-hand side of the equation, containing the electrodynamic terms. Thus a natural separation is possible along the following lines: To treat dynamical effects one ignores the slow variation of f from collisional effects and therefore sets the right-hand side of the equation equal to zero to this order. The remaining equation, being a statement of continuity of phase volume, is merely Liouville's equation for the plasma. In this context it is usually called the "collisionless Boltzmann equation" or the "Vlasov equation," after a theorist who was one of the first to apply this equation to plasmas.

On the other hand, if one is primarily interested in collisional effects, and not in short time scale dynamical phenomena, then on the collisional time scale the dynamical terms can be taken to average to zero or to damp away (provided the plasma is stable), leading to the equation

$$\frac{\partial f_j}{\partial t} = \left(\frac{\delta f_j}{\delta t}\right)_{coll} \tag{48}$$

The detailed evaluation of the collision integrals required to solve this equation was first carried out by Rosenbluth and co-workers.[13] Their vehicle was the Fokker-Planck equation, an equation from classical kinetic theory giving the transport in velocity space caused by interparticle collisions.

The Fokker-Planck equation for a plasma takes to form of a complicated integrodifferential equation involving moments of the (evolving) distribution functions, together with derivatives in velocity space. Except in very simple cases it has not been solved analytically; in general it must be tackled by numerical methods, using digital computers. We shall not repeat the equation here but refer the reader to the literature,[11,13,14] and will here only describe the salient features of the results.

As might be expected, considering the nature of the random processes involved, the effect of collisions as predicted by the Fokker-Planck equation is to attempt to drive the plasma, whatever its initial velocity distribution, toward an isotropic Maxwellian velocity distribution. In many cases boundary conditions in velocity space, arising from magnetic confinement, for example, may prevent the attainment of a Maxwellian distribution. In this case the solutions will approach a "compromise" between the two competing effects—i.e., disordering by collisions, and ordering through selective processes arising from the boundary conditions. A case in point is mirror confinement. Here the particle dynamics of the situation do not allow the containment of particles with velocity vectors too nearly parallel to the magnetic field. This circumstance produces a "loss cone" in velocity space and introduces a persistent degree of anistropy into the Fokker-Planck solutions.[15,16] Furthermore, since the coulomb cross section increases inversely with the square of the relative particle energy in a collision, the low energy particles are therefore the most rapidly scattered into the loss cone. Thus not only are the Fokker-Planck solutions for mirror confinement anisotropic, but they must also vanish at zero particle velocity. This is to be contrasted with the Maxwellian, which has its maximum value at $v = 0$. Figures 5 and 6 illustrate the types of asymptotic solutions obtained in the case of mirror confinement, for various values of the "mirror ratio" R, i.e., the ratio of the field at the mirror to that at the minimum between the mirrors. The variables used in the plots are the scalar velocity v and $\mu = \cos \theta = v_{\parallel}/v$, i.e., the pitch angle of the particle orbits with respect to the field lines at the midplane. It can be seen that the solutions for large mirror ratio approach isotropy and a Maxwellian shape.

As an illustration of the manner in which an initially non-Maxwellian distribution will evolve toward a Maxwellian with the passage of time, Fig. 7 reproduces results obtained by Rosenbluth et al.,[13] showing the collisional relaxation of an electron plasma.

2. Collisional Relaxation of Test Particles

There exists a simpler and very useful method for estimating the time scale for collisional relaxation phenomena, without recourse to the

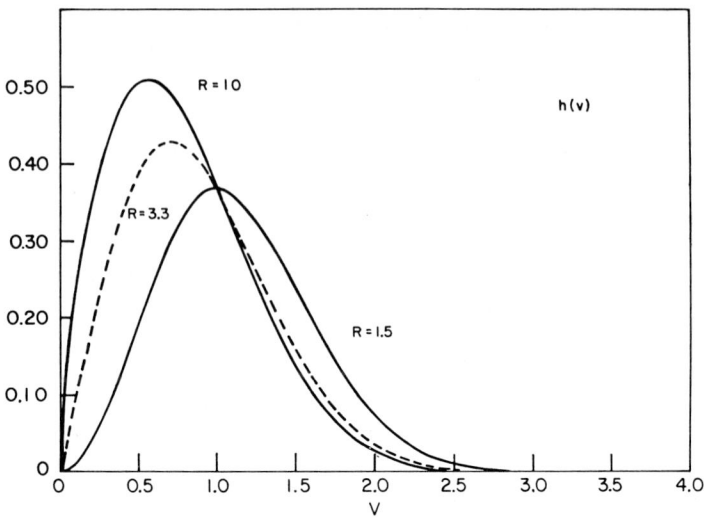

Fig. 5. Distribution function in particle speed, $h(v)$, arising from particle–particle collisions in a mirror-confined plasma.

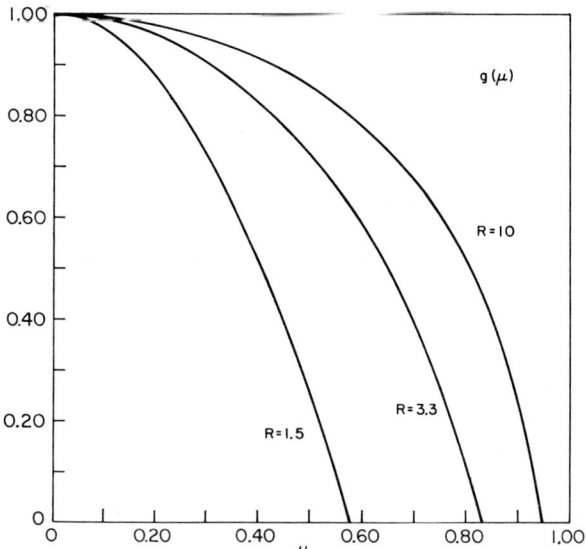

Fig. 6. Distribution function $g(\mu)$, $\mu = \cos \theta$, arising from particle–particle collisions in a mirror-confined plasma. Note effect of loss cone in cutting off $g(\mu)$ beyond the loss-cone angle $\mu_c = (1 - 1/R)^{1/2}$.

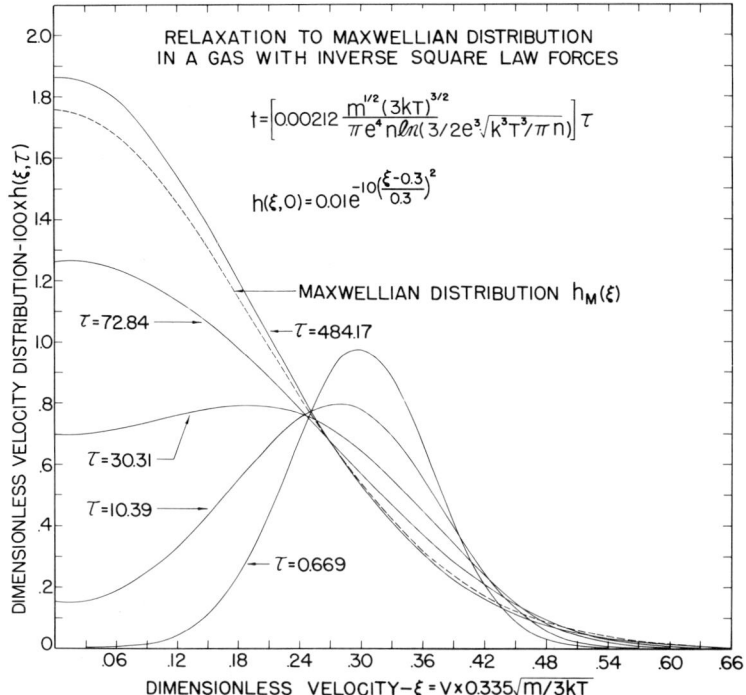

Fig. 7. Solutions of Fokker-Planck equation showing relaxation of peaked distribution function toward a Maxwellian distribution as a result of particle–particle collisions.

Fokker-Planck equation. One considers a "test particle" of a given energy and evaluates the collisional interaction integrals for it, while it is moving in, say, a Maxwellian plasma. Chandrasekhar[17] and Spitzer[1] have performed extensive calculations following up this idea.

One particularly useful result is obtained by evaluating the integrals giving the expectation value of the increase in kinetic energy transverse to the original direction of the test particle. It is clear that when this increase equals the original energy, the particle will have essentially "relaxed," i.e., forgotten its original motion as far as initial anisotropy is concerned. In evaluating this quantity we can make more meaningful the definition of "collisionless" or "high temperature" as applied to a plasma. Plasmas whose dimensions are small compared to the mean free path for relaxation would generally satisfy such a definition.

The quantity we desire may be derived from the expectation value of the square of the rate of increase perpendicular velocity, $v_\perp$, of a test particle of charge Z moving in a Maxwellian field of particles of charge

Z_1 and density n_1. Spitzer finds for this quantity the expression

$$\langle (\Delta v_\perp)^2 \rangle = \frac{8\pi e^4 n_1 Z_1{}^2 Z^2 \ln \Lambda}{M^2 v} F(x) \tag{49}$$

$F(x)$ is a function expressing the effect of the relative velocities of the test and field particles; $x = (v/\bar{v}_1)$, where $\frac{1}{2}M_1\bar{v}_1{}^2 = \kappa T_1$. In the limit $x \gg 1$, $F(x) \to 1$. At $x = 1$, $F(x) = 0.63$ and at $x = 0.1$, $F(x) = 0.075$.

If we determine the time at which $\langle (\Delta v_\perp)^2 \rangle$ becomes equal to v^2 we may calculate an expression for the "relaxation time" we desire, i.e., we set

$$\langle (\Delta v_\perp)^2 \rangle \cdot \tau_d = v^2 \tag{50}$$

The result can be expressed in terms of an effective coulomb scattering cross section σ_d, through

$$\tau_d = (n_1 \sigma_d v)^{-1} \tag{51}$$

where

$$\sigma_d = 2\pi Z^2 Z_1{}^2 e^4 (\ln \Lambda) F(x) / W^2 \text{ cm}^2$$

and

$$W = \tfrac{1}{2} M v^2$$

Note that σ_d is simply the classical Rutherford scattering cross section for large-angle scattering, multiplied by the terms $(\ln \Lambda) F(x)$. The increase, as noted before, arises from the inclusion of multiple small-angle scattering arising from the myriad "distant collisions" with particles over all collision impact parameters less than λ_D.

With W expressed in electron volts,

$$\sigma_d = 1.3 \times 10^{-13} \frac{Z^2 Z_1{}^2}{W^2{}_{\text{ev}}} (\ln \Lambda) F(x) \text{ cm}^2 \tag{52}$$

Alternatively, the value for τ_d can be expressed, with W in eV, as

$$\tau_d = 0.55 \times 10^7 \frac{W^{3/2}}{n_1 \ln \Lambda F(x)} \left(\frac{A^{1/2}}{Z^2 Z_1{}^2} \right) \tag{53}$$

where $A = M/M_p$, the mass relative to the proton.

Though the basic scattering cross sections depend only on energy, it is important to notice the large differences in rates of the above relaxation processes for ions and electrons. Because of their higher velocity, electrons undergo collisional relaxation at a rate $(M/m)^{1/2}$ faster than that for ions. To take an illustrative example, consider the case of a 10-keV proton moving in a proton–electron plasma of density 10^{14} ions cm^{-3} and a kinetic temperature of 10^4 eV. Here $x = 1$, $F(x) = 0.63$, and $\ln \Lambda \approx 20$. In this case one finds $\tau_d \approx 4.4 \times 10^{-3}$ sec.

Contrast this case with that for the electrons in the same plasma. Here $x \approx 1$, $F(x) \simeq 1$, and $\tau_d \approx 60$ μsec.

The significance of the above examples is twofold. First, it shows that ion–ion relaxation times are likely to be rather long in a high temperature plasma, corresponding to a collision mean free path, from Eq. (51), of some 6 km in the example. Second, in any nonequilibrium plasma, such as one confined in an open-ended system, the electrons will tend to relax toward isotropy much more rapidly than the ions. In fact it is frequently the case that they may exist in a state close to an isotropic Maxwellian, while at the same time the ions of the same plasma may be decidedly non-Maxwellian. As we have mentioned earlier, this circumstance leads to positive ambipolar potentials in mirror confined plasmas, is important in analyzing certain types of plasma instabilities, and is relevant to understanding some of the processes occurring in space plasmas.

In addition to the relaxation processes just discussed, significant energy exchange processes between different components of the plasma can take place. For example, a low energy ion moving in a sea of hot electrons will gain energy from the electrons until its mean energy equals the mean energy of the electron distribution, $\frac{3}{2}\kappa T_e$. Conversely, an energetic ion will be "cooled" in moving through a cloud of lower temperature electrons, even though its mean velocity may be substantially less than that of the electrons. If the electron kinetic temperature is substantially lower than the ion mean energy, this "dynamical friction" effect may dominate all other collision processes in its effect on the ion's energy. Accompanying the transfer of energy from hot ions is, of course, a heating of the electrons.

A concise analytical expression for the dynamical friction processes we have just described is the following:[18]

$$\left(\frac{dW}{dt}\right)_{ei} = 4\pi \sqrt{2} \ln \Lambda \, \frac{n_e Z^2 e^4}{(\pi m \kappa T_e)^{1/2}} \left(\frac{m}{M}\right) \left(1 - \frac{W}{\frac{3}{2}\kappa T_e}\right) \text{ erg sec}^{-1} \quad (54)$$

For ion energies small compared to electron mean energies, ion heating will proceed at a rate given by (54) in the limit $W \ll \frac{3}{2}\kappa T_e$. Taking $\ln \Lambda \approx 20$, and with T_e in electron volts, (54) becomes

$$\frac{dW}{dt} = 0.88 \times 10^{-7} \left(\frac{Z^2}{A}\right) \frac{n_e}{T_e^{1/2}} \text{ eV sec}^{-1} \quad (55)$$

To take an example, if $n_e = 10^{14}$ cm^{-3} and $T_e = 10^4$ eV (about 10^{8}°K), then low energy protons will be heated at a rate of 88 keV sec^{-1}.

In the other limit $\frac{3}{2}\kappa T_e \ll W < (M/m)(\frac{3}{2}\kappa T_e)$, where the electrons are cooler than the ions, but their mean velocity is still larger than that of the ion, cooling of the ion will proceed, at a rate given from Eq. (54), of (T_e in eV)

$$\frac{1}{W}\frac{dW}{dt} = -0.48 \times 10^{-7} \frac{n_e(Z^2/A)}{T_e^{3/2}} \sec^{-1} \tag{56}$$

This circumstance leads to an ion energy decreasing exponentially with time at a rate found by integrating Eq. (56)

$$W = W_0 \exp\left[-t/\tau_{ie}\right] \tag{57}$$

where the time constant is

$$\tau_{ie} = 1.7 \times 10^7 \frac{T_e^{3/2}}{n_e}\left(\frac{A}{Z^2}\right) \sec \tag{58}$$

If the electron temperature is rather low, this time constant can be very short. For example, for $n_e = 10^{14}$ cm^{-3} and $T_e = 10$ eV, one finds $\tau_{ie} = 5.5$ μ for protons. The time constant, of course, increases rapidly with electron temperature, so that in the example above, at $T_e = 10^4$ eV, it has increased to 0.17 sec.

It should be clear from the previous discussion that the various collisional relaxation and energy transport processes in a plasma proceed at rates which may vary by many orders of magnitude between different situations. While laboratory plasmas might exhibit characteristic collisional times ranging between fractions of a microsecond and seconds, in interplanetary plasmas these times might be as long as years (or even millenia!). Furthermore, the ratios appropriate to electrons and ions more often than not differ widely within the same plasma, so that processes in one species (the electrons) may early achieve collisional equilibrium while the ions may exist for the entire duration of the observation in a state far from collisional equilibrium.

One consequence of situations such as we have described is that within a plasma the various degrees of freedom of the particles are often decoupled from each other. Thus the compression of a plasma in a direction along the field lines may find the ions behaving as a one-dimensional gas, while the electrons respond as an ordinary gas would. At the same time, compression of the plasma in the direction transverse to the magnetic field would proceed (for the ions) as in a two-dimensional gas. In general one has for such situations (adiabatic compressions) that the appropriate temperature varies as

$$T \sim C^{\gamma-1} \tag{59}$$

where C is the geometric compression ratio (ratio of initial to final length along the field line for longitudinal compression, or ratio of initial to final radius for transverse compression). Here $\gamma = (f+2)/f$ from kinetic theory, where f is the number of degrees of freedom.

In the case of longitudinal compression $f = 1$ so that

$$T_\parallel \sim C^2 \tag{60}$$

Perpendicular to the field $f = 2$ and

$$T_\perp \sim C \tag{61}$$

If the compression proceeds slowly compared to relaxation times (for the electrons of the plasma, for example), then $f = 3$, $\gamma = \frac{5}{3}$ as in an ordinary gas and

$$T \sim C^{2/3} \tag{62}$$

Among the circumstances where these various compressional effects may appear are: (*a*) in laboratory experiments involving rapidly varying magnetic fields for plasma compression and heating, (*b*) in certain types of plasma instabilities, and (*c*) in cosmical situations involving moving regions of plasma containing trapped magnetic fields (as in Fermi's proposed mechanism to explain the origin of cosmic rays).[19]

3. Particle Dynamics and the Vlasov Equation

(*a*) *Particle Drifts.* We have described how it may often be possible in analyzing plasma phenomena to perform a natural separation of the Boltzmann equation into a slowly varying part associated with collisional effects and a collision-free part (the Vlasov equation) describing higher frequency phenomena arising from noncollisional dynamics. What are these "noncollisional dynamical effects" to which we allude? They are simply the collective continuations of the elementary dynamics of charged particles moving in response to electromagnetic or gravitational fields.

The starting point for our discussion is therefore the equation of motion for a charged particle, given earlier (Eqs. (2)) in relativistic form. For our purposes here it will be sufficient to discuss the nonrelativistic limit, for which the equations take the form

$$M\ddot{\mathbf{r}} = e\left[\mathbf{E} + \frac{\dot{\mathbf{r}}}{c} \times \mathbf{B}\right] + \mathbf{G} \tag{63}$$

The solution of even this simple-appearing equation in the general case is very difficult, because of the presence of velocity-dependent forces. However, most of the important features of the motion necessary to understand the physics involved can be seen from some simple cases.

Consider first the case **G** = 0 (i.e., ignore gravity). We now introduce a new coordinate system moving at the velocity

$$\mathbf{v}_E = c\,\frac{\mathbf{E} \times \mathbf{B}}{B^2} \; ; \quad |\mathbf{v}_E| \ll c \tag{64}$$

We have met this velocity already in the development of the MHD equations (see Eq. (36)). In the new frame of reference defined by (64) it is readily verified that the equations of motion take the form

$$M\ddot{\mathbf{r}}' = e\left[\frac{\dot{\mathbf{r}}'}{c} \times \mathbf{B}\right] + e\,\frac{\mathbf{E} \cdot \mathbf{B}}{B} \tag{65}$$

Parallel to **B** the equation is simply

$$M\ddot{z} = eE_\| \tag{66}$$

i.e., accelerated motion in the presence of an electric field. Perpendicular to **B** the motion predicted depends on the details of the spatial and temporal variation of **B**. In the special case **B** = constant, the solutions are circles of arbitrary center, as we shall verify directly. In the motion perpendicular to **B**, let $r = x + iy$ and substitute in the two coupled equations for the motion derived by taking the x and y components of Eq. (65). In this way one obtains a single equation in the complex variable r:

$$\dot{r} = -i\Omega r \tag{67}$$

where $\Omega = eB/Mc$. This equation is solved by the rotating solution

$$r = ae^{i\Omega t} \tag{68}$$

where $a = Mv_\perp c/eB$. The motion described is, of course, simply a rotation at the cyclotron or gyrofrequency of the particle at a radius equal to a.

For future reference we give the magnitude of the frequencies and orbit radii.

For electrons:

$$\Omega_e = -1.76 \times 10^7 B \text{ radians/sec} \tag{69}$$

For ions:

$$\Omega_i = 0.965 \times 10^4 (Z/A)B \text{ radians/sec} \tag{70}$$

For example, in a field of 10^4 G, $v_e = |\Omega_e|/2\pi = 28{,}000$ MHz and $v_i = 15.4$ MHz for protons.

Similarly, the orbit radii are given by:

For electrons:

$$a_e = 3.3\,\frac{W^{1/2}}{B} \text{ cm, } (W \text{ in electron volts}) \tag{71}$$

For ions:

$$a_i = 144\left(\frac{A^{1/2}}{Z}\right)\frac{W^{1/2}}{B} \text{ cm} \qquad (72)$$

To recapitulate, in the preceding discussion we have shown that in the presence of an electric field and a uniform magnetic field, the orbit of a charged particle can be depicted as a helical motion along the field lines upon which a transverse electric drift, v_E, has been superposed;

$$v_E = c\left[\frac{\mathbf{E} \times \mathbf{B}}{B^2}\right] = c\left(\frac{E_\perp}{B}\right) \qquad (73)$$

We note that this drift is the same one met in the discussion of the hydromagnetic equations. Note also, in that context, that v_E is in the same direction for ions and electrons, and thus does not lead to separation of charge within the plasma, at least to the order of approximation discussed here.

The "drift" picture of plasma particle motions just described has been taken as the starting point for the treatment of more complicated particle dynamic and plasma effects, those arising in non-uniform fields, by Alfvén,[20] by Spitzer,[1] and by others. The discussion to follow will be based on the same viewpoint.

Returning to the original equation of motion, Eq. (63), we note that the effect of gravity, represented by the term **G**, appears in nearly the same way as does **E**. It follows that the effect of gravity is also to introduce a transverse drift of magnitude

$$v_g = g_\perp/\Omega \text{ cm/sec} \qquad (74)$$

Note that v_g is opposite in sign for electrons and ions (Ω is taken as negative for electrons) and thus leads to a tendency toward charge separation in a plasma. In the presence of magnetic fields usually employed in laboratory experiments, v_g is negligibly small compared to other drifts. In the earth's magnetic fields outside the atmosphere it is small but possibly significant for some processes; in a field of 0.01 G (the value of the earth's field at 3 earth radii) a proton will drift transversely due to gravity at about 10 cm/sec.

A drift closely related to the gravitational drift, but of much greater practical importance, arises if a drift velocity (for example v_E) is itself a function of time, i.e., if the drift frame is accelerated. In such cases the value of the acceleration of the drifts, $d\mathbf{v}/dt$, takes the place of **g** in the gravity drift equation.

In the presence of magnetic field gradients, very important drift motions arise. We will here use a perturbation approach to derive expressions for these drifts, applicable in the limit of "weak" gradients—that is, ones which satisfy the restriction

$$a\left|\frac{\nabla B}{B}\right| \ll 1 \qquad (75)$$

implying small variation in the strength of the magnetic field across an orbit. This restriction is more often than not well satisfied in magnetically confined plasmas.

Consider first a field with a transverse gradient, one where we may expand the field as $B = B_0 (1 + \varepsilon x + \cdots)$, where

$$\varepsilon = \frac{1}{B}\frac{dB}{dx}\bigg|_{x=0}$$

and $\varepsilon a \ll 1$. We may now write the equations of motion (in a frame where $E_\perp = 0$) in cartesian coordinates as

$$\ddot{x} = \Omega \dot{y}$$
$$\ddot{y} = -\Omega \dot{x} \qquad (76)$$

where $\Omega = \Omega_0 (1 + \varepsilon x)$.

In the limit $\varepsilon \to 0$, the solutions to these equations are of the form

$$x = a \sin \Omega_0 t$$
$$y = a \cos \Omega_0 t \qquad (77)$$

In the case $\varepsilon \neq 0$ let us attempt the following change of variables: let $x' = x$, $y' = y + bt$, $b \ll \Omega$. In other words, we transform to a drift frame moving perpendicular to ∇B. In these variables the equations of motion become

$$\ddot{x}' = \Omega_0(1 + \varepsilon x')(\dot{y}' - b)$$
$$\ddot{y}' = \Omega_0(1 + \varepsilon x')\dot{x}' \qquad (78)$$

Expanding, we obtain (dropping the prime)

$$\ddot{x} = \Omega_0 \dot{y} + \Omega_0(\varepsilon x \dot{y} - b) - \varepsilon b x$$
$$\ddot{y} = \Omega_0 \dot{x} + \Omega_0 \varepsilon x \dot{x} \qquad (79)$$

The last term in the equation for $\ddot{x}$ is of second order in the perturbation and may be dropped. We may now determine the drift frame velocity, b, by noting that in this frame $\dot{x}$ and $\ddot{x}$ together with $\dot{y}$ and $\ddot{y}$ must have zero time average. By inspection of the equations we see that this requirement can be satisfied if we set $b = \varepsilon x \dot{y}$, and to this order insert the unperturbed solutions for x and $\dot{y}$, finding

$$b = \varepsilon a^2 \Omega_0 \overline{\sin^2 \Omega_0 t} = \tfrac{1}{2}\varepsilon a^2 \Omega_0 \qquad (80)$$

THE PHYSICS OF HIGH TEMPERATURE PLASMA 175

In the equation for $\ddot{y}$, the perturbation terms are of the form $x\dot{x} = \sin \Omega t \cos \Omega t$, so that $\overline{x\dot{x}} = 0$ to first order in ε. We have therefore shown that the transverse drift frame velocity $v_{d\perp}$ is equal to b. It may be rewritten in the form

$$v_{d\perp} = \frac{1}{2}\varepsilon \frac{a^2 \Omega_0^2}{\Omega_0} \tag{81}$$

i.e.,

$$v_{d\perp} = \frac{1}{2}\left(\frac{\nabla B}{B}\right)_\perp \cdot \frac{v_\perp^2}{\Omega}$$

Note that $v_{d\perp}$ is perpendicular to both B and $(\nabla B)_\perp$ and is oppositely directed for electrons and ions.

A related and similar drift arises from the parallel velocity component of the particle as it moves along a field line which is curved with radius of curvature R (not to be confused with the mirror ratio R_M). As discussed by Spitzer, this drift may be evaluated by introducing a new coordinate system rotating with the angular velocity $v_\parallel/R$ about the center of curvature. In this frame the particle possesses no longitudinal velocity but instead is subjected to a centrifugal force $(mv_\parallel^2/R)$ that produces a transverse drift as though the particle were moving under an equivalent gravitational acceleration. Thus from (74) this drift is

$$v_{d\parallel} = v_\parallel^2/R\Omega \tag{82}$$

If electric currents in the plasma are weak, so that the magnetic field within the plasma has its vacuum value, then $(\nabla B/B)_\perp = 1/R$ and one obtains for the sum of the two magnetic drifts

$$v_d = v_{d\perp} + v_{d\parallel} = (1/\Omega R)(\tfrac{1}{2}v_\perp^2 + v_\parallel^2) \tag{83}$$

Note again that v_d is oppositely directed for electrons and ions.

For reference, we summarize the drifts found thus far in Table I.

TABLE I

Drift	Equation	Relative sense for electrons and ions
Electric	$v_E = c[(\mathbf{E} \times \mathbf{B})/B^2] = c(E_\perp/B)$	same
Gravitational	$v_g = g_\perp/\Omega$	opposite
Gradient B	$v_{d\perp} = \tfrac{1}{2}(\nabla B/B)_\perp (v_\perp^2/\Omega)$	opposite
Gradient B	$v_{d\parallel} = v_\parallel^2/\Omega R$	opposite

The importance of the drifts is that they represent the particulate mechanisms that operate in the attainment of equilibrium, in the propagation of certain plasma waves, and in the driving of many types of plasma instabilities.

As in example of the role of particle drifts in magnetostatic equilibrium situations, consider again the problem of lack of equilibrium in a simple torus. In such a torus (see Fig. 2), the field intensity varies inversely with radius, i.e., at every point in the cross section ∇B points inward, toward the polar axis of the toroid. In this gradient the electrons and ions undergo the $v_{d\perp}$ and $v_{d\|}$ drifts, the electrons up and the ions down in the figure. The immediate effect of these drifts will be to cause negative charge to be displaced upward and positive charge downward, leading to a vertical electric field. This electric field, now perpendicular to **B**, will lead in turn to an outward electric drift, $v_E = c(E/B)$ for both ions and electrons, destroying confinement. According to this same picture we can understand the role of rotational transform in promoting equilibrium in a torus. When rotational transform is present, uncanceled charges accumulating in the upper part and lower part of the torus have an opportunity to neutralize each other by flowing along the lines of force, which no longer are coplanar circles closing on themselves, but instead spiral helically around the interior at the torus. But even in this case there exists the possibility of losing a special group of drifting particles—those that are moving too slowly along the field lines as compared to their velocity perpendicular to the lines. These particles may actually drift to the walls before they have made a full transit around the torus (which would average out their drift when rotational transform is present). They would then be lost unless some other mechanism caused them to precess more rapidly around the magnetic axis. It is believed that the radial electric field accompanying the equilibrium state in the presence of radial diffusion losses may provide this mechanism, but there is as yet no conclusive proof that this amelioration will be adequate.

Before leaving the subject of particle drifts we mention a concept useful in the discussion of drifts and in the solution of certain plasma problems. This concept is the *guiding center*. We have already noted that particle drifts can be thought of as a circling motion upon which transverse motion has been superposed. This picture can be equally well represented as the motion of a translating "guiding" center around which the particle rotates. When particle orbit diameters are small and drifts are relatively slow, this picture is both realistic and useful.

(*b*) *The Magnetic Moment—Adiabatic Invariants.* Stimulated in large part by the fusion confinement problem and by space research, some remarkable advances have been made in recent years in the understanding of charged particle orbits in complex magnetic fields. A major aid to this understanding has been the appreciation of the importance of the adiabatic invariants that characterize various periodicities of the particle motions in a confining field. As defined in classical physics, an "adiabatic invariant" is a particular function (of the variables of an almost periodic motion) that approaches a constant in the limit of adiabatic (here, slow) changes of one of the parameters of the system. Since the existence of adiabatic invariants represents such an important facet of treating the confinement of particles in mirror systems, we shall first discuss the general nature of such invariants before considering specific examples. The derivation to follow is based on a discussion by W. Thompson.[21]

Consider an almost periodic system where some quantity, say the amplitude of oscillation, can be described as a series in a small parameter ε.

$$r = r_0 + \varepsilon r_1 + \varepsilon^2 r_2 + \cdots \qquad (84)$$

and where the zeroth order solution, r_0, is periodic, so that r satisfies the equation

$$\ddot{r} + \Omega^2 r = \varepsilon g(r, t) \qquad (85)$$

Given such a system the statement is made that the function $\alpha(r_0)$ of the zeroth order solution defined through the integral

$$\alpha(r_0) = \oint \dot{r} \, dr \qquad (86)$$

taken over a period of the motion, is constant to at least the second order in ε, i.e., it is asymptotically constant for sufficiently small ε. Note that the integral is just the area in the phase space r, $\dot{r}$, i.e., it is the "action" integral of classical mechanics. According to classical theory this integral represents an extremum among possible trajectories.

A sufficient proof of the constancy of $\alpha(r_0)$ to at least second order in ε is to show that the time derivative of $\alpha(r_0 + \varepsilon r_1)$ vanishes to this order. This we may readily do for the problem at hand. We have

$$\frac{d}{dt}[\alpha(r_0 + \varepsilon r_1)] = \frac{d}{dt}\oint (\dot{r}_0 + \varepsilon \dot{r}_1) \, dr = \frac{d}{dt}\oint (\dot{r}_0 + \varepsilon \dot{r}_1)^2 \, dt \qquad (87)$$

Expanding, we find

$$\frac{d}{dt}[\alpha(r_0 + \varepsilon r_1)] = \frac{d}{dt}\oint \dot{r}_0^2 \, dt + 2\varepsilon \int (r_0 \dot{r}_1 + \dot{r}_0 r_1) \, dt + 0(\varepsilon^2) \qquad (88)$$

The first term vanishes because of the periodicity of $r_0(t)$. We may then integrate the first term of the second integral finding for the second integral expression the result

$$2\varepsilon \left\{ \dot{r}_0 \dot{r}_1 \Big|_a^b - \oint (\ddot{r}_1 \dot{r}_0 - \dot{r}_0 \ddot{r}_1) \, dt \right\} \tag{89}$$

Since $\dot{r}_0$ is periodic, the first term vanishes when evaluated at the end points ($\dot{r}_0 = 0$ at both limits); the second term is identically zero. Thus da/dt must be at least not greater than order ε^2, completing our proof.

We may rewrite $a(r_0)$ in terms of the frequency parameter, Ω, for the motion through the relation

$$a(r_0) = \oint \dot{r}_0 \, dr_0 = \oint (\dot{r}_0)^2 \, dt \tag{90}$$

Since r_0 is periodic we may write $\theta = \Omega t$, so that

$$a(r_0) = \Omega \oint \left(\frac{dr}{d\theta}\right)^2 d\theta = \tfrac{1}{2}\Omega a^2 \tag{91}$$

where a is the amplitude of the oscillation of r. From what has preceded it can be seen that $\Omega(t)$ need not be constant; it is only required that $d\Omega/dt = 0(\varepsilon)$, i.e. that the change in Ω per period is small.

We have shown that the quantity $a = \tfrac{1}{2}\Omega a^2$ is an adiabatic invariant of quasi-periodic motions with frequency Ω and amplitude a. The quantity Ωa^2 is of course proportional to the product of frequency and energy (kinetic or potential) of the oscillation.

In detailed discussions of particular invariants it has been shown that the constancy of such invariants is extremely well preserved as long as ε is indeed small. In fact, it has been shown that the variations in a from period to period of the motion become exponentially small in the parameter ε^{-1}, i.e., the variations can be represented by functions of the form

$$\left|\frac{\delta a}{a}\right| \sim e^{-1/\varepsilon} \tag{92}$$

As can be readily seen, this bound rapidly becomes very small in the limit $\varepsilon \to 0$. Furthermore, it can be shown that for ε reasonably small the variations of a may take on a periodic character themselves, oscillating between definite limits,[22] rather than growing with the square root of the number of periods, as would be the case if the fluctuations were uncorrelated.

The first adiabatic invariant of the particle motions in a magnetic field that we will consider is the simplest and most important one—the adiabatic invariant associated with the periodic circling of particles about their guiding centers.[20] Here the frequency, Ω, is simply the cyclotron frequency, and the amplitude, a, is the orbit radius. The frequency will be subject to slow variations as the particle moves from place to place in an inhomogeneous field, or as it responds to a slow variation in magnetic field with time. Here, therefore, he expression for the invariant is

$$a = \tfrac{1}{2}\Omega a^2 = \tfrac{1}{2}\Omega \left(\frac{v_\perp^2}{\Omega^2}\right) = \frac{c}{Ze}\left(\frac{\tfrac{1}{2}Mv_\perp^2}{B}\right) \tag{93}$$

We note that a is proportional to the ratio of the energy of rotation of the particle, divided by the local value of the magnetic field. Note now that the electric current loop represented by the circularly moving charge is given by

$$\dot{q} = Ze\Omega/2\pi \tag{94}$$

and the magnetic moment represented by such a circular current is

$$\mu = \dot{q}A = \frac{Ze\Omega}{2\pi}\pi a^2 = \frac{Ze}{2}\frac{Mv_\perp^2}{ZeB} = \frac{\tfrac{1}{2}Mv_\perp^2}{B} \tag{95}$$

It follows from (93) that $\mu = W_\perp/B$ is an adiabatic invariant of the particle motion,* and that it is numerically equal to the magnetic moment generated by the circulating charge.

In a static magnetic field not only is the magnetic moment an adiabatic invariant but the total particle energy is an exact invariant. This fact has an important consequence—it provides the rationale for the magnetic mirror effect, and for mirror confinement.[4] In particular, since $W_\perp/B$ and $W = W_\perp + W_\parallel$ are both constants of the motion, it follows that when a particle moves along a field line into a region of increased magnet field, its rotational energy must increase at the expense of the energy of motion parallel to the field, i.e., the particle moves more slowly in this direction, and may even be stopped and "reflected." This mirroring effect is easy to calculate. We have, at any point, z, along a field line that

$$W_\parallel(z) + W_\perp(z) = W = \text{constant} \tag{96}$$

But $W_\perp = \mu B$, so that

$$\frac{\partial W_\parallel}{\partial z} = -\frac{\partial W_\perp}{\partial z} = -\mu\frac{\partial B}{\partial z} = -\frac{\partial}{\partial z}(\mu B) \tag{97}$$

since $\mu = $ constant. Thus the quantity μB acts as a sort of magnetic potential for motion along the field lines, so that we may write

$$F = -\nabla_\parallel(\mu B) \tag{98}$$

* Actually an approximation to the exact adiabatic invariant, but one which is increasingly good as $\varepsilon \to 0$.

This result implies that a particle moving in its helical path toward an increasing field will feel a retarding force proportional to $\nabla_\parallel B$. The particle will be turned around (reflected) by this magnetic mirror action before it reaches the mirror peak, provided the mirror field B_{MAX} satisfies the condition $\mu B_{MAX} > W$.

Since $\mu = W_\perp/B$ and $W_\perp/W = v_\perp^2/v^2 = \sin^2\theta$ (where θ is the pitch angle of the helical motion; $\theta = 0$ corresponds to motion exactly along the field lines), the constancy of μ implies that a simple transformation law exists for the pitch angles. We see that between any two points z_1, z_2 along a field line the pitch angle will transform according to the relation

$$\sin\theta\,(z_2) = [B(z_2)/B(z_1)]^{1/2} \sin\theta\,(z_1) \tag{99}$$

i.e., $B^{1/2}$ acts something like an "index of refraction" for the pitch angle, analogous to Snell's law for optical media.

It follows from (99) that the mirror trapping condition can also be written in terms of a "loss cone" in velocity space (pitch-angle space), by defining a critical pitch angle for reflection. For particles launched at the point of minimum field between the mirrors, one therefore has a critical loss cone angle, θ_c, defined through

$$\theta_c = \sin^{-1}\left(\frac{B_0}{B_{MAX}}\right)^{1/2} = \sin^{-1}\left(\frac{1}{R_M}\right)^{1/2} \tag{100}$$

where R_M is the mirror ratio, (B_{MAX}/B_0).

Once trapped between mirrors, particles will oscillate between the mirroring points while slowly drifting transverse to the field. The existence of the oscillatory motion along the field lines between the mirroring points suggests the existence of another adiabatic invariant, that associated with the parallel motion. Indeed one exists, which we may deduce immediately from the general definition (86). If z represents position along the field lines then

$$J = \oint \dot{z}\,dz \tag{101}$$

is an adiabatic invariant. The quantity J is recognized as the *action integral* of classical mechanics.

The existence of adiabatic invariants in the motion of particles trapped between magnetic mirrors has enabled the proof of some powerful principles for such containment. The classic paper discussing this equation has been published by Northrop and Teller.[23] Prior to the elucidation of those questions, one might have wondered whether the precessional drifts of particles trapped in complex fields such as magnetic wells (see Fig. 3) would cause the particles to "walk" across the confining field after precessing around the system a few times. That this circumstance cannot occur for adiabatically confined particles can be proved readily from the constancy of μ and J[22,28]:

First, the constancy of μ implies that whatever lines of force a particle happens to be moving on when it reaches the endpoint and is reflected, the value of B at that point must always be the same. Second, the constancy of J implies that as the particle is drifting slowly in the azimuthal direction while oscillating between its reflection points, it must choose those lines for which J as well as μ is constant. Thus when it has finally precessed once around the field and comes back to the vicinity of its starting line it must again return exactly to that line (since any other line at a larger or smaller radius than the precession surface would have a different value of J between the endpoints B.

That adiabatic particles (i.e., ones whose orbits are small in radius compared to the characteristic dimensions of the confining field) are trapped essentially indefinitely in mirror systems of even very complex geometry has been abundantly proved by experiment. For example, relativistic electrons trapped in the earth's field many years ago in the ARGUS experiment are still in evidence—after traveling in their trapped orbits over distances equal to many light years! In the laboratory, experiments with energetic electrons have also demonstrated extremely long confinement (minutes), limited only by scattering of the electrons against residual gas atoms.[24,25] The only requirement that this kind of confinement be achieved is that the orbit radius of the particle be small compared to the shortest characteristic gradient distance of the confining mirror field. As a rule of thumb, meeting the condition $a|\nabla B/B| \lesssim 0.1$ is generally adequate to insure adiabaticity. However, in keeping with the exponential nature of the convergence to adiabaticity, even a relatively small increase in this parameter soon leads to the destruction of particle confinement.

As a practical matter, the adiabaticity condition implies that in confinement experiments the size and strength of the confining field should be adjusted to be consistent with the type and energy of the particle that is to be confined. In the context of space research this adiabaticity requirement is abundantly well satisfied—often by orders of magnitude!

The quantities μ and J, in addition to being invariants for particles moving within inhomogeneous but static fields, are also invariant in time-varying fields, provided this variation is not too rapid. For μ the adiabaticity requirement is that the field should vary only slightly in one cyclotron period, i.e., that $\Omega \tau_B \gg 1$, where τ_B is the characteristic time constant for variation of the field. A similar condition applies for J.

Consider first the invariance of μ. In a time-varying magnetic field a particle will experience an accelerating force owing to the existence of the electric field associated with $\partial B/\partial t$. From Maxwell's equations and Stokes' theorem, the total emf around the orbit will be

$$\text{emf} = \oint \mathbf{E} \cdot d\mathbf{l} = -\frac{1}{c}\frac{\partial}{\partial t} \oint \mathbf{B} \cdot d\mathbf{A} \qquad (102)$$

This quantity will in turn be closely approximated by $-(1/c)(\partial/\partial t)(\pi a^2 B)$. This emf produces a torque on the particle and thus a time rate of increase of angular momentum, given by

$$\frac{d}{dt}(Ma^2\Omega) = eaE_\phi = \frac{e}{2\pi}\oint \mathbf{E} \cdot d\mathbf{l} \qquad (103)$$

But from Eq. (102) this must equal $-\tfrac{1}{2}(d/dt)(Ma^2\Omega)$. But these two requirements can clearly only be simultaneously satisfied if

$$\frac{d}{dt}(a^2\Omega) = 0, \quad \text{i.e.,} \quad \frac{d}{dt}\left(\frac{W_\perp}{B}\right) = \frac{d\mu}{dt} = 0 \qquad (104)$$

Thus in both time- and space-varying fields μ is an adiabatic invariant.

Note that the constancy of μ in a time-varying field implies that particles will gain perpendicular energy in an increasing field, and vice versa ($W_\perp \sim B$).

Note also that the constancy of μ implies that the magnetic flux through the orbit is a constant of the motion:

$$\phi = \pi B^2 a = \pi B \frac{v_\perp^2}{\Omega^2} = \frac{2\pi Mc^2}{e^2}\left(\frac{W_\perp}{B}\right) = \text{constant} \qquad (105)$$

We see therefore that as a particle moves from place to place in either a static or a slowly varying magnetic field, it maintains constant flux through its orbit, the diameter shrinking or expanding depending upon the local value of B. Not only does the flux through individual orbits remain constant, but as we shall see, the flux through the entire ensemble (i.e., the plasma itself) also has an invariant property. We therefore recover from the drift picture the result from hydromagnetics: plasma considered as a fluid moves as though it were a compressible perfect conductor, preserving flux through each element of the fluid as it moves in the magnetic field.

To prove the existence of the flux invariant we need only to invoke the constancy of μ and J. These two invariants insure that a third nearly periodic quantity exists, the precessional drift motion around the axis of the field, arising from its radial gradient. In simple situations the flux invariance of this motion is easy to demonstrate directly. Consider, for example, the equation of motion of a particle in an axially symmetric time-varying magnetic field. As in the discussion of the particle drifts, we may analyze this problem by introducing the complex coordinate $r = x + iy$, then expressing the induction electric field in terms of the time rate of change of B, through Maxwell's equations, finding

$$\ddot{r} = i\frac{e}{2mc}[\dot{B}r + 2\dot{r}B] \qquad (106)$$

Multiplying by r we find

$$r\ddot{\mathbf{r}} = \frac{ie}{2mc}\frac{d}{dt}[Br^2] \qquad (107)$$

Consider now a particle moving in a small near-circular orbit at radius r_0. We may determine the frame in which its guiding center of rotation remains fixed by requiring that $\langle r_0 \dot{r}\rangle = 0$. From Eq. (107) we see that this frame is the one in which $\langle Br^2\rangle = $ constant, i.e., that one in which the flux enclosed by a circle centered on the origin and passing through the guiding center is a constant of the motion to this order. This same result is found if one evaluates the frame moving at the electric field drift velocity $v_E = c(E_\perp/B)$ in the field; motion at this velocity corresponds to motion on a flux surface.

Summarizing, provided adiabaticity requirements are satisfied, particles once trapped in a magnetic mirror field will oscillate between the mirrors while their guiding centers precess on closed drift surfaces. As long as these drift surfaces do not come within an orbit radius of intersecting physical surfaces, the particles will remain trapped indefinitely if the invariants μ and J remain undisturbed. Any process which is to interfere with this situation must do so nonadiabatically, i.e., it must occur on a time scale at least as short as the periodicities involved. In a magnetic well with a sufficiently intense static magnetic field only two processes can lead to nonadiabatic behavior: (1) particle–particle collisions and (2) high frequency plasma instabilities. On the basis of all these considerations it can now be understood why the existence of magnetostatic equilibria rests on such a solid experimental and theoretical base in the case of open-ended confinement systems. The existence of the μ, J, and ϕ invariants for the particles of plasmas once trapped between magnetic mirrors implies strong constraints on the subsequent behavior of that plasma.

4. Guiding Center Equations

When the details of the high frequency motion of the particles of a plasma are unimportant (as in magnetostatic equilibrium or in low frequency wave or instability problems), a useful simplification of the collisionless Boltzmann equation is the use of guiding center equations. Here one writes a continuity equation for the density of guiding centers of each particle species, subject to the guiding center drifts and to the requirement for neutrality.

If we therefore take $g(\mathbf{x}, \mathbf{v})$ to be the density of guiding centers, the continuity equation for the guiding center density becomes simply

$$\frac{\partial g}{\partial t} + \nabla \cdot (\mathbf{v}_D g) = 0 \tag{108}$$

where $\mathbf{v}_D$ represents the totality of the particle guiding center drifts relevant to the problem at hand. This equation, coupled with Maxwell's equations, then defines the set of equations to be solved. We shall later give an example of the use of the guiding center equations to solve a problem in plasma stability.

Before leaving the subject of guiding center equations, however, we note an important representation for the use of these equations. If the guiding center distribution function is specified as a function of the particle magnetic moment μ and the particle energy W, constants of the motion in the guiding center approximation, then it follows that $g(\mu, W)$ will be a constant everywhere within the plasma. This circumstance leads to major simplification in the analysis of some types of problems. An example is the calculation of the spatial dependencies of the plasma density and pressure, now to be discussed.

It is first convenient to change from the μ, W representation to a closely related one, $\nu = \frac{1}{2}(v_\perp^2/B)$ and $\varepsilon = \frac{1}{2}v^2$, the magnetic moment and energy per unit mass, respectively. The phase space volume element in ν, ε coordinates is now related to that in the usual $v_\perp, v_\parallel$ coordinates through

$$dV = \pi dv_\perp^2 \, dv_\parallel = 2\pi B/q \, d\nu \, d\varepsilon \tag{109}$$

where $q = v_\parallel = \sqrt{2}(\varepsilon - \nu B)^{1/2}$.

Thus in these units the plasma density can be expressed as

$$n(B) = \sqrt{2}\pi B \int_\nu \int_\varepsilon g(\nu, \varepsilon)(\varepsilon - \nu B)^{-1/2} \, d\nu \, d\varepsilon \tag{110}$$

where $g(\nu, \varepsilon)$ need only be specified as a constant on any given flux tube element passing through the plasma. Expressed in this way the density is clearly a function of B only on any given flux surface, leading to the consideration of confined distributions appropriate to magnetic wells (where ν and ε represent meaningful constants of the motion).

Similarly,

$$p_\perp = \frac{\pi}{\sqrt{2}} B^2 \int_\nu \int_\varepsilon g(\nu, \varepsilon)[\mu(\varepsilon - \nu B)^{-1/2}] \, d\nu \, d\varepsilon \tag{111}$$

$$p_\| = 2\pi B \int_\nu \int_\varepsilon g(\nu, \varepsilon)(\varepsilon - \nu B)^{1/2} \, d\nu \, d\varepsilon \tag{112}$$

By differentiation of the expression for $p_\|$ it will be seen that these definitions for $p_\|$ and $p_\perp$ automatically satisfy the mirror pressure balance of hydromagnetics, Eq. (16), along any line of force.

As a specific example of the use of ν, ε coordinates we will calculate the ion density distribution of a plasma confined between magnetic mirrors. For a distribution function we will choose an analytic form which approximately represents a collisional "loss-cone" solution of the type discussed in connection with Figs. 5 and 6, here for the case of a mirror ratio $R_M = 2$.

In $v_\perp^2, v_\|^2$ coordinates, g is given at the midplane ($B = B_0$) as:

$$\begin{aligned} g(v_\perp^2, v_\|^2) &= (v_\perp^2 - v_\|^2)^{1/2} \exp(-v_\perp^2) & v_\perp > |v_\|| \\ &= 0 & v_\perp < |v_\|| \end{aligned} \tag{113}$$

In ν, ε coordinates this becomes (independent of position within the plasma)

$$\begin{aligned} f(\nu, \varepsilon) &= \sqrt{2}(2\nu B_0 - \varepsilon)^{1/2} \exp(-2\nu B_0) & \varepsilon < 2\nu B_0 \\ &= 0 & \varepsilon > 2\nu B_0 \end{aligned} \tag{114}$$

Thus we have for the plasma density

$$n(B) = \int_0^\infty d\nu \int_0^{2\nu B_0} (2\nu B_0 - \varepsilon)^{1/2}(\varepsilon - \nu B)^{-1/2} \exp(-2\nu B_0) \, d\varepsilon \tag{115}$$

Making the change of variables $2\nu B_0 - \varepsilon = u^2$, the integrations may be readily performed, leading to

$$n(B) = n(B_0)(2 - B/B_0) \tag{116}$$

Thus the density is a maximum at $B = B_0$ and falls to zero at $B = 2B_0$ (at the mirrors). Of course the actual spatial dependence of n depends on the spatial dependence of B.

The calculation of the spatial dependences of n, $p_\perp$, and $p_\|$ is important when one considers the stability properties of bounded plasmas and the detailed calculation of ambipolar electric fields.[8-10]

5. The Vlasov Equation

We return now to the use of the Boltzmann equation for a plasma in the collisionless limit. To repeat, the resulting equation, which has come to be known as the Vlasov equation, takes the form, for each particle species in the plasma,

$$\frac{df}{dt} = \frac{\partial f}{\partial t} + v \cdot \nabla f + \frac{e}{M}\left(\mathbf{E} + \frac{1}{c}\mathbf{v} \times \mathbf{B}\right) \cdot \nabla_v f = 0 \tag{117}$$

Note that this equation is Liouville's equation for a phase volume element in the plasma. In other words it simply states that $f(\mathbf{v}, \mathbf{r})$ remains constant with time when evaluated along the trajectories of the infinitesimal group of particles lying initially within $d\mathbf{r}$ and $d\mathbf{v}$ of $\mathbf{r}$ and $\mathbf{v}$. Macroscopic quantities, such as the plasma density, are then evaluated as a function of time by integration over the velocity coordinates, together with introducing the constraints imposed by Maxwell's equations, charge neutrality, and the boundary conditions. Since the equations thus obtained are nonlinear in character, only in special cases has it been possible to obtain exact solutions. However, perturbation theoretic methods have been developed for the solution of a wide variety of problems. In fact in the last few years a prodigious effort has gone into solving certain classes of problems in plasma physics via the Vlasov equation. Generally this effort has been aimed at the problem of discovering plasma instabilities or of investigation the nonlinear consequences of such instabilities. The motivations for this effort have been first the controlled fusion problem, and more recently, the search for an understanding of phenomena observed in plasmas in space.

We shall later discuss a specific example of the use of the Vlasov equation for the analysis of a high frequency instability. At this point we shall outline an important perturbation method of solving this equation for low β plasma situations, i.e., in plasmas where the perturbation in the magnetic field is small compared to its static value.

The method described here is known as "the method of characteristics." It was first applied to plasma problems by Rosenbluth and co-workers.[26] Following their lead others have successfully tackled many otherwise very difficult problems using this method. To set up the method one first assumes that the distribution function can be given in the form $f = f_0 + f_1$, $f_1 \ll f_0$, i.e., a static equilibrium state plus a time-varying perturbation. (Here the subscript "1" denotes a perturbed quantity.) In equilibrium we then take $f = f_0$ and $\mathbf{E} = 0$, which may involve going to a moving frame where $\mathbf{E} = 0$ (the electric field drift frame for the dc electric field, if any is present). In the perturbed state $\mathbf{E} = \mathbf{E}_1 = -\nabla \phi$ (the low β assumption), and $\nabla \cdot E_1 = 4\pi \rho_1$, so that from Poisson's equation

$$\nabla^2 \phi_1 = -4\pi \rho_1 = -4\pi \sum_j e_j \int f_{1j} \, d^3 v \tag{118}$$

If we insert $f = f_0 + f_1$ into the Vlasov equation and linearize—i.e., neglect quantities quadratic in the perturbed quantities—there results

$$\frac{\partial f_1}{\partial t} + \mathbf{v} \cdot \nabla f_1 + \frac{e}{M}(\mathbf{v} \times \mathbf{B}_0) \cdot \nabla_v f_1 = -\frac{e}{M} \mathbf{E}_1 \cdot \nabla_v f_0 \qquad (119)$$

We now recognize the fact that the left-hand side of the equation is simply a statement of (df_1/dt), the total time derivative of f_1, as determined by following an infinitesimal group of particles along the *unperturbed* trajectories in $\mathbf{r},\mathbf{v}$, i.e., the linearized equation becomes

$$\frac{d}{dt} f_1 = -\frac{e}{M} \mathbf{E}_1 \nabla_v f_0 \qquad (120)$$

We may now determine f_1 formally simply by integrating forward in time along the unperturbed trajectories:

$$f_1(\mathbf{r}, \mathbf{v}, t) = -\frac{e}{M} \int_{-\infty}^{t} [\mathbf{E}_1(\mathbf{r}', t')] \cdot \nabla_{v'} f_0 \, dt' \qquad (121)$$

or, finally,

$$f_1(\mathbf{r}, \mathbf{v}, t) = \frac{e}{M} \int_{-\infty}^{t} \nabla \phi_1(\mathbf{r}', t') \cdot \nabla_{v'} f_0 \, dt' \qquad (122)$$

Here $\mathbf{r}$, $\mathbf{v}'$, t' are the coordinates of an infinitesimal particle group which is at $\mathbf{r}$, $\mathbf{v}$, t at time t. We tacitly assume the disturbance to vanish at $t = -\infty$ and look for the behavior of f_1 for $t > 0$. In this way we have found an equation for f_1, in terms of ϕ_1, that can be solved by Fourier analysis. To complete the solution one assumes the existence of a perturbed potential in the form of a wave motion of arbitrary frequency and wavenumber. Inserting this into (122) determines f_1 in terms of ϕ_1. Integration of f_1 over velocity and summation over particle species then yields another expression for ϕ_1, through Poisson's equation. Elimination of ϕ_1 then leads to the *dispersion equation* for the wave; i.e., a relationship of the form $D(\omega, k) = 0$ between the frequencies and the wavenumbers of allowed waves. But the assumed wave was taken to have a time dependence of the form $\exp(i\omega t)$ (or the complex conjugate). It follows that solutions of $D(\omega, k) = 0$ that have a *positive* imaginary component for ω correspond to decaying waves; those with Im $\omega < 0$ correspond to unstable waves, i.e., ones that grow exponentially in time. Solutions giving real ω values usually correspond either to undamped waves or to marginally stable ones, depending on circumstances. The classic paper which represented the first correct analysis of the growth and damping of elementary plasma waves was written by Landau in 1946.[27] It is in large part for this reason that the collisionless wave damping mechanism that plays such an important role in the behavior of high temperature plasmas has been named "Landau damping."

We pause for a moment in our headlong flight into the complex plane to recapitulate the Pilgrim's Progress to this point. We have dwelt at length on some of the basic concepts and methods of analysis that are needed to gain an understanding of modern plasma physics. We have seen that plasma has a many-sided personality—sometimes behaving more like compressible conducting fluid—sometimes more like a congenial group of drifting individual particles, each carrying out slow drifts in the magnetic and electric fields within which it finds itself. We have thus far seen only hints of the complex and perverse side of the plasma personality, those where collective effects appear full-blown and the problems of prediction are more like those of mob psychology than of individual analysis. We will now therefore leave these more casual encounters with the plasma state and begin the discussion of those twin problem children—plasma waves and plasma instabilities. Good luck! We both need it.

III. Waves in High Temperature Plasmas

The question of what waves can be propagated in high temperature plasma is a very important one, and one having virtually no analogy with the propagation of waves in an ordinary gas. In such gases, as the collisional mean free path is increased beyond the size of the system all sound wave propagation ceases. Not so with high temperature plasma. Here the electrostatic and the electromagnetic fields provide a means for coupling between the particles so that they can participate in, and influence, coherent wave propagation, even when collisional processes are totally unimportant. If one adds to this picture the coupling between the particle velocity components provided by any static magnetic field in which the plasma is immersed, it still further complicates an already complicated situation. We cannot hope here to give an adequate treatment of waves in plasmas, but will have to be content with discussing a few of the most important classes. In following this discussion it should be kept in mind that the subject of plasma waves is intimately related to the question of plasma instabilities. Many, if not most, plasma instabilities can be traced to mechanisms within the plasma that permit a resonant feed of energy to a plasma wave from some ordered plasma motion—for example, systematic particle drifts or the orbiting of the particles at their cyclotron frequencies. In such "wave–particle" instabilities the plasma wave grows in amplitude at the expense of the free energy reservoirs represented by the ordered (i.e., nonthermal) components of the particle motions.

A. Electron Plasma Oscillations

The most elementary wave in a high temperature plasma is that associated with so-called "plasma oscillations," the periodic oscillations of the electron component of the plasma about the mean position of the ions of the plasma. Here the restoring force, once a displacement is produced, is the electric field associated with the fractional separation of the charges. The simplest form of plasma oscillation is that which occurs in a "cold" plasma (zero electron temperature), a case that is useful to discuss for heuristic reasons. Consider a slab of plasma of uniform density, sharply bounded in x. The magnetic field is either taken to be zero or to be perpendicular to the faces of the slab. If now the electrons of the plasma are uniformly horizontally displaced to the right, surface charges will result, proportional to the displacement.

$$\sigma = e n_e x \tag{123}$$

So that $E_x = 4\pi\sigma = 4\pi n_e e x$. But the equation of motion for the electrons is

$$e E_x = m_e \ddot{x} \tag{124}$$

resulting in the differential equation

$$\ddot{x} = -\left(\frac{4\pi n_e e^2}{m_e}\right) x = -\omega_{pe}^2 x \tag{125}$$

where $\omega_{pe} = (4\pi n_e e^2/m_e)^{1/2}$, the electron plasma frequency. The solution to Eq. (125) is, of course, simple harmonic motion, and may be taken to be the form $x = x_0 \sin \omega_{pe} t$. It follows that the dispersion equation $D(\omega, k) = 0$ takes the simple form

$$D(\omega, k) = (\omega^2 - \omega_{pe}^2) = 0 \tag{126}$$

Here the waves exhibit no "dispersion"; in this zero temperature approximation all wavelengths of the wave in a homogeneous plasma are characterized by the same frequency. It follows that in this case the group velocity, $\partial\omega/\partial k$, is zero so that the wave is nonpropagating.

When finite temperature effects are introduced into the equations, dispersion appears, the group velocity is nonzero, and propagation is possible over a finite band of wavelengths. The class of electrostatic plasma oscillations that can occur in this case is of fundamental importance in plasma physics, and they are peculiar to the plasma state. They bear no relation to classical electromagnetic waves, being longitudinal, and are furthermore characterized by a damping that persists even in the limit of vanishing collision frequencies. As we have noted earlier, this damping, one of the fundamental processes in high temperature plasma, was first discussed rigorously by Landau in his classic paper, and thus has come to be known as "Landau damping."

To understand how a wave can be damped in the absence of collisional dissipative effects we need to consider what happens when electrostatic waves propagate through a plasma composed of particles with a distribution of velocities (for example, a Maxwellian distribution). Since these waves are longitudinal it follows that if the wave phase velocity happens to match with the speed and direction of some group of particles selected from the distribution, then a resonant exchange of energy between the wave

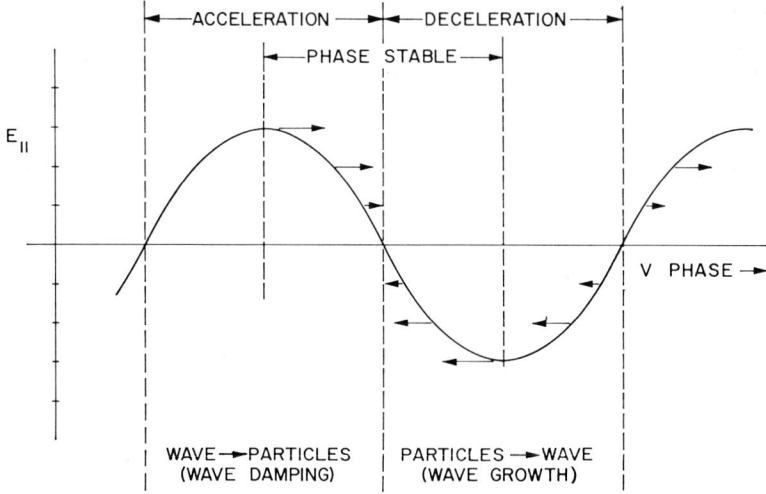

Fig. 8. Schematic of mechanisms of wave growth and wave damping processes arising from wave–particle interactions in a plasma.

and these particles can take place. Depending on the relative phases, the interaction can be one in which energy is fed from wave to particles as in a linear accelerator, or the reverse can occur. However, again as in an accelerator, one will find that certain phases are stable, while others are not. Referring to Fig. 8 we see that particles whose instantaneous phase lies close to the stable phase tend to be bunched at this phase—those moving slightly slower are speeded up; those slightly faster are slowed down. In the first case energy is extracted from the wave; in the second, particle energy is delivered to the wave. Suppose now that at the phase velocity in question the distribution function in that velocity component has a negative slope, i.e., $(\partial f/\partial v) < 0$ at $v = \omega/k$. In this case there will always be more particles with velocities slightly less than the phase velocity, ω/k, than there are with phase velocities slightly greater than the wave phase velocity.

It follows that the energy loss from the wave required to speed up the more populated laggard group will always exceed that fed to the wave by the less populous faster moving group; ergo, the wave is damped. In the process the coherent wave fields will be transformed into presumably incoherent particle motion. Note, however, that in special circumstances the phenomenon of Landau damping can operate in reverse, that is, waves can *grow* at the expense of the particle motion. This situation can arise if the distribution function of the particle has a region of positive slope, $\partial f/\partial v > 0$, that occurs at velocities which resonate with the wave.

Returning to the analysis of plasma oscillations in a hot plasma, the problem can be treated via the Vlasov equation and the method of characteristics, as described in Eq. (122).

As a test wave we consider a longitudinal plane electrostatic wave, describable by a potential of the form

$$\phi_1 = \tilde{\phi}_1 \exp\left[i(k_\| z - \omega t)\right] \tag{127}$$

This assumed disturbance is to be inserted into Eq. (122) to find the perturbed distribution function f_1,

$$f_1 = \frac{e}{M} \int_{-\infty}^{t} \nabla \phi_1 \cdot \nabla_v f_0 \, dt' \tag{128}$$

The integration is to be performed parametrically along the unperturbed trajectories in the relevant coordinates (here only z, because of the plane-wave nature of the assumed perturbation). Thus we take $z = v_\| t$ as the unperturbed trajectory and insert this into Eq. (128). Having calculated f_1 we may then evaluate the perturbed charge density, ρ_1, and use this to evaluate ϕ_1, through Poisson's equation $\nabla^2 \phi_1 = -k_\|^2 \phi_1 = -4\pi\rho_1$, with

$$\rho_1 = \sum_j e_j \int f_1 \, d^3v \tag{129}$$

The summation being carried out over particle species. There results an integral form of the dispersion equation:

$$-k^2 \tilde{\phi}_1 = -4\pi \sum_j \frac{e_j^2}{M_j} \int d^3v \int_{-\infty}^{t} dt' \, \tilde{\phi}_1 \left[ik_\| \frac{\partial f_0}{\partial v_\|} \right] \exp\left[i(k_\| v_\| - \omega)t'\right] \tag{130}$$

At this point we will take $d^3v = 2\pi v_\perp \, dv_\perp \, dv_\|$, that is, we shall specialize to distribution functions that can be described in cylindrical coordinates in velocity space. Such a representation is, for example, appropriate in the presence of a uniform magnetic field oriented parallel to the z axis.

Upon performing the integration over t' and canceling $\tilde{\phi}_1$, which appears on both sides of the equation, the wave dispersion relation is obtained:

$$D(\omega, k) = k_\parallel^2 - 2\pi \sum_j \omega_{pj}^2 \int_0^\infty v_\perp \, dv_\perp \int_{-\infty}^\infty \frac{\partial f_0/\partial v}{v_\parallel - \omega/k_\parallel} \, dv_\parallel = 0 \quad (131)$$

This dispersion equation provides us with a means for determining the propagation characteristics of the plasma waves. That is to say, for any given frequency ω we may use (131) to determine the asymptotic value of the propagation wavenumber, $k_\parallel$. But it is apparent from the singular nature of the integrand in (131), showing a "wave resonance" at $k_\parallel v_\parallel = \omega$, that the dispersion equation cannot be satisfied in general with real values of both k and ω. If, for example, we use (131) to investigate the propagation constants for a traveling wave with real frequency ω, then $k_\parallel$ must be complex: $k_\parallel = k_r + ik_i$. Depending on the sign of k_i, either wave damping ($k_i > 0$) or wave growth ($k_i < 0$) will occur,

$$\phi \sim \exp\,[i(k_r z - \omega t)] \cdot \exp\,[-k_i z] \quad (132)$$

To evaluate $D(\omega, k)$ we must, of course, first perform the indicated integrals over $v_\parallel$, presumably along any convenient contour between the limits. But we see immediately that the answer thus obtained would then depend on the contour chosen, i.e., whether it was one which passed above or below the pole at $v_\parallel = \omega/k_\parallel$ (considering $v_\parallel$ to be a complex variable for the integration). In attempting to resolve this ambiguity, Vlasov supposed that one should take the principal value of the integral, thereby obtaining a purely real result. It was Landau who later showed that Vlasov's assumption was wrong, and that causality dictated the use of a contour integral, the contour being one which rendered $D(\omega, k)$ a continuous function of ω and k, as it must be from physical considerations. The contour consists of one which at all times passes below the Landau pole at $v_\parallel = \omega/k_\parallel$. Thus if Im $\omega/k_\parallel > 0$, the contour is simply the real $v_\parallel$ axis, but if Im $\omega/k_\parallel < 0$, the contour must be deformed into the lower half-plane so as to pass below the pole. In this way one defines the proper function of $\omega/k_\parallel$, through analytic continuation into the lower half-plane.

The exact form of $D(\omega, k)$ thus obtained will of course depend on the details of the equilibrium distribution function, f_0. We shall later see that Eq. (131) will predict either damping (stable waves) or growth (unstable waves) depending on the sign of $\partial f_0/\partial v_\parallel$ for velocities $v_\parallel$ near the "resonant" value $k_\parallel v_\parallel = \omega$.

An important case exhibiting stable waves is provided by the Maxwellian distribution

$$f_0 = \exp\,[-(v_\perp^2 + v_\parallel^2)/v_0^2]; \qquad \langle v^2 \rangle = \tfrac{3}{2} v_0^2 \quad (133)$$

Inserting this distribution function into (131), the dispersion equation takes the form

$$D(k, \omega) = k_\parallel^2 + \sum_j \frac{\omega_{pj}^2}{v_{0j}^2} \left\{ 2\left[1 + \sigma_j \frac{1}{\pi^{1/2}} \int_{-\infty}^{\infty} \frac{e^{-x}\,dx}{x - \sigma_j}\right] \right\} = 0 \quad (134)$$

where

$$\sigma_j = \omega/k_\parallel v_{0j}$$

The function

$$Z(\sigma) = \frac{1}{\pi^{1/2}} \int_{-\infty}^{\infty} \frac{e^{-x}\,dx}{x - \sigma} \quad (135)$$

the so-called "plasma dispersion function" has been tabulated.[28] For Im $\sigma < 0$, appropriate to damped waves as in the present case, the asymptotic form for $Z(\sigma)$ is

$$Z(\sigma) = i2\pi^{1/2}e^{-\sigma^2} - \frac{1}{\sigma}\left[1 + \frac{1}{2\sigma^2} + \frac{3}{4\sigma^4} + \cdots\right] \qquad \sigma \gg 1 \quad (136)$$

Thus the term $2[1 + \sigma Z(\sigma)]$ in (134) becomes, asymptotically,

$$2[1 + \sigma Z(\sigma)] \approx \left[i \cdot 4\pi^{1/2}\sigma e^{-\sigma^2} - \frac{1}{\sigma^2} - \frac{3}{2\sigma^4} - \cdots\right] \quad (137)$$

We now see from the form of the variable σ that when it is evaluated for the ions of the plasma at frequencies $\omega \approx \omega_{pe}$ appropriate to electron plasma oscillations, $\sigma_{\text{ion}} \gg 1$ so that the contribution of the heavier mass ions will generally be negligible (i.e., they remain essentially motionless under these conditions). Thus we may ignore the ion terms in this limit and obtain as the asymptotic form of the dispersion equation

$$D(k, \omega) = k_\parallel^2 - \frac{\omega_{pe}^2}{\bar{v}_{oe}^2}\left\{\left[\frac{1}{\sigma^2} + \frac{3}{2\sigma^4}\right] - i4\pi^{1/2}\sigma e^{-\sigma^2}\right\} = 0 \quad (138)$$

It is apparent from the form of this equation that for real ω the solutions must yield complex $k_\parallel$ values, corresponding to propagation with damping. To demonstrate this explicitly we expand (138) in the limit Im $k_\parallel \ll |k_\parallel|$, finding (the subscript on k will be dropped)

$$|k|^2 + 2ik_i k_r + \frac{\omega_{pe}^2}{\bar{v}_{oe}^2}\left\{i4\pi^{1/2}\frac{\omega}{|k|\,v_{oe}} \exp\left[-(\omega^2/|k|^2 v_{oe}^2)\right]\right.$$
$$\left. - \frac{|k|^2 v_{oe}^2}{\omega} - \frac{3}{2}\frac{|k|^4 v_{oe}^4}{\omega^4}\right\} = 0 \quad (139)$$

Separating this equation into its real and imaginary parts we find, first for the real part:

$$|k|^2 - \frac{\omega_{pe}^2}{v_{oe}^2}\left\{\frac{|k|^2 v_{oe}^2}{\omega^2}\left[1 + \frac{3}{2}\frac{k^2 \bar{v}_{oe}^2}{\omega^2}\right]\right\} = 0 \quad (140)$$

But

$$\frac{\bar{v}_{oe}^2}{\omega_{pe}^2} = \frac{m\bar{v}_{oe}^2}{4\pi n_e^2} = 2\lambda_D^2 \qquad (\lambda_D = \text{Debye length})$$

so that we finally obtain

$$|k|^2 \lambda_D^2 = \frac{1}{3}\left(\frac{\omega^2}{\omega_{pe}^2}\right)\left(\frac{\omega^2}{\omega_{pe}^2} - 1\right) \tag{141}$$

Note that in the limit $T_e \to 0$, $\lambda_D \to 0$, so that we recover the previously derived simple dispersion equation: $\omega^2 - \omega_{pe}^2 = 0$, Eq. (126).

For $(\omega - \omega_{pe}) \ll \omega_{pe}$, $k\lambda_D \to 0$, i.e., the wavelength becomes very long compared to the Debye length. However, as $\omega/\omega_{pe} \to 1.5$, $|k|\lambda_D \to 1$, i.e., the wavelengths become very short. As we shall show, in this limit the wave damping becomes very large, so that for all practical purposes these plasma waves will only propagate for a relatively narrow range of values of ω, namely for $\omega_{pe} \lesssim \omega \lesssim 1.1\omega_{pe}$. To calculate the Landau damping of the waves we have just described, we return to the imaginary part of the dispersion equation and solve for the ratio Im k/Re k, that is for α, the damping constant per wavelength, finding

$$\frac{k_i}{k_r} = \alpha = -\left(\frac{\pi}{2}\right)^{1/2} \frac{1}{k^3\lambda_D^3}\left(\frac{\omega}{\omega_{pe}}\right) \exp\left[-\frac{1}{2k^2\lambda_D^2}\left(\frac{\omega^2}{\omega_{pe}^2}\right)\right] \tag{142}$$

We see immediately that as long as $k^2\lambda_D^2 \ll 1$, the damping will be exponentially small per wavelength, but if λ becomes comparable to λ_D the damping will be very large. This result therefore provides again a verification that λ_D defines the limit of "fine-grainedness" of plasma phenomena. Figure 9 shows the effect of finite temperature ($\lambda_D \neq 0$) on the propagation and attenuation of the waves.

Using (141) we may rewrite (142) in the form

$$\alpha = 3^{3/2}\left(\frac{\pi}{2}\right)^{1/2}\left(\frac{\omega_{pe}}{\omega}\right)^2 \frac{1}{(\omega^2/\omega_{pe}^2 - 1)^{3/2}} \exp\left(-\frac{\frac{3}{2}}{\omega^2/\omega_{pe}^2 - 1}\right) \tag{143}$$

We see that in the limit $\omega \to \omega_{pe}$ the damping rate does indeed become exponentially small, but that as ω increases it rapidly becomes large (corresponding to the decrease of λ toward the Debye length). Since the wave phase velocity is given by $v_p = \omega/k$, we may also write the result in the simple, and important, form

$$\alpha = -2\pi^{1/2}\left(\frac{v_p}{v_{oe}}\right)^3\left(\frac{\omega_{pe}}{\omega}\right)^2 \exp\left[-v_p^2/v_{oe}^2\right] \tag{144}$$

This equation shows that the attenuation per wavelength varies exponentially with the *square* of the ratio of the phase velocity to the mean electron velocity. The origin of this dependence is the fact that the wave

THE PHYSICS OF HIGH TEMPERATURE PLASMA

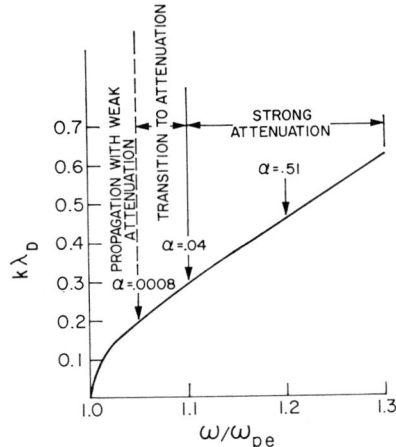

Fig. 9. Plot of dispersion function for longitudinal electrostatic plasma waves. Approximate attenuation constants are shown, indicating regions of weak, intermediate, and strong attenuation by Landau damping.

damping is being produced by the resonant absorption of wave energy by those electrons moving nearly at the phase velocity of the wave. In a Maxwellian plasma the number of such electrons decreases exponentially with the ratio v^2/v_{oe}^2. This simple but very significant result has only recently been verified experimentally, in one of the classic experiments of modern plasma physics.[29] Figure 10 shows a comparison between (a) theoretical and (b) experimentally determined wave attenuations at two different electron temperatures, showing close agreement with theory. It is also clear from the form of Eq. (144) that when $v_p^2/v_{oe}^2 \gg 1$ the attenuation will depend sensitively on the details of the distribution function at its high energy tail end. This result is also borne out by the experiments, in which it was shown that minor modification of the distribution function by means of electron streams had a large effect on the damping.

In the case we have here considered, the wave–particle resonance effects produce damping. As we have noted, however, the original dispersion equation also predicts wave growth, in those cases where $\partial f_0/\partial v_\parallel$ is positive when evaluated at the wave phase velocity. While a Maxwellian distribution always produces damping, nonequilibrium distributions such as those likely to be met in magnetically confined plasmas or in other circumstances may produce wave growth, that is, instability. A simple example is that of an electron beam sent through a plasma. Here the beam represents a "bump on the tail" of the plasma distribution function, a region where $\partial f_0/\partial v_\parallel$ may be positive. When this

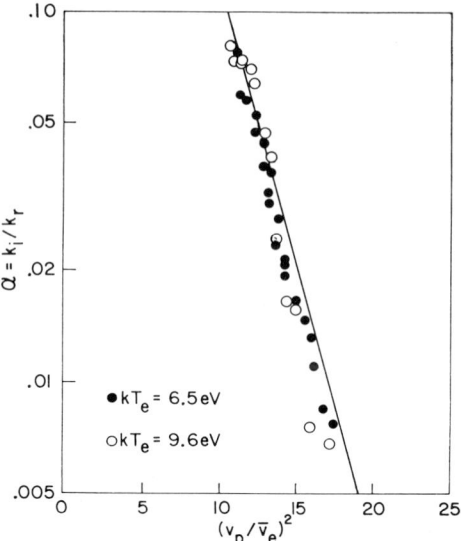

Fig. 10. Comparison of experimental and theoretical attenuation coefficients for longitudinal electrostatic plasma waves propagated along a plasma column (after Malmberg and Wharton, Ref. 29).

is the case, electrostatic plasma waves may grow at an exponential rate in space, rapidly reaching large amplitudes, until their effect in turn destroys the locally peaked nature of the distribution function, meanwhile heating the rest of the plasma electrons. Beam instabilities of this general type have been observed in many experiments. The basic mechanism that they exemplify is one common to many other types of plasma instabilities.

Before leaving the subject of electrostatic plasma waves, we should note an additional important wave of this type, the so-called "ion acoustic wave"[30] that occurs at frequencies much lower than the electron plasma frequency. It will be remembered that the ionic terms were, appropriately, neglected in discussing waves at frequencies of order ω_{pe}. If we now reexamine the general dispersion equation, Eq. (134), in the lower frequency limit and in the limit of small ion thermal velocities, we may readily obtain a dispersion relation for the phase velocity in the form (neglecting small damping terms)

$$v_p^2 = \omega^2/k^2 = \tfrac{3}{2}(\kappa T_e/M)[1/(1 + \tfrac{3}{2}k^2\lambda_D^2)] \qquad (145)$$

In the limit $k^2\lambda_D^2 \ll 1$ the phase velocity is seen to be a velocity equal to the thermal velocity the ions would have if they possessed the temperature of the electrons; thus the expression for the wave velocity resembles that giving the velocity of an acoustic wave in an ordinary medium. Here the electrons provide the coupling between the ions through their electrostatic forces. As $k^2\lambda_D^2 \to 1$ the frequency of the ion acoustic wave tends to approach the value, $\omega \approx \omega_{pi} = (m/M)^{1/2} \omega_{pe}$.

Summarizing the discussion of longitudinal electrostatic plasma oscillations in a plasma, we have seen that the electronic oscillations of this type are highly dispersive, existing only over a relatively narrow band of frequencies above the plasma frequency, $\omega_{pe} = (4\pi n_e e^2/m_e)^{1/2}$ radians/sec. Furthermore, these waves may exhibit either Landau damping or unstable growth, depending on the details of the electron distribution functions. The electron plasma frequency $f_{pe} = \omega_{pe}/2\pi \approx 9 \times 10^3 \, n_e^{1/2}$ Hz lies in the microwave frequency range for most laboratory plasmas, but may be as low as a few kiloHertz for space plasmas.

In addition to the electronic oscillations we have seen also that the ions can undergo electrostatic oscillations at lower frequencies down to $\omega \approx \omega_{pi} = (m/M)^{1/2}\omega_{pe}$, in a wave known as the "ion acoustic wave."

B. Electromagnetic Waves in a Plasma

Although the longitudinal electrostatic plasma waves just discussed have no counterpart in ordinary media, under the proper circumstances transverse electromagnetic waves may be propagated through a plasma much as in other media. In the absence of a magnetic field it is relatively simple to describe their properties. The dispersion equation for these waves, in cases where thermal effects are unimportant, may be derived from simple equations of motion and Maxwell's equations.

From Maxwell's equations we may readily obtain the appropriate wave equation in the form

$$\nabla^2 \mathbf{E} = \frac{1}{c^2} \frac{\partial^2 \mathbf{E}}{\partial t^2} + \frac{4\pi}{c^2} \frac{\partial \mathbf{j}}{\partial t} \tag{146}$$

In the plasma $\mathbf{j} = e(n_i \mathbf{v}_i - n_e \mathbf{v}_e)$ (taking $Z = 1$), so that

$$\frac{\partial \mathbf{j}}{\partial t} = e\left(n_i \frac{\partial \mathbf{v}_i}{\partial t} - n_e \frac{\partial \mathbf{v}_e}{\partial t}\right) \tag{147}$$

But we also have, from the equations of motion,

$$m_e \dot{\mathbf{v}}_e = -e\mathbf{E} \quad \text{and} \quad M_i \dot{\mathbf{v}}_i = e\mathbf{E} \tag{148}$$

Therefore for the average motion of the charge elements we have

$$\frac{\partial \mathbf{j}}{\partial t} = e^2 \left(\frac{n_i}{M_i} + \frac{n_e}{m_e}\right) \mathbf{E} \approx \frac{n_e e^2}{m_e} \mathbf{E} \tag{149}$$

since $M_i \gg m_e$. Substituting this result into (146) the wave equation takes the form

$$\nabla^2 \mathbf{E} - \left(\frac{4\pi n_e e^2}{m_e c^2}\right)\mathbf{E} = \frac{1}{c^2}\frac{\partial^2 \mathbf{E}}{\partial t^2} \tag{150}$$

Now consider the propagation of a transverse plane wave E_y, propagating in the z direction,

$$E_y = E_0 e^{i(kz-\omega t)}$$
$$E_x = E_z = 0 \tag{151}$$

Substitution of this wave into the wave equation yields a dispersion equation,

$$D(\omega, k) = k^2 - \frac{\omega^2}{c^2}\left(1 - \frac{\omega_{pe}^2}{\omega^2}\right) = 0 \tag{152}$$

containing the ubiquitous plasma frequency.

Propagation (i.e., $k^2 > 0$) therefore only occurs for $\omega > \omega_{pe}$. Furthermore, it can be seen that the plasma behaves as though it had a dielectric constant less than unity, since the phase velocity is greater than c

$$v_p = \frac{\omega}{k} = \frac{c}{(1 - \omega_{pe}^2/\omega^2)^{1/2}} \tag{153}$$

becoming very large as $(\omega - \omega_{pe}) \to 0$.

The group velocity, $\partial\omega/\partial k$, is of course less than c.

$$v_g = \partial\omega/\partial k = c(1 - \omega_{pe}^2/\omega^2)^{1/2} \tag{154}$$

vanishing as $(\omega - \omega_{pe}) \to 0$. For $\omega < \omega_{pe}$, as seen from the dispersion equation the wave becomes evanescent, no real propagation being possible. This situation resembles closely the case of wave behavior below "cutoff" in a microwave waveguide.

In the limit $\omega \ll \omega_{pe}$ the "wave" penetrates into the plasma a distance of order $|k|^{-1} \approx c/\omega_{pe}$. This distance $\delta = c/\omega_{pe}$ is known as the "collisionless skin depth." In a plasma of density 10^{12} electrons/cm³, for example, $\delta \approx 0.5$ cm.

1. Effect of Magnetic Field

The propagation of transverse waves in a plasma is greatly altered by the presence of a magnetic field. The effect of the magnetic field is to couple the transverse degrees of freedom of the particle motions and to introduce a marked anisotropy into the wave propagation characteristics, resembling somewhat the situation of light propagation in an optically active crystal.

To evaluate the wave dispersion equation for the case at hand it is necessary to replace the simple equations of motion for the electrons and ions as used above by ones appropriate in a magnetic field. As noted, the magnetic field introduces an essential anisotropy into the problem so that the wave propagation characteristics depend on the direction of the wave vector **k** with respect to **B**. However, for propagation either parallel or perpendicular to **B** the results are relatively simply obtained and stated.[31] They are most easily given in terms of an effective index of refraction for the waves $N = kc/\omega$, or what is equivalent, in terms of the dielectric constant, $K = N^2$. We quote the results here without derivation.

For propagation parallel to **B**, there are two oppositely rotating circularly polarized waves. The first of these, the "electron cyclotron wave," rotating in the same sense as the electron cyclotron motion, has the dielectric constant

$$K_e = 1 - 2 \frac{\bar{\omega}_p{}^2/\omega^2}{(1 + \Omega^+/\omega)(1 - \Omega^-/\omega)} \tag{155}$$

The second, or "ion cyclotron wave" rotating in the ion direction, has the dielectric constant

$$K_i = 1 - 2 \frac{\bar{\omega}_p{}^2/\omega^2}{(1 - \Omega^+/\omega)(1 + \Omega^-/\omega)} \tag{156}$$

Here we have introduced the notation $\bar{\omega}_p{}^2 = (\omega_{pe}{}^2 + \omega_{pi}{}^2)/2 \approx \omega_{pe}{}^2/2$ (i.e., $\bar{\omega}_p{}^2$ is the average squared plasma frequency) and $\Omega^- = |\Omega_e|$, $\Omega^+ = \Omega_i$, the absolute values of the electron and ion cyclotron frequencies (Ω_e is negative by convention).

In the limit $B \to 0$, Ω^+ and Ω^- vanish and we recover the previously derived result for the dielectric constant, allowing propagation ($K > 0$) only for $\omega > \omega_{pe}$. However, when $B \neq 0$ we see that propagation can reappear for special bands of frequencies below ω_{pe}. For example, in the case of the electron cyclotron wave, Eq. (155), consider cases where $\omega_{pe}{}^2 \gg \Omega_e{}^2$. Here we can see directly from the equation that propagation will be possible for $\omega > \omega_{pe}$ (the usual case), will vanish for $\omega_{pe} > \omega > |\Omega_e|$, but will reappear for $\omega < |\Omega_e|$. This particular wave is well known in ionospheric studies, where it has come to be known as the "whistler" mode. This appellation came about because of the discovery that lightning storms could launch such waves into the ionosphere, guided by the earth's magnetic field. Where these same magnetic lines return to earth, a great distance away, a low frequency radio receiver will detect the wave as a descending audio tone—a cosmic "whistler." The detailed theory of this mode not only explains the reason for the downward tone, but also explains the efficient guiding of the waves by the earth's field.[30] The descending frequency character of the detected signals is readily explained in terms of a frequency-dependent group velocity. Since the launching impulse (a lightning bolt) is short in duration it will initiate a broad spectrum of wave frequencies. But if the group velocity is a function of frequency, the signals first arriving at the receiver will correspond to the fastest waves, here observed to be of higher frequency. Later waves will be of a progressively lower frequency, giving rise to a descending tone.

We may directly demonstrate the validity of the above interpretation by calculating the low frequency propagation characteristics of the mode. In the limit $\Omega^- \gg \omega > \Omega^+$, Eq. (155) reduces to

$$N^2 = k^2 c^2 / \omega^2 \approx \omega_{pe}^2 / \omega \Omega^- \tag{157}$$

From this expression we find for the group velocity,

$$v_g = \frac{d\omega}{dk} = 2 \frac{(\omega \Omega^-)^{1/2}}{\omega_{pe}} c \tag{158}$$

Thus $v_g \sim \omega^{1/2}$, so that as observed at the receiving point $\omega \sim 1/T^2$, that is, ω decreases with increasing time after the first signal appears. Furthermore, $v_g \ll c$ under these conditions, so that the propagation delay time may be fairly long compared to ordinary radio waves.

The "ion cyclotron wave" described by (156) is also of considerable interest. This wave can be propagated for $\Omega^+ > \omega > 0$ (as well as for $\omega > \omega_{pe}$). For $\omega \to \Omega^+$ this wave becomes highly dispersive, and the group velocity approaches zero, a circumstance that has been put to practical use: Suppose that circularly polarized ion cyclotron waves are launched, by magnetic induction, in a region of relatively high magnetic field, where $\Omega^+ > \omega$. As the wave propagates down the field lines toward a weaker field region, Ω^+ will decrease toward ω, the group velocity will tend toward zero (resulting in increased wave energy density), and a resonant transfer of energy from the wave to the ions will occur, "heating" the plasma ions. This process has been likened to the breaking up of ocean waves on a beach, and the physical arrangement we have described has been dubbed a "magnetic beach." This idea provides a very efficient and practical way to heat the ions of a high temperature plasma.

At very low frequencies $\omega \ll \Omega^+$, both the cyclotron waves take on a different character, becoming the now famous Alfvén or hydromagnetic waves. In this limit the wave is not dispersive and the dielectric constant assumes a constant value, which from (155) or (156) is found to be

$$K_e = K_i = K_H = 1 + 2\bar{\omega}_p^2 / \Omega^+ \Omega^- \tag{159}$$

Since $m_e \ll M_i$, this becomes

$$K_H = 1 + 4\pi n M c^2 / B^2 = 1 + 4\pi \rho c^2 / B^2 \tag{160}$$

the "hydromagnetic dielectric constant." Note that $K_H = 1 + 2\times$ (mass energy density/magnetic energy density). In all but low density plasmas or plasmas immersed in very strong fields, $K_H \gg 1$. The corresponding wave velocity is the Alfvén velocity[20]

$$v_A = \frac{c}{(K_H)^{1/2}} = c\frac{c}{(1 + 4\pi\rho c^2/B^2)^{1/2}} \tag{161}$$

Since K_H is independent of frequency in this limit, the group and phase velocities of these Alfvén waves are the same. As ω is increased the Alfvén waves will split into two circularly polarized waves, merging continuously into the ion and electron cyclotron waves we have discussed.

We have thus far been considering waves propagating generally parallel to **B**. The picture for waves propagating perpendicular to **B** is rather different.

Consider first the case of linearly polarized electromagnetic waves perpendicularly incident on the plasma, but with their electric vector parallel to **B**. In this case the electrons are free to move along the field lines (in a constant B field) in response to the wave electric field and we recover the simple result obtained in the absence of a field, namely

$$k^2c^2 = \omega^2(1 - \omega_{pe}^2/\omega^2) \tag{162}$$

Waves of this orientation, which propagate freely transversely through the plasma provided $\omega > \omega_{pe}$, have been put to use to measure the plasma density. A beam of microwaves, polarized as above, is transmitted through the plasma. The presence of the plasma causes a phase shift (advancing the phase) in the transmitted wave that can be detected by interferometric techniques. Knowing the thickness of the plasma this phase information can be converted to a measurement of the plasma electron density.[32] As $f_{pe} \approx 9 \times 10^3 n_e^{1/2}$, the microwave technique is generally useful only up to plasma densities of order 10^{13}, where $f_p \approx 3 \times 10^7$ MHz so that microwaves of $\lambda < 1$ cm are required. At higher densities lasers have been successfully employed to perform the same type of measurement.

When the wave is perpendicularly incident but with its electric vector transverse to **B**, the propagated wave exhibits a complicated "hybrid wave" character. In this case the appropriate dielectric constant is[31]

$$K_\perp = 2K_e K_i/(K_e + K_i) \tag{163}$$

Here propagation may occur only for certain discrete bands of frequencies, those where K_e and K_i (see Eqs. (155) and (156)) have signs and relative magnitudes such that $K_\perp > 0$. The conditions for propagation are thus a mixture of those appropriate to plasma electrostatic waves and cyclotron waves. An example is the case of propagation at the "hybrid frequency" $\omega = (\Omega^+\Omega^-)^{1/2} = \Omega_i(M/m)^{1/2}$. Here the expression for $K_\perp$ reduces to

$$K_\perp \approx 1 - (\omega_{pe}/\Omega_e)^4 \cdot |\Omega_e/\Omega_i| \qquad (164)$$

so that propagation is possible provided the plasma density is such as to satisfy the inequality

$$\omega_{pe}^2/\Omega_e^2 < (|\Omega_e/\Omega_i|)^{1/2} = (M_i/m_e)^{1/2} \qquad (165)$$

Note that in this case the plasma would be beyond "cutoff" for normally incident waves of the other polarization (parallel to **B**) at the same frequency. For such waves, as seen from (162), evaluated at the hybrid frequency, the cutoff density is much lower, propagation only occurring for densities such that

$$\omega_{pe}^2/\Omega_e^2 < |\Omega_i/\Omega_e| = m_e/M_i \qquad (166)$$

a factor of about 80,000 lower for the case of protons.

The electrostatic plasma waves, the electron and ion cyclotron waves, the Alfvén waves, and the various hybrids of all these comprise the main spectrum of waves that can be propagated in a homogeneous plasma immersed in a uniform magnetic field. With the exception of the electrostatic plasma waves, where the effects of finite temperature are crucial, we have not discussed the influence of thermal particle motions on the other types of waves. However, these waves, contrary to the electrostatic plasma waves, propagate in the absence of thermal effects. Consequently, it can be expected that the inclusion of thermal motions in the wave equation will not drastically alter the character of these waves. This is indeed the case; resonances are weakened and quantitative changes in the character of the wave dispersion appear but the basic picture remains very similar. What is new is that, as in the electrostatic plasma waves, collisionless damping or growth can appear, arising from resonant energy exchanges between waves and particles. We will not discuss such cases here but will later discuss some examples that arise in connection with plasma instability.

C. Drift Waves

The already complex situation, of wave propagation in a homogeneous plasma immersed in a uniform magnetic field, is made even more complex when density gradients or magnetic gradients are present. In such situations "drift waves" may exist, arising from differential motions of ions and the electrons in these gradients. If the phase velocity of one of these waves happens to correspond to some mean drift velocity, the wave may be strongly damped or it may grow unstably. Most drift waves, in common with the propagating electrostatic waves, depend on finite temperature effects for their existence, but some persist even in the limit of vanishing orbit sizes. Drift waves may be purely transverse, $\mathbf{k} \cdot \mathbf{B} = 0$, or may propagate at a slight angle to the field, $k_\parallel \ll k_\perp$. In the latter case, electron Landau damping effects along the field lines play a dominant role in determining the angle of propagation of the wave.

In addition to distinctions with respect to direction of propagation, drift waves fall into two general classes as far as their frequency spectrum is concerned: (1) Low frequency waves, associated with hydromagnetic effects; these generally have wavelengths large compared to particle orbit diameters. (2) High frequency waves, occurring at or near ion or electron cyclotron frequencies or their harmonics; waves characterized by very short wavelengths, often small even compared to orbit diameters. This latter circumstance arises because for a drift wave to exist its phase velocity must be comparable to the drift that sustains it. But in this case these drifts, as our previous discussion has shown, are always slow compared to the ion or electron velocities themselves. Since $\omega \sim \Omega_i$ or greater and since the phase velocity ω/k must be comparable to the drift velocity, it follows that $\Omega_i/k \approx v_d$, i.e., $(\Omega_i/k\bar{v}_i) = v_d/\bar{v}_i \ll 1$ so that $k(\bar{v}_i/\Omega_i) = k\bar{a}_i \gg 1$.

The low frequency drift waves can be treated adequately through the use of either hydromagnetic or guiding center equations. However, to deal with high frequency drift waves requires the Vlasov equation and entails very sophisticated analytical techniques. We will here limit the discussion to a single example of each type of drift wave, and refer the reader to the extensive literature for additional details.

As we have suggested, a prime requirement for a drift wave to appear is that there exist a gradient in some plasma or environmental parameter. Among the parameters whose gradients have been considered are (1) density, (2) magnetic field, (3) temperature, and, even, (4) impurity concentration. Associated with each gradient or combination of gradients there will exist differential drift motions within the plasma that can support waves. These waves may propagate with either positive damping or negative damping, that is, unstable growth. As the parameters are changed, stable drift waves may become unstable, and vice versa. In other words these waves almost always exhibit some kind of initial threshold condition. One of the tasks of the plasma theorists is to delineate those regimes for which each drift wave is stable. This task is one closely related to the problem of plasma confinement, that monomania of the fusion researcher, since unstable drift waves can cause anomalously rapid plasma transport out of the confining field. Their efficacy in causing such transport, however, depends very much on the nature of the wave considered. Generally speaking, long-wavelength, low frequency waves, of which the prime example is the hydromagnetic instability, cause the most rapid transport, whereas the high frequency waves, being of very short wavelength, cause much slower transport, as far as cross-field motion is concerned. This matter will be mentioned again in connection with the discussion of plasma instabilities.

1. Hydromagnetic Drift Waves

Plasma cannot be confined magnetically without giving rise to the possibility of hydromagnetic drift waves—stable or unstable. An obviously necessary ingredient of confinement is that a negative density gradient should exist at the plasma boundaries. A second necessary ingredient is that the lines of force of the confining magnetic field possess curvature—whether it be the curvature associated with toroidal systems or the curvature present in magnetic mirror fields. Field curvature produces drifts; particle drifts in the presence of a density gradient can support drift waves. To facilitate the discussion of these waves we will introduce a convenient fiction, often employed in analyzing this type of problem. Since the particle drifts produced by a gravitational field are very similar in character to those resulting from field curvature, we will replace the curvature drifts in the problem with ones associated with a fictitious "g" operating in a constant B field, later identifying this "g" with the real drifts.[2,33] The rationale for the use of a fictitious gravity is that it avoids the difficult problem of working with the curvilinear coordinates of the actual B field.

In keeping with the low frequency nature of hydromagnetic disturbances, and with the fact that thermal motions of the particles play a secondary role in such phenomena, we may employ guiding center equations (108) to treat the problem.

The basic equations are (1) the equation of continuity for the density of guiding center for each species (ions and electrons);

$$\frac{\partial N}{\partial t} + \nabla \cdot (N\mathbf{v}_D) = 0 \tag{167}$$

together with (2) the equation for the particle drift velocities $\mathbf{v}_D$

$$\mathbf{v}_D = \frac{\mathbf{g} \times \mathbf{B}}{\Omega B} + c\frac{\mathbf{E} \times \mathbf{B}}{B^2} + \frac{(d\mathbf{v}_D/dt) \times \mathbf{B}}{\Omega B} \tag{168}$$

(gravitational drift) (electric field drift) (acceleration drift)

and (3) Poisson's equation for the electrostatic potential ($Z = 1$ assumed)

$$\nabla^2 \phi = 4\pi(N_i - N_e)e \tag{169}$$

In addition we shall assume that β is small (plasma pressure small compared to $B^2/8\pi$) so that the electric field is given by the gradient of a potential, $\mathbf{E} = -\nabla \phi$, and therefore $\nabla \times \mathbf{E} = 0$. We also take $\mathbf{E} = 0$ in the equilibrium state (or transform to a frame where this is the case). We will be satisfied to consider the characteristics of small amplitude waves and will therefore work with linearized equations, neglecting quantities that are of second order in the wave amplitudes. The plasma will be assumed to have its density gradient in the y direction, with $\mathbf{B}$ being parallel to the z axis (Fig. 11).

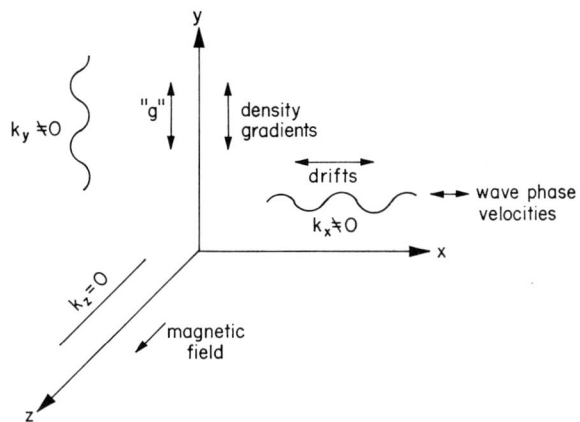

Fig. 11. Coordinate system for hydromagnetic drift wave calculation.

Expanding the continuity equation we obtain

$$\frac{\partial N}{\partial t} + \mathbf{v}_D \cdot \nabla N + N \nabla \cdot \mathbf{v}_D = 0 \qquad (170)$$

for each species. This is the equation that is to be linearized.

Consider now the acceleration drift, containing the term

$$\frac{d\mathbf{v}_D}{dt} \times \mathbf{B} = \frac{c}{B^2}\left(\frac{d\mathbf{E}}{dt} \times \mathbf{B}\right) \times \mathbf{B} \qquad (171)$$

$\mathbf{g}$ is assumed constant in time; it does not contribute to this term. Expanding we find

$$\frac{d\mathbf{v}_D}{dt} \times \mathbf{B} = \frac{c}{B^2}\left\{\frac{d\mathbf{E}}{dt}(\mathbf{B} \cdot \mathbf{B}) - \mathbf{B}\left(\frac{d\mathbf{E}}{dt} \cdot \mathbf{B}\right)\right\}$$

$$= c\frac{d\mathbf{E}}{dt} \qquad (172)$$

since $\mathbf{E} \cdot \mathbf{B} = 0$ at hydromagnetic frequencies, as required by the condition $(\mathbf{E} + (1/c)\mathbf{v} \times \mathbf{B}) = 0$. Now

$$\frac{d\mathbf{E}}{dt} = \frac{\partial \mathbf{E}}{\partial t} + (\mathbf{v}_D \cdot \nabla)\mathbf{E} = \frac{\partial \mathbf{E}}{\partial t} + \frac{\mathbf{g} \times \mathbf{B}}{\Omega B} \cdot \nabla \mathbf{E} \qquad (173)$$

The second term is negligible compared to the first, so that the final expression for the totality of the drift velocities reduces to

$$\mathbf{v}_D = \frac{\mathbf{g} \times \mathbf{B}}{\Omega B} + c\frac{\mathbf{E} \times \mathbf{B}}{B^2} + \frac{c}{\Omega B}\frac{\partial \mathbf{E}}{\partial t} \qquad (174)$$

We now linearize the continuity equation by writing $N = N_0(y) + N_1(x, y, t)$, $N_1 \ll N_0$, neglecting terms of order N_1^2 or $N_1\mathbf{E}$ (since $\mathbf{E} \sim N_1$). We also neglect terms of order εN_1 as compared to N_1, i.e., assume that the characteristic length associated with the density gradient is large compared to the wavelengths of the wave as measured in the y direction. With these linearizations the guiding center equation becomes

$$\frac{\partial N_1}{\partial t} + \frac{g}{\Omega}\left(\frac{\partial N_1}{\partial x}\right) = \frac{c}{B} E_x \left(\frac{\partial N_0}{\partial y}\right) + \frac{c}{B} N_0\left[\frac{\partial E_y}{\partial x} - \frac{\partial E_x}{\partial y}\right]$$
$$+ \frac{c}{\Omega B}\left[\frac{\partial \dot{E}_x}{\partial x} + \frac{\partial \dot{E}_b}{\partial y}\right] N_0 = 0 \qquad (175)$$

The next-to-the-last term vanishes because $\nabla \times \mathbf{E} = 0$ by assumption. We now take N_1 to be a test wave of the form

$$N_1 = \tilde{N} \exp\left[i(k_x x + k_y y - \omega t)\right] \qquad (176)$$

and substitute in Eq. (175), obtaining

$$N_1 = \frac{i}{(\omega - gk_x/\Omega)} \frac{c}{B} \left\{ E_x \frac{\partial N_0}{\partial y} + \frac{N_0}{\Omega}\left[\frac{\partial E_x}{\partial x} + \frac{\partial E_b}{\partial y}\right]\right\} \qquad (177)$$

Now $\mathbf{E} = -\nabla \phi$, and we may take ϕ also to be of the form of the test wave

$$\phi_1 = \tilde{\phi} \exp\left[i(\omega t + k_x x + k_y y)\right] \qquad (178)$$

so that we obtain an expression for N_1 in terms of ϕ_1

$$N_1 = \frac{1}{\omega - gk_x/\Omega} \frac{c}{B}\left\{k_x \frac{\partial n_0}{\partial y} - \frac{\omega}{\Omega}(k_x^2 + k_y^2)N_0\right\}\phi_1 \qquad (179)$$

We will have one such equation for each species. Since we have here taken $k_z = 0$, this automatically insures that $\mathbf{E} \cdot \mathbf{B} = 0$, as required by the hydromagnetic equations. It follows from the fact that $k_z = 0$ that Landau damping effects from electron thermal motions along the field lines can play no role for these waves and need not be included, there being no component of **E** along **B** to provide the necessary coupling. We shall later see that if k_z is nonvanishing, important new effects can arise not included in the hydromagnetic wave picture here being discussed.

The perturbed density N_1 may now be eliminated from the equation through Poisson's equation

$$\nabla^2 \phi_1 = -(k_x^2 + k_y^2)\phi_1 = -k^2 \phi_1 = -4\pi(N_{1i} - N_{1e})e \quad (180)$$

In the expression for the charge density contributed by the electrons, N_{1e}, we may neglect Ω_e^{-1} as compared to Ω_i^{-1}, obtaining finally

$$k^2 \phi = \frac{-4\pi N_0 ec}{\Omega_i B} \left\{ \frac{\omega^2 k^2 + gk_x^2 \varepsilon}{\omega^2 - (\omega/\Omega_i)gk_x} \right\} \quad (181)$$

where $\varepsilon = (1/N_0)(\partial N_0/\partial y)$. Making the substitution $b^2 = B^2/4\pi N_0 mc^2$, $\Omega_i = eB/MC$, we obtain the dispersion equation in the form

$$D(k, \omega) = \omega^2(1 + b^2) - \omega \left[\frac{gk_x b^2}{\Omega_i} \right] - \varepsilon g \left(\frac{k_x^2}{k^2} \right) = 0 \quad (182)$$

This dispersion equation, quadratic in ω, predicts stable waves if its solutions are real, unstable if complex. Noting that $(1 + b^2)/b^2 = 1 + 1/b^2 = 1 + 4\pi\rho c^2/B^2 = K_H$, the hydromagnetic dielectric constant, we may rewrite the dispersion equation in the form

$$\omega^2 - \left(\frac{\omega}{\Omega_i}\right)\frac{gk_x}{K_H} - \varepsilon g \left(\frac{k_x^2}{k^2}\right)\left(\frac{K_H - 1}{K_H}\right) = 0 \quad (183)$$

Solving this equation for ω we obtain

$$\omega = \frac{1}{2}\left\{\frac{gk_x}{\Omega_i K_H} \pm \left[\left(\frac{gk_x}{\Omega_i K_H}\right)^2 + 4\varepsilon g\left(\frac{k_x^2}{k^2}\right)\left(\frac{K_H - 1}{K_H}\right)\right]^{1/2}\right\} \quad (184)$$

We see first that when g is positive (g in the same direction as the density gradient) the solutions correspond to two stable waves with oppositely directed phase velocities:

$$v_H = \frac{\omega}{k_x} = \frac{g}{2\Omega_i K_H} \pm \frac{1}{2}\left[\left(\frac{g}{\Omega_i K_H}\right)^2 + \frac{4\varepsilon g}{k^2}\left(\frac{K_H - 1}{K_H}\right)\right]^{1/2} \quad (185)$$

At low densities $K_H \to 1$ and one phase velocity approaches 0, while the other approaches g/Ω_i, the gravitational drift velocity. At high densities, $K_H \gg 1$, and

$$v_H = \pm(\varepsilon g/k^2)^{1/2} \quad (186)$$

independent of density and varying as $\varepsilon^{1/2}$.

Our main interest in these waves arises when we make the identification of the ficticious $g \approx \bar{v}^2/R$, as mentioned earlier, so that we are in effect treating hydromagnetic drift waves as they would appear in nonuniform magnetic fields. Noting also that $\bar{v}^2/\Omega_i^2 = a_i^2$, the mean square ion orbit radius, we may rewrite the dispersion equation (183) for ω in the form

$$\omega^2 - \omega\Omega_i\left[\frac{k_xR}{K_H}\left(\frac{a_i^2}{R^2}\right)\right] + \varepsilon R\left(\frac{a_i^2}{R^2}\right)\left(\frac{k_x^2}{k^2}\right)\left(\frac{K_H - 1}{K_H}\right)\Omega_i^2 = 0 \quad (187)$$

Being a quadratic equation, this equation yields two values for ω, which are given by the complicated expression

$$\omega = \frac{\Omega_i}{2}\left\{\frac{k_xR}{K_H}\left(\frac{a_i^2}{R^2}\right) \pm \left[\left[\frac{k_xR}{K_H}\left(\frac{a_i^2}{R^2}\right)\right]^2 + 4\varepsilon R\left(\frac{a_i^2}{R^2}\right)\left(\frac{k_x^2}{k^2}\right)\left(\frac{K_H - 1}{K_H}\right)\right]^{1/2}\right\}$$
(188)

Since $a_i \ll R$ by assumption, it follows that $\omega \ll \Omega_i$, as is required for hydromagnetics to apply. Note also that the sign of the phase velocity ω/k_x depends on the direction of the density gradient relative to the curvature of the field lines; $\varepsilon R > 0$ corresponds to field lines convex toward the plasma surface and $\varepsilon R < 0$ corresponds to field lines bulging away from the plasma.

As long as $\varepsilon R > 0$ it can be seen from the form of the solution that ω will always be real, corresponding to stable waves. If $\varepsilon R < 0$, the second term under the square root becomes negative and under the proper conditions may dominate, so that ω becomes complex, with two complex conjugate roots, $\omega = \omega_r \pm i\omega_i$. Since the original test wave was postulated to vary as $e^{-i\omega t}$, the existence of a positive imaginary component in ω implies that one of the two wave solutions is unstable, growing exponentially with time with an e-folding time ω_i^{-1}. We shall discuss the instability represented by this case at greater length in the section on plasma instabilities. It represents one of the most important and earliest predicted types in plasma research—the hydromagnetic "flute" or "interchange" instability.

2. *"Finite-Orbit" Effects*

The drift waves just described have been treated only in the hydromagnetic limit $ka_i \ll 1$, $a_i/R \ll 1$, and for the case $k_z = 0$ (infinite wavelength along the field lines). By so doing we have ignored two important effects. The first of these is the so-called "finite-orbit effect" arising from the fact that in a magnetically confined plasma where the electrons and the ions have finite and comparable temperatures the mean diameters of the orbits of the ions and electrons are quite different. For example, if $T_e = T_i$, then $a_i/a_e = (M/m)^{1/2}$ (assuming $Z = 1$). This circumstance implies that in the non-uniform electric field associated with the plasma wave the ions and electrons will not undergo **E** × **B** drifts at exactly the same rate, since the average electric field experienced by the ions as they circle in their orbits will differ from that experienced by the electrons, even for an electron whose guiding center happens to coincide at the same instant with the ion's guiding center. These finite-orbit effects have been treated rigorously by use of the Vlasov equation.[27,34] However, they can also be incorporated as a correction to the guiding center equations that we have employed here. To do this one need only evaluate the average electric field experienced by the particle in question as it moves within the test wave field, and then use this averaged field in the drift velocity equation. Thus one takes

$$v_E = c \frac{\langle \mathbf{E} \rangle \times \mathbf{B}}{B^2} \tag{189}$$

where $\langle \mathbf{E} \rangle$ is to be calculated by taking a time-averaged value of **E** around the particle orbit. In a plane wave field it is not difficult to show that for waves with $k_z = 0$,

$$\langle \mathbf{E} \rangle = J_0(ka)\mathbf{E} \tag{190}$$

giving rise to substantial changes in the electric drifts if wavelengths approach the orbit radius, a. For our purposes it will in some cases be sufficient to consider only terms of order k^2a^2 or lower, so that we then take

$$\langle \mathbf{E} \rangle \approx (1 - k^2a^2/4)\mathbf{E} \tag{191}$$

from the first two terms of the expansion of the Bessel function J_0.

In addition to correcting the electric drifts an additional finite orbit correction needs to be made to the guiding center equations. This correction simply allows for the fact that in the region of a density gradient the actual plasma density and the density of guiding centers are not quite identical, since the actual particle density at a point is composed of the averaged sum of the contributions from all particles whose guiding centers lie within an orbit radius of the point in question.

The overall effect of adding the two corrections described here to the guiding center equations and then recalculating the dispersion equation for hydromagnetic drift waves is to yield a modified equation, of the form (compare Eq. (182))

$$\omega^2[J_0^2(k_x\bar{a}_i) + b^2] - \omega\left\{\frac{gk_x}{\Omega_i}b^2 + \Omega_i\left(\frac{\varepsilon}{k_x}\right)[1 - J_0^2(k_x\bar{a}_i)]\right\} - \varepsilon g = 0 \tag{192}$$

where we have neglected the electron orbit size, a_e, compared to a_i and k_y relative to k_x. In the limit $ka_i \to 0$, this equation of course reverts to the previously derived one, Eq. (182).

The dispersion equation obtained above now includes most of the important features of low frequency transverse drift waves in an inhomogeneous plasma, i.e., one with density gradients transverse to the magnetic field. When g is replaced by $\bar{v}^2/R$, as before, the equation also gives a reasonable representation of the effects of magnetic field curvature. As we shall later show, the inclusion of finite orbit effects has a profound influence on the stability of these waves. At hydromagnetic frequencies, the case here considered, finite orbit effects may stabilize otherwise unstable waves; at higher frequencies they may actually lead to the stimulation of new types of instabilities. Indeed they point up the fact that to view a high temperature plasma as a structureless fluid medium is to ignore most of the complexity of its behavior.

3. Low Frequency Drift Waves with $k_\| \neq 0$

There remains one more type of low frequency drift wave that it is important to discuss. This is the case which arises when one considers the propagation of test waves with the wave **k** vector not exactly perpendicular to **B**, so that though $k_\| \ll k_\perp$, it is still finite. In this case $\mathbf{E} \cdot \mathbf{B} \neq 0$ for the electric field of the wave so that the effect of electron motion along **B** plays an important role. In the low β limit, $\beta < m/M$, we may analyze this type of drift wave by an extension of the guiding center analysis first discussed. We need only recognize that with frequencies sufficiently low to satisfy the requirement $\omega \ll k\bar{v}_e$ one can assume that the electrons will maintain their thermal distribution in the presence of the component of wave electric field along **B** so that the perturbed electron density is given by the simple Boltzmann relation

$$n_e = n_e \exp[e\phi/\kappa T_e] \approx n_0(1 + e\phi/\kappa T_e) \tag{193}$$

since $e\phi \ll \kappa T_e$ for the test wave. The drift wave that we shall find in this case requires only a density gradient and a finite electron temperature for its existence; ion thermal velocities and magnetic curvature play no essential role and appear as corrections, electric drifts being the only ones essential to the problem. Here, however, test waves are to be taken to be of the form

$$\exp[i(k_x x + k_z z - \omega t)]$$

We may assume $k_y = 0$ for simplicity. Following essentially the same procedure as before we will find the dispersion equation for these waves takes the form[35]

$$\omega^2 - k_x\left(\frac{\kappa T_e c}{eB}\varepsilon\right)\omega - k_z^2\left(\frac{\kappa T_e}{M}\right) \quad (194)$$

In the limit of zero density gradient ($\varepsilon \to 0$), this equation has the two solutions

$$\omega = \pm k_z(\kappa T_e/M)^{1/2} \quad (195)$$

i.e., "ion sound" waves with the phase velocity along the field lines equal to

$$c_i = \omega/k_z = (\kappa T_e/M)^{1/2} \quad (196)$$

the same waves that we have earlier discussed (Eq. (145)).

When $\varepsilon \neq 0$, Eq. (194) predicts that two drift waves with electric field components perpendicular to **B** can exist, even in the limit $k_z = 0$. In this limit the dispersion equation (194) becomes

$$\omega[\omega - k_x(\kappa T_e c/eB)^n] = 0 \quad (197)$$

corresponding to one solution of zero frequency, and another propagating transversely with the phase velocity

$$\omega/k_x = (\kappa T_e c/eB)\cdot\varepsilon = (\kappa T_e/m_e)(1/\Omega_e)\cdot\varepsilon \quad (198)$$

This velocity will be recognized as simply equal to the mean thermal drift velocity of the electrons arising from the density gradient.

When $k_z \neq 0$ the wave frequencies as predicted by the dispersion equation are

$$\omega = \frac{\varepsilon k_x}{2}\left(\frac{c_i^2}{\Omega_i}\right)\left\{1 \pm \left[1 + \left(\frac{k_z^2}{k_x^2}\right)\frac{\Omega_i^2}{\varepsilon^2 c_i^2}\right]^{1/2}\right\} \quad (199)$$

The waves represented by these solutions may be of low frequency, and depend for their existence on finite electron temperature. The reasons for discussing them at all at this point may seem obscure, but these waves are indeed very important to the subject of high temperature plasma physics. The reason: such drift waves may be driven unstable by finite orbit effects, even at very low densities and in a uniform magnetic field; all that is required is that a density gradient be present. For this reason the instability thus found has been dubbed a "universal instability," an appellation that certainly deserves attention if one is going about the business of magnetic confinement. We shall again defer treatment of these sinister entities to the upcoming section on plasma instabilities.

D. Summary

Perhaps the best single word to describe the wave propagation problem in plasmas is "bewildering." Yet from the complexities of the situation a few simple points may be distilled. We have first seen that plasmas, because of the fundamental role that the tendency toward space charge neutrality plays, may propagate a purely longitudinal electrostatic wave not possible in ordinary media. We have further seen that this wave propagates and is damped by an important process, Landau damping, that does not depend on collisional effects but is rooted in the effect of the thermal velocities of the electrons.

In addition to the longitudinal waves, we have found that transverse electromagnetic waves may propagate through a plasma, but that their propagation is also strongly affected by the same plasma wave resonances found above, as well as by resonances with the cyclotron frequencies of the ions and the electrons. At low frequencies these waves become either the fundamentally important hydromagnetic Alfvén wave, in which the plasma exhibits an effective dielectric constant $K_H = 1 + 4\pi\rho c^2/B^2$, or else electrostatic "ion sound" waves propagating at the phase velocity $c_i = (\kappa T_e/M)^{1/2}$.

Upon considering inhomogeneous plasmas, "drift waves" were found propagating generally transverse to the direction of density gradients, and influenced by (or even arising only in the presence of) magnetic gradients.

The need to include the effects of finite orbit size in discussing plasma waves was stressed, and the new waves arising when the nonhydromagnetic effect of electric fields parallel to **B** were discussed.

Besides their importance to the general understanding of plasma behavior, plasma waves bear a fundamental relationship to the problem of plasma instability, since they are the carriers for this disease of the plasma state, a subject the discussion of which may unfortunately no longer be deferred.

IV. Plasma Instabilities

Preoccupation with the subject of plasma instabilities has been the most pronounced characteristic of high temperature plasma research, both in the laboratory and in space research. We have already indicated the reasons for this circumstance: In space, almost every plasma phenomenon observed owes its origin to some interaction between plasma and magnetic fields—effects that we would describe on earth as plasma instabilities. In fusion research, where magnetic confinement is a prerequisite to the solution of the fusion power problem, understanding and controlling plasma instabilities is essential to achieving a degree of confinement adequate to permit self-sustaining thermonuclear reactions.

In the present context what do we really mean when we refer to "plasma instability"? We mean any cooperative plasma motion that can regenerate itself, starting from the normal level of random fluctuations in a plasma, in a time short compared to that of collisional processes in the plasma. Thus "plasma instability" can refer to motions that range all the way from a gross velocity of the plasma as a whole across a confining magnetic field, to high frequency, short-wavelength oscillations of the plasma accompanied by intense fluctuating electric fields, but perhaps by relatively little transport across the magnetic field. Thus we will be concerned not only with the existence of instabilities but also with their consequences. When instabilities cause a too rapid loss of plasma from a magnetic confining field they become of special interest to fusion researchers. When they strongly influence energy transport between the particles of a plasma and give rise to heating effects they may be of special interest to space researchers and astrophysicists attempting to understand the aurorae or the origin of cosmic rays. Of all the hallmarks of the high temperature plasma state, the complexity of its unstable behavior is foremost. Yet despite this complexity it has been found possible to analyze such problems in great detail, to predict means for the manipulation and control of plasma instabilities, and in many cases to verify these predictions with exactitude by experimental observations.

The division between gross instabilities and the fine-scale "microinstabilities" that we have suggested is a useful one to adopt. By their nature, gross instabilities are slow ($\omega \ll \Omega_i$), and involve wavelengths that are generally large compared to orbit diameters. Whether they appear as unstable drift waves, or simply as a gross sideways motion of the plasma, they owe their origin to a very simple circumstance: If a magnetically confined plasma can exchange some of its internal energy for directed motion by distorting or by starting to move in some direction across the confining field, such an exchange will occur. (As fusion researchers well know, plasmas are keenly aware of Finagle's law: if something can go wrong—it will!) The simplest examples of gross instabilities driven in such a manner are those where the plasma is able to expand in the course of moving transversely. Here the free energy driving the motion is simply the $p\,dV$ work available when any gas expands. But on the hydromagnetic time scale, and in a high temperature plasma, we have seen that if a plasma moves across a field it must move in such a way as to preserve constant magnetic flux through its interior. Thus the question as to whether a plasma is incipiently grossly instable or not depends simply on the question as to whether a transverse perturbation exists which allows the plasma to achieve a net expansion in volume, at the same time enclosing a constant amount of magnetic flux. The simplest example of such a flux-preserving unstable expansion would be the behavior of a plasma in a magnetic field with a transverse gradient in one direction only. In such a case, as illustrated in Fig. 12, a simple sideways drift toward the region of weaker field would permit the plasma to expand, at the same time preserving constant flux. Conversely, a displacement toward the region of stronger field is a stable one, i.e., the plasma is compressed, creating a

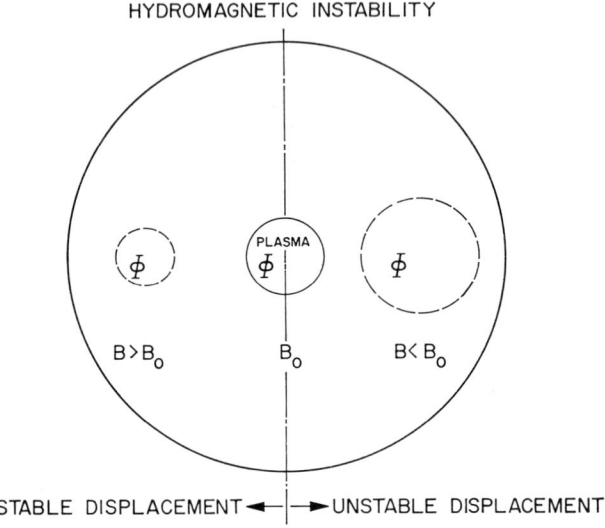

Fig. 12. Schematic illustrating hydromagnetically stable and unstable displacements of plasma column in a magnetic field, the intensity of which varies with position transversely.

restoring force. It follows from this simple picture that a plasma column located at the "bottom" of a magnetic well field (see Fig. 3c) is stable against any form of flux-preserving gross instability. Any displacement from the equilibrium will in this case result in forces tending to restore the plasma to the bottom of the well. Conversely, because the field decreases radially, plasmas confined in ordinary mirror fields (see Fig. 3b) will exhibit a tendency for gross instability, unless other constraints, such as finite orbit effects, inhibit the instability.

It is now easy to see why plasma cannot be contained in a simple torus, but instead rapidly expands toward the outer wall, i.e., in the direction of weaker field. In more complicated toroidal fields, such as those in the stellarator, the situation is not so easy to discern. We may, however, derive a simple stability criterion which helps to assess the stability of such systems. To derive this criterion we need only invoke the expansion energy argument, used above to explain the origin of gross instability.

Consider a flux-preserving displacement of an element of the plasma, as shown in Fig. 13. Comparing the situation in the two portions, we see that the volume enclosed is in the two cases given by

$$\text{Vol}_1 \sim \int (\text{Area})_1 \, dl$$

$$\text{Vol}_2 \sim \int (\text{Area})_2 \, dl$$

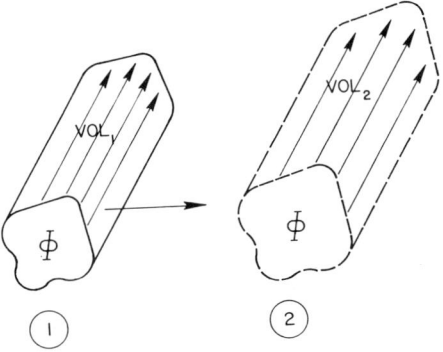

Fig. 13. Schematic illustrating the origin of the $\int dl/B$ stability criterion for hydromagnetic stability.

If $\text{Vol}_2 < \text{Vol}_1$ the displacement will be a stable one, and vice versa. On the other hand, the magnetic flux $\phi = B \cdot (\text{Area})$, so that

$$\int \text{Area} \, dl \sim \int \phi \frac{dl}{B} \sim \int \frac{dl}{B}$$

since $\phi = $ constant. It follows, therefore, that if the quantity

$$I = \int \frac{dl}{B}$$

is a maximum at the location of the plasma, the plasma will be stable against gross instabilities on the hydromagnetic time scale.

Before discussing the application of the above criterion to toroidal systems we will return to our earlier discussion of the hydromagnetic drift waves. These waves represent the normal modes through which the plasma will respond to perturbations. They are therefore intimately related to the question of gross instabilities, which are nothing more than unstable hydromagnetic wave modes.

Recalling Eq. (188) for the frequency of the drift waves, we found that in a magnetic field the lines of which have a radius of curvature R, the frequency of the drift wave is given by

$$\omega = \frac{\Omega_i}{2} \left\{ \frac{k_x R}{K_H}\left(\frac{a_i^2}{R^2}\right) \pm \left[\left[\frac{k_x R}{K_H}\left(\frac{a_i^2}{R^2}\right)\right]^2 + 4\varepsilon R\left(\frac{a_i^2}{R^2}\right)\left(\frac{k_x^2}{k^2}\right)\left(\frac{K_H - 1}{K_H}\right) \right]^{1/2} \right\} \ll \Omega_i \quad (200)$$

When $\varepsilon R > 0$ (field lines convex toward the plasma), this equation has only unstable (real) solutions. This situation corresponds to the case of the magnetic well field (Fig. 3c) with plasma concentrated near the bottom of the well. On the other hand, if $\varepsilon R < 0$, lines concave toward the plasma, the wave may become unstable. This occurs when the terms inside the square root become negative, i.e., if

$$\varepsilon > \frac{k^2 \bar{a}_i^2}{2|R|} \frac{1}{K_H(K_H - 1)} \quad \text{unstable} \quad (201)$$

If k could become arbitrarily small (λ very large), the plasma would be instable at arbitrarily small densities or density gradients. However, in a confined plasma λ cannot exceed a maximum value given by $\lambda_{max} \approx 2\pi r$, with r being the mean plasma radius. If we then also estimate the minimum value of the density gradient parameter by

$$\varepsilon_{min} \approx 1/r$$

an appropriate condition for stability is

$$K_H(K_H - 1) < \frac{a_i^2}{2r|R|} \ll 1 \quad \text{stable} \quad (202)$$

Thus while this condition cannot be satisfied for ordinary densities where $K_H \gg 1$, it shows that there does exist a density threshold for the hydromagnetic drift instability, below which it is stable.

The above circumstance illustrates an important point: while simple energetic arguments such as the dl/B criterion may provide sufficient conditions for stability, not all plasma systems are unstable that do not satisfy such criteria. To sustain an instability, not only must the unstable motion be energetically favorable (free energy available), but also the conditions must be such that an allowed mode of plasma motion can tap this energy reservoir through "resonant" energy exchange. Thus the existence of a threshold density for instability is a common feature of drift wave instabilities of all types, both low and high frequency. The origin of the threshold effect can in each case be traced to the fact that below the threshold the wave in question cannot achieve resonance with the specific source of plasma free energy that provides the reservoir for its growth.

With this concept in mind we return to our consideration of the simple hydromagnetic drift instability. Let us examine the expression for the phase velocity of the wave in the x direction, given by ω/k_x. From Eq. (200) this velocity is

$$v_H = \left\{ \frac{g}{2\Omega_i K_H} \pm \frac{1}{2}\left[\left(\frac{g}{\Omega_i K_H}\right)^2 + \frac{4\varepsilon g}{k^2}\left(\frac{K_H - 1}{K_H}\right) \right]^{1/2} \right\} \quad (203)$$

At the threshold for instability the second term vanishes and the two velocities converge to the value

$$v_H \text{ (threshold)} = g/2\Omega_i K_H \quad (204)$$

Thus at low densities, where $K_H \to 1$, the wave phase velocity at threshold approaches its maximum value, $g/2\Omega_i$. It follows that when instability just appears it will appear as a slow growth superposed on a relatively rapidly propagating wave. In a cylindrical plasma column at low densities, one would therefore always expect to see the hydromagnetic drift instability accompanied by strong rotational effects. This is indeed what is observed. On the other hand, at high densities, where $K_H \gg 1$, the phase velocity becomes small, so that here one would expect the unstable plasma to develop exponentially growing perturbations with little accompanying rotation. In this limit we see from Eq. (184) that the growth rate becomes much more rapid, reaching the value

$$\omega_H = \sqrt{\varepsilon g} \text{ sec}^{-1} \quad (205)$$

This growth rate is the same as the classical result for the growth rate of Rayleigh-Taylor instabilities in a dense fluid supported against gravity by a light fluid (here the magnetic field plays that role).

If we again make the replacement $g \to \bar{v}_i^2/R$, the above growth rate becomes

$$\omega_H = \bar{v}_i(\varepsilon/R)^{1/2} \approx \bar{v}_i/(rR)^{1/2} \quad (206)$$

i.e., the instability grows on the time scale it would take for an ion to traverse the distance $(rR)^{1/2} \approx L$, where L is the length of the plasma region as measured along the field lines. In a confined high temperature plasma this time scale is very short, of the order of a few microseconds, indicating the catastrophic nature of gross instabilities.

The simplest unstable perturbation of a plasma column would be a gross sideways displacement. Here the azimuthal "wavelength" is equal to the mean plasma circumference. Depending on circumstances and on the initial perturbation, short-wavelength disturbances can also appear, leading to an azimuthal rippling of the plasma surface, so that the plasma would begin to resemble a fluted column. This circumstance explains the origin of the name "flute instability" often applied to the hydromagnetic drift instability.

The picture we have just described applies in the limit where ion orbit sizes are sufficiently small compared to all other dimensions. When this is not the case, the "finite orbit effects" earlier alluded to will modify the situation, leading to stabilization or partial stabilization of the flute instability. Finite orbit stabilization[27,34] thus provides another example of stability achieved in the face of situations energetically favorable for instability. Recalling the modification to the hydromagnetic drift wave equation given earlier, it was found that the effect of finite orbit corrections was to modify the simple dispersion equation to the form

$$\omega^2[J_0^2(k_x \bar{a}_i) + b^2] - \omega\{gk_x b^2/\Omega_i + \Omega_i(\varepsilon/k_x)[1 - J_0^2(k_x \bar{a}_i)]\} + \varepsilon g = 0$$
(207)

If again we make the correspondence $g \to v_i^2/R$ and expand the Bessel functions, retaining terms of order $k^2 a^2$, we obtain the equation

$$\omega^2 - \omega\Omega_i\left\{\frac{k_x R}{K_H}\left[1 + \frac{\varepsilon R}{2}(K_H - 1)\right]\frac{a_i^2}{R^2}\right\} + \varepsilon R\left(\frac{a_i^2}{R^2}\right)\Omega_i^2\left(\frac{K_H - 1}{K_H}\right) = 0$$
(208)

This dispersion equation is almost identical to the one earlier found (Eq. 187), except for the expression

$$(\varepsilon R/2)(K_H - 1)$$

that appears in the second term of the equation. But in the solution of the dispersion equation for its roots, it is this term, squared, that provides stabilization when the third term is negative ($\varepsilon R < 0$). Without the finite orbit correction term we found only a lower density threshold for instability; with this term included an *upper* threshold may appear—in this case for the reappearance of stability.

We may evaluate this stability criterion directly from Eq. (208). For simplicity we consider the high density limit, $K_H \gg 1$, finding

$$\varepsilon R(k_x^2 a_i^2) > 16 \quad \text{stable} \quad (209)$$

Stability here depends on satisfying a minimum condition on the orbit radius, given the density gradient, curvature, and wavelength parameters. Long wavelengths are least well stabilized, so that we may derive an approximate condition for the stability of a plasma column against fluting by making the identifications

$$k_x = \varepsilon \approx \frac{1}{r} \qquad L^2 = Rr$$

where L is the length of the plasma column (appropriate to mirror-confined plasmas). We then find the approximate condition, valid at high densities,

$$(La_i/r^2) > 4 \qquad \text{stable} \tag{210}$$

The significance of this result is that if $L \gg r$ (a long, thin plasma), then stability can still be achieved without violating the requirement $a_i \ll r$. Physically, this simply means that when the field curvature is weak, even the small relative electric drifts associated with the difference in ion and electron orbit sizes are sufficient to provide a charge separation and thus a restoring force sufficient to stabilize the drift waves. Since this mechanism is most effective against short wavelength perturbations (larger ka) it follows that there will be limitations on the wavelengths observed in even an unstable plasma.

We have earlier seen that when plasma confined in magnetic fields with lines concave toward the plasma is subject to the flute instability, stability against this mode may be guaranteed by use of magnetic wells, whose field lines are everywhere convex toward the plasma. But what of toroidal systems? Here it is patently impossible to arrange matters so that the field lines are everywhere convex toward the plasma. In fact, under normal circumstances in a torus such as the stellarator the field lines will almost everywhere be *concave* toward the plasma, owing to the necessity for rotational transform in which the lines spiral helically around the magnetic axis. What, therefore, can be done to suppress the drift instability in the torus? It turns out that the finite orbit effects just considered are inadequate to stabilize long-wavelength modes in toroidal systems, so that one must turn to other possibilities. Two methods have been proposed to deal with this problem. The first of these, proposed for stabilization of the stellarator, is the use of magnetic shear. If the helical rotational transform windings on a stellarator are properly arranged they will produce a rotational transform which increases with distance away from the magnetic axis. In this case the helical pitch of the field lines will increase as one approaches the chamber walls starting from the magnetic axis. The effect of this will be to create a basketweave-like pattern of field lines, as shown in Fig. 14. In such a field it is not possible for a simple flute-like perturbation to develop and propagate unchanged across the field. In fact, in the limit of infinite plasma conductivity, no such perturbation can cross the magnetic field without distorting it; to cross would require currents and additional energy. That this situation is theoretically stable at high plasma temperatures can be shown by an extension of the max dl/B principle discussed earlier.[36] At lower temperatures finite resistivity can introduce slow-growing modes, but these should become innocuous as the temperature is increased. Shear stabilization therefore stands as one of the fundamental approaches to the stabilization of closed systems.

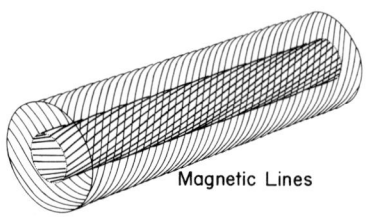

Fig. 14. Illustration of magnetic field with shear. Helical pitch of field lines increases with radius, so that the field lines of successive flux surfaces are skewed with respect to inner ones.

The second idea for stabilizing toroidal systems is of more recent origin. When the power of the magnetic well for the stabilization of flute instabilities was proved, the question was soon raised as to whether this same idea had application to the torus. Although it was soon appreciated that no possible combination of external field windings could produce a magnetic-well type field throughout the length of a torus, the idea emerged that such a situation might be created on the average. In such an "average well" it would be arranged geometrically that particles moving along the field lines would, by virtue of the field geometry, spend more of their time in field regions characterized by "good" curvature than they would in regions of "bad" curvature. In this way, provided the particle mean free paths are very long, so that the average particle "remembers" its traversals of the good curvature regions while whizzing through the bad ones, the overall effect will be a stabilizing one. In such a situation the max $\int dl/B$ criterion for stability can be satisfied, leading to stability against fluting.

One of the first proposed[37] and first tested average well systems is the so-called "toroidal octupole," shown schematically in Fig. 15. In this device "floating" internal conductors are used to create the pattern of magnetic lines shown. Here the field lines are seen to close on themselves the short way around the interior of the torus, since they encircle the internal conductors. Tracing around a typical magnetic line within the confinement region, one sees that the field curvature toward the plasma

THE PHYSICS OF HIGH TEMPERATURE PLASMA

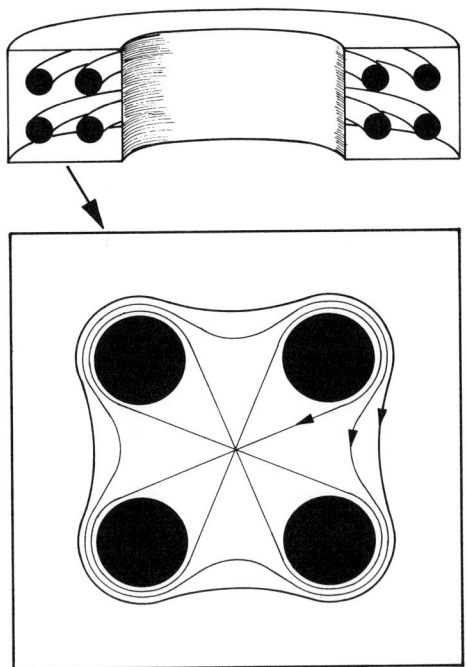

Fig. 15. Cutaway schematic drawing of the toroidal octupole. Section shows configuration of field lines around conductors. Line defining ψ (critical) is outermost (heavy) line. Inside ψ (critical) lines possess favorable average curvature.

will alternate between good and bad. Note, however, that in the bad curvature regions B is large, whereas in the good curvature regions B is weaker, in fact vanishing on the interior axis of the torus. It follows that when evaluated along such lines $\int dl/B$ will have its maximum value for those lines passing through the plasma near to the interior axis, diminishing monotonically outward until a critical flux surface, ψ (critical), is reached as shown in Fig. 15. Beyond this surface $\int dl/B$ increases again (B becomes weaker and the path length increases). This circumstance implies that plasma contained on lines within ψ (critical) will be hydromagnetically stable, while that residing on lines outside ψ (critical) will be unstable. Multipole fields of the toroidal octupole type are among the simplest examples of average magnetic wells, but many other types have been proposed, including ones which use no internal conductors.[38] However, those with internal conductors, such as the multipole fields, have a marked advantage: they can be made axially symmetric. In such cases it can be shown rigorously that magnetostatic equilibrium exists for plasma immersed in such fields. The disadvantage of these multipole fields is the technical difficulty, sometimes verging on impossibility, of supporting or floating the internal conductors within the hot plasma without compromising the confinement. Toroids with only external conductors do not suffer from this difficulty, but cannot be axially symmetric, so that it is much more difficult to prove that they permit equilibria. Indeed, if equilibrium states exist in such systems they are likely to be sensitively dependent on errors in the field. This latter circumstance arises because in a nonaxially symmetric toroidal field it is much more difficult to insure that magnetic lines nominally contained within the plasma do not find their way out to the chamber wall, as they course around the torus, through the effect of geometrical perturbations and mechanical tolerances. The problem is a bit analogous to the task of herding a group of water snakes around and around in a toroidal fish bowl, when one must do this entirely from the outside, and when no snake must be allowed to touch the walls at any time! To adequately solve the confinement problem for a high temperature plasma, particles must remain trapped within the torus for thousands of round trips. To the extent that they are guided mainly by the field lines, these same lines must not lead the plasma particles to the chamber wall during a comparable number of revolutions.

In summary, in our discussion of hydromagnetic instabilities to this point we have shown that wherever a magnetically confined plasma state exists together with curved magnetic lines, hydromagnetic drift waves can be expected. Depending on the sign of this curvature as seen by the plasma boundary, these waves may be stable or incipiently unstable. Uncontrolled, they would rapidly destroy confinement. They are most effectively controlled in open-ended magnetic well fields, where the field curvature is everywhere favorable. Failing this, one must appeal to finite-orbit effects, magnetic shear, or the use of average magnetic wells to control these instabilities. In the search for effective means to accomplish stability in closed systems many specific ideas have been proposed and are under test, but the issue is not yet satisfactorily resolved. We shall return to this question in a later section.

A. The "Firehose" and the "Mirror" Instabilities

Before leaving the subject of hydromagnetic instabilities, two additional types should be mentioned that have been theoretically predicted[39] and observed experimentally. These are called, respectively, the "firehose" instability and the "mirror" instability. While the hydromagnetic drift instability requires a density gradient for its existence, these instabilities do not; they may occur even in an infinite homogeneous plasma. But, like the hydromagnetic drift instability, these also do not require or involve any direct resonant interaction with the individual particle motions for their stimulation; their free energy reservoir is pressure anisotropy, i.e., departure from equality of $p_\perp$ and $p_\parallel$. Both of these instabilities represent unstable Alfvén waves feeding from the plasma streaming motions implied by pressure anisotropy.

The easier of the two instabilities to understand is the firehose instability, as it is closely analogous to the familiar instability of a high pressure flexible hose carrying a large flow of water. A local kinking or bending of the hose, once initiated, will grow from the transverse momentum imparted to the hose by the streaming water as it flows around the bend. In the same way, when $p_\parallel \gg p_\perp$ in a plasma, the counterstreaming flow implied by this condition will tend to create growing kinks in the flux tubes of the confining field along which the plasma is streaming.

The critical pressure anisotropy condition for the firehose instability is most simply derived from the hydromagnetic equations that were earlier presented. The relevant one is Eq. (14) for the force balance perpendicular to the magnetic field lines. This may be written in the form

$$\rho \left[\frac{\partial \mathbf{v}}{\partial t}\right]_\perp + \nabla_\perp \left(p_\perp + \frac{B^2}{8\pi}\right) - \left[\left(\frac{B^2}{8\pi}\right) + (p_\perp - p_\parallel)\right]\frac{(\mathbf{B} \cdot \nabla)\mathbf{B}}{B^2} = 0 \quad (211)$$

We consider now a homogeneous plasma in magnetostatic equilibrium and evaluate the effect of a small departure from equilibrium leading to a slight perturbation of the magnetic field, the perturbed B being given by $\mathbf{B} = \mathbf{B_0} + \hat{\mathbf{B}}$, with $\hat{\mathbf{B}} \ll \mathbf{B_0}$. Linearizing Eq. (211) and differentiating with respect to time, we find the equation

$$\rho \left[\frac{\partial^2 \mathbf{v}}{\partial t^2}\right]_\perp - \frac{B_0}{4\pi}\left[1 + \frac{p_\perp - p_\parallel}{B_0^2/8\pi}\right]\frac{\partial^2 \hat{\mathbf{B}}}{\partial z \, \partial t} = 0 \quad (212)$$

But for a high temperature homogeneous plasma we must have (see Eq. (32))

$$\mathbf{E} + \frac{\mathbf{v}}{c} \times \mathbf{B} = 0 \quad (213)$$

which simply implies that resistive effects may be ignored on the hydromagnetic time scale. Also, from Maxwell's equations

$$\nabla \times \mathbf{E} = -\frac{1}{c}\left(\frac{\partial \mathbf{B}}{\partial t}\right) \quad (214)$$

so that when (214) is combined with (213) we obtain

$$\nabla \times (\mathbf{v} \times \mathbf{B}) = \frac{\partial \mathbf{B}}{\partial t} \quad (215)$$

Expanding and linearizing this equation (i.e., dropping terms of second order in the perturbed quantities), we find for the perpendicular component

$$B_0 \left[\frac{\partial \mathbf{v}}{\partial z}\right]_\perp = \frac{\partial \mathbf{\dot{B}}}{\partial t} \tag{216}$$

If this relation is now differentiated with respect to time and substituted into (212), we obtain a wave equation for the perpendicular component of the fluid velocity:

$$\left[\frac{\partial^2 \mathbf{v}}{\partial t^2}\right]_\perp = \frac{B_0^2}{4\pi\rho}\left[1 + \frac{p_\perp - p_\parallel}{B_0^2/4\pi}\right]\left[\frac{\partial^2 \mathbf{v}}{\partial z^2}\right]_\perp \tag{217}$$

From its form we can see immediately that this equation represents the propagation of a modified Alfvén wave along the field lines, since when $p_\perp = p_\parallel$ (scalar pressure) it reduces to the wave equation for Alfvén waves with phase velocity $v_A = (B^2/4\pi\rho)^{1/2}$.

To obtain a dispersion equation from the wave equation we assume transverse test waves of the form

$$[\mathbf{v}]_\perp = v_0 \exp[i(k_z z - \omega t)]$$

and substitute into the wave equation, obtaining

$$\omega^2 = \frac{B^2}{4\pi\rho}\left[1 + \frac{p_\perp - p_\parallel}{B_0^2/8\pi}\right]k_z^2 \tag{218}$$

This equation represents stable waves (ω real) only as long as the quantity

$$\left[1 + \frac{p_\perp - p_\parallel}{B_0^2/8\pi}\right] > 0 \tag{219}$$

The critical condition for unstable wave growth is therefore

$$(p_\parallel - p_\perp) > B_0^2/4\pi \quad \text{unstable} \tag{220}$$

Thus in any plasma where $p_\parallel > p_\perp$ the firehose instability may occur provided the plasma pressure is high enough. Written in terms of $\beta = \{p_\perp/(B_0^2/8\pi + p_\perp)\}$, the above condition for instability is simply

$$\beta > 2p_\perp/(p_\parallel + p_\perp) \quad \text{unstable} \tag{221}$$

consistent with the limiting value of $\beta = 1$ for magnetostatic equilibrium in an isotropic plasma. Note that if $p_\parallel \gg p_\perp$, the firehose instability may occur even for $\beta \ll 1$, i.e., under conditions where the equilibrium state of the plasma produces a negligible perturbation in the vacuum magnetic field. The existence of the firehose instability demonstrates that even in the absence of direct wave–particle resonance effects, streaming motion such as that implied when $p_\parallel > p_\perp$ in a magnetically confined plasma can represent a potent source of instability.

The growth rate for the firehose instability, obtainable from the dispersion equation, is

$$\omega_i = k_z v_A \left[\frac{p_\parallel - p_\perp}{B_0^2/8\pi} - 1 \right]^{1/2} \sec^{-1} \quad (222)$$

Short wavelength perturbations are seen to grow most rapidly. This will remain true up the the point where finite-orbit effects step in and modify the equations, reducing the growth rate. We may estimate an upper limit to the growth rate by setting $k_z \approx a_i^{-1}$ in the expression for the growth rate and taking the limit $(p_\parallel - p_\perp) \gg B_0^2/4\pi$. The result is

$$\omega_i \approx \Omega_i (\bar{v}_\parallel^2 / \bar{v}_\perp^2)^{1/2} \quad (223)$$

i.e., a rate of exponentiation of the order of the ion cyclotron frequency, a rapid growth rate indeed. Thus in any plasma where the critical pressure for the firehose instability is appreciably exceeded, one would expect to observe a spectrum of unstably growing Alfvén waves, with wavelengths down to the order of ion orbit radii. The growth of these unstable waves would be expected to lead to "rippling" of the magnetic field lines together with rapid diffusion of the plasma in velocity space. Though we have not as yet discussed this aspect of the effect of plasma instabilities, it should be clear that in the present example the end result of the instability will be to tend to destroy the plasma pressure anisotropy that stimulated it in the first place, leading to a new state that would be either stable or only marginally unstable.

Our second example of a hydromagnetic instability feeding on pressure anisotropy is the "mirror" instability. It can also be looked on as the unstable Alfvén wave and its analysis may be carried forward in a way similar to that employed for the firehose instability, except that one must consider waves propagating perpendicular to the magnetic field. In this case for instability the pressure anisotropy must be such that $p_\perp > p_\parallel$. The physical mechanisms involved in the growth of this instability are relatively simple. Suppose, for example, that $p_\perp \gg p_\parallel$. In this circumstance the particles of the plasma have relatively low velocities along the field lines; most of their motion consists of circling motion around the lines of force of the field, which we shall assume is originally uniform and parallel in the region of interest. If now a localized upward fluctuation in the particle density occurs, the diamagnetic effect of the added particles $(p_\perp + B^2/8\pi = \text{constant})$ will locally depress the field. But now if the total plasma pressure exceeds a critical value this diamagnetic depression of the field will be sufficient to lead to trapping even more particles in the miniature mirror field thus created, digging a still deeper hole, and so on and so on. Thus the "mirror instability" has nothing directly to do with mirror confined plasmas, but can occur in any situation where the relevant critical condition is exceeded.

Since the mirror instability involves perturbed velocity components parallel to **B**, thermal plasma wave effects will enter into an exact analysis of the problem in addition to the Alfvén waves. Thus in the general case where electron and ion temperatures are comparable but not equal, the critical condition for onset of this instability is a somewhat complicated one.[30] But in the interesting special cases where $T_{\|e} = T_{\perp i}$ and $T_{\perp e} = T_{\perp i}$, or where $T_e \ll T_i$, the condition simplifies to

$$\frac{p_\perp}{p_\|} > \frac{B^2/8\pi + p_\perp}{p_\perp} \quad \text{unstable} \tag{224}$$

i.e.,

$$\beta > \beta_M = p_\|/p_\perp \quad \text{unstable} \tag{225}$$

Thus if $p_\perp \gg p_\|$ this instability may occur even when $\beta \ll 1$. However, as the pressure approaches isotropy, $\beta_M \to 1$, consistent with the limiting value of β required by considerations of magnetostatic equilibrium.

An interesting and amusing feature of the mirror instability is that it could in theory be initiated by the *loss* of plasma from a previously stable system. Suppose that the conditions are such that β is finite but that $p_\perp/p_\|$ lies just below β_M. If now any process (for example, end losses through mirrors) leads preferentially to the loss of particles moving parallel to the field lines, the critical ratio of $p_\perp/p_\|$ may be exceeded and instability will ensue. Presumably a similar effect could occur with the firehose instability if $p_\perp$ is preferentially depleted. Such effects, and other similar ones, may be involved in the complicated plasma instabilities that occur in the earth's magnetosphere during magnetic storms.

The hydromagnetic instabilities we have been considering—namely, the hydromagnetic drift or flute instability, the firehose, and the mirror instability—are all examples of instabilities fed from gross free energy reservoirs—expansion, rotational effects, or pressure anisotropy. All involve some type of unstable hydromagnetic wave motion, drift waves in a density gradient or Alfvén or electroacoustic waves, and all occur at frequencies small compared to ion cyclotron frequencies, and with phase velocities small compared to particle velocities. None, therefore, involve direct resonant interactions between individual particles and the waves so that knowledge of the detailed structure of the particle distribution functions is normally unnecessary in discussing them. From the standpoint of seeking for ways to avoid these instabilities clear prescriptions can be given from theory: Use of magnetic wells, of average magnetic wells, or of magnetic shear should stabilize the flute instability; control of pressure anisotropy and limitation of β should eliminate the firehose and the mirror instabilities. From a practical standpoint it is indeed fortunate that the gross instabilities can be cured by these straightforward prescriptions, for they (particularly the flute instability) are the most disruptive of confinement of all. It follows that in the search for stable plasma confinement, elimination of gross instabilities is the first order of business. Once this has been accomplished, attention may be turned to the less catastrophic, but more pernicious, wave–particle "resonance" instabilities, which we shall now discuss.

B. *Wave–Particle Instabilities*

Once one begins to consider unstable waves fed by direct resonant interactions with particle motions—that is by streaming, cyclotron motions, or gradient drifts—the catalog of possible instability modes becomes enormous. In the last few years a prodigious theoretical effort has gone into the analysis of these instabilities and their consequences. We cannot hope here to give a detailed account of all the various instability modes that have been predicted or analyzed, but will instead discuss a few important examples. Fortunately, theory has shown that the number of nonhydromagnetic instability modes that may appear in a given situation depends critically on the details of the particle distribution functions, the number decreasing rapidly as the degree of randomization is increased. Furthermore, the study of these instabilities in the laboratory is facilitated by the fact that they are not capable of producing the extremely rapid plasma transport and loss that characterizes the gross hydromagnetic instabilities. A simple way to understand this important difference in loss rates is to note that in cases where enhanced diffusion is caused by unstable fluctuations within the plasma, the diffusion velocity can be described via a diffusion coefficient, so that the mean transverse diffusion velocity is given by

$$\langle v_\perp \rangle = D_\perp \left(\frac{\nabla n}{n}\right)_\perp \qquad (226)$$

The diffusion coefficient, $D_\perp$, is on the other hand determined by the rate of transverse "random walking" caused by the fluctuations. That is to say, it is proportional to the rate of making these steps, ω, multiplied by the square of the "step length" (i.e., the transverse correlation length, $\lambda_\perp$). Thus

$$D_\perp \approx \omega \lambda_\perp^2 \qquad (227)$$

In wave–particle instabilities $\lambda_\perp$ must be identified with the perpendicular wavelengths, and the frequency ω with the frequency spectrum of the instability. But because of the phase-resonance requirements for wave–particle instabilities, $\omega - kv = 0$, there exists a reciprocal relationship between ω and $\lambda_\perp$; $\omega \sim \lambda_\perp^{-1}$. Thus we find $D_\perp \sim \omega \lambda_\perp^2 \sim \omega^{-1}$. Because of this inverse dependence on frequency, high frequency instabilities where $\omega \approx \Omega_i$ or higher are generally found to be incapable of producing transverse plasma transport at rates comparable to those caused by hydromagnetic instabilities, where $\omega \ll \Omega_i$. Even when one considers open-ended confinement systems, where losses may occur because of pitch angle diffusion, it is generally found that the average rates of particle losses caused by nonhydromagnetic instabilities are much slower than those stemming from hydromagnetic instabilities.

The two necessary ingredients for any wave–particle instability to occur are (1) that there should exist a "carrier" plasma wave to which particle energy can be fed, and (2) that a free energy reservoir of directed or ordered periodic motion particle motion exists with particle velocities that can match that of the wave. The simplest example of such a free energy reservoir is that associated with counterstreaming motion within a plasma. One of the most elementary waves which can be driven unstable by this motion is the electrostatic plasma oscillation. The "two-stream" instability that arises in this case is of fundamental importance, being an example of a wider class of instabilities.

The critical conditions for the two-stream instability and its general characteristics may be readily deduced from the dispersion equation for electrostatic plasma waves that we earlier derived (Eq. (131)):

$$D(\omega, k) = k_\parallel^2 - 2\pi \sum_j \omega_{pj}^2 \int_0^\infty v_\perp \, dv_\perp \int_{-\infty}^\infty \frac{\partial f_0/\partial v_\parallel}{v_\parallel - \omega/k_M} \, dv_\parallel = 0 \quad (228)$$

Let us apply this equation to the case of two equal-density groups of electrons counterstreaming through each other. An example would be the case of two colliding and interpenetrating plasma "blobs." In the rest frame of each blob we shall assume that the electron distribution is Maxwellian. Thus in the laboratory frame the distribution functions for the electrons can be represented in the form

$$f_0 \sim \exp\{-[|v_\parallel \pm v_s|^2 + v_\perp^2]/v_0^2\}$$

We will ignore small corrections associated with the motion of the ions, because of their much larger mass. If we now insert the above f_0 into the dispersion equation, Eq. (228), we are led to the result

$$k_\parallel^2 + \frac{1}{2}\frac{\omega_p^2}{v_0^2}\{2[1 + \sigma_1 Z(\sigma_1)] + 2[1 + \sigma_2 Z(\sigma_2)]\} = 0 \quad (229)$$

where ω_p is the plasma frequency for the combined plasma, $\sigma_1 = \omega/k_\parallel v_0 - v_s/v_0$ and $\sigma_2 = \omega/k_\parallel v_0 + v_s/v_0$. The term v_s represents the streaming velocity of each plasma with respect to the laboratory (see Eq. (134) for comparison, with $v_s = 0$).

Let us first consider the asymptotic limit σ_1 and $\sigma_2 \gg 1$, appropriate when the relative streaming velocity v_s is larger than the thermal velocity v_0. In this limit we find (see Eq. (136) for the asymptotic form of the Z function):

$$k_\parallel^2 - \frac{1}{2}\frac{\omega_p^2}{v_0^2}\left[\frac{1}{\sigma_1^2} + \frac{1}{\sigma_2^2}\right] = 0 \qquad (230)$$

which may be written as

$$1 - \frac{1}{2}\left[\frac{1}{(\nu - \nu_0)^2} + \frac{1}{(\nu + \nu_0)^2}\right] = 0 \qquad (231)$$

with $\nu = \omega/\omega_p$ and $\nu_0 = k_\parallel v_z/\omega_p$. This equation has four roots, found by solving for ν^2:

$$\nu^2 = \tfrac{1}{2}\{1 + 2\nu_0^2 \pm (1 + 8\nu_0^2)^{1/2}\} \qquad (232)$$

First note that if $\nu_0 \geq 1$, all four roots of the equation are real, corresponding to stability. This implies that $\nu_0 = k_\parallel v_s/\omega_p \geq 1$ is a sufficient condition for stability, i.e.,

$$k_\parallel \geq \omega_p/v_s \qquad \text{stable}$$

Instability can therefore only occur if the wavelength exceeds a critical value; not all wavelengths are unstable. The physical origin of this condition is, of course, traceable to the requirement for phase velocity resonance.

There also exists a critical streaming velocity, below which the system becomes stable, for any wavelength. This critical value may be determined exactly from the characteristics of the Z function. We shall here obtain an approximate expression for it from the lowest order terms of the power series expansion of $Z(\sigma)$,

$$Z(\sigma) \approx i\pi^{1/2} - 2\sigma, \qquad \sigma \ll 1 \qquad (233)$$

Expanding the dispersion equation to first order in σ we find

$$k_\parallel^2 + \frac{2\omega_{pe}^2}{v_0^2}\left\{1 + \left(\frac{\omega}{k_\parallel v_0}\right)i\pi^{1/2} - 2\left(\frac{\omega}{k_\parallel v_0}\right) - 2\left(\frac{v_s}{v_0}\right)^2\right\} = 0 \qquad (234)$$

We now recall the result previously obtained, that for $v_s = 0$ the solutions to the dispersion equation corresponded to heavily damped stable waves, the imaginary term provided the damping. Marginal stability will therefore occur in the limit $\omega \to 0$ (vanishing imaginary term) and $k_\parallel \to 0$

(remembering that long wavelengths are the most unstable). In this limit we find for marginal stability

$$1 - 2(v_s/v_0)^2 = 0$$

i.e., $v_s \leq 0.7\, v_0$, stable.

We therefore find that the counterstreaming velocity must exceed a critical value compared to the natural spread in velocities of the distribution functions before instability can occur. Below this critical relative velocity the plasma system becomes stable against electron plasma oscillations at any density. If one examines this same situation as to electron–ion oscillations, a somewhat different criterion for stability emerges, but the qualitative result is similar. Concise ways of analyzing these types of problems using the so-called Nyquist diagrams have been worked out and published.[40]

C. Loss-Cone Instabilities

An instability closely related to the two-stream instability is predicted theoretically to occur in plasmas whose distribution functions are of the "loss-cone" type. As the name implies, a loss-cone distribution is one which is deficient in particles of low rotational velocity, $v_\perp$, as compared to $v_\parallel$, the velocity along the field lines. This is the situation that arises naturally in the confinement of particles between magnetic mirrors. In such cases the mirror loss cone enforces the condition $f_0(v_\perp{}^2, v_\parallel{}^2) = 0$ for $v_\perp < \gamma |v_\parallel|$, where γ is a function of the mirror ratio R_M. Such distribution functions are illustrated in Figs. 5 and 6. Figure 16 shows a

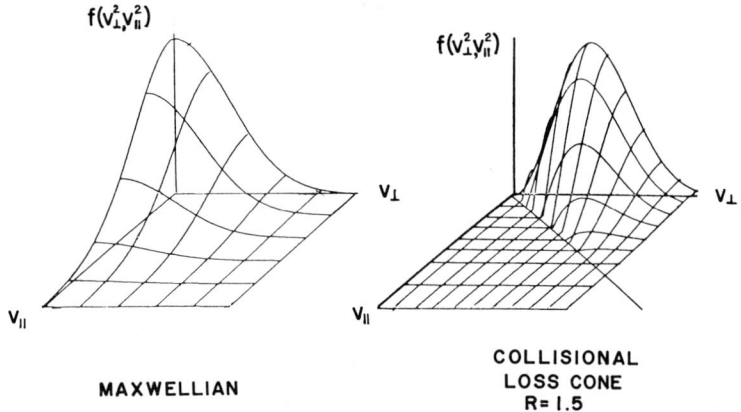

Fig. 16. Comparison between a collisional loss-cone distribution for mirror ratio 1.5 with a Maxwellian distribution.

comparison between a Maxwellian and a collisional equilibrium loss-cone distribution appropriate to a mirror system with a mirror ratio of 1.5. Since there is no obvious source of counterstreaming motion in a plasma in magnetostatic and collisional equilibrium between mirrors, we need to see how this streaming can in fact arise. We note that counterstreaming is signalled by the appearance of "double humping" in the distribution function in some direction in space. If we now consider the distribution

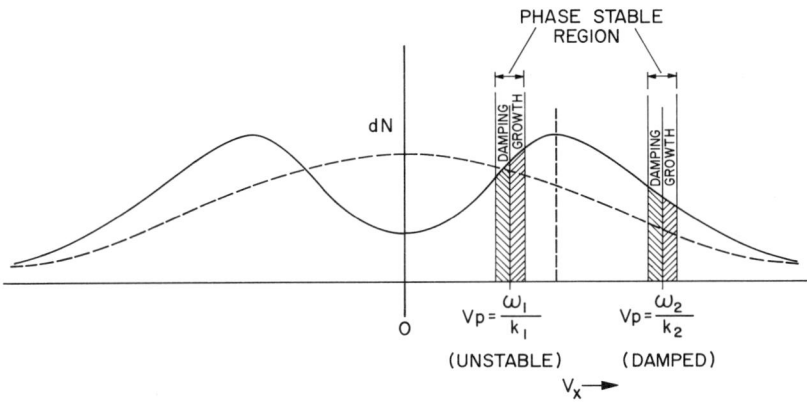

Fig. 17. Inverted population in direction perpendicular to field for collisional loss-cone distribution, showing creation of a "double-humped" distribution owing to anisotropy in velocity-space. Dotted line indicates Maxwellian distribution for comparison. Regions of possible wave growth and wave damping are shown (see Fig. 8 for wave mechanisms involved).

function of the ions contained between mirrors as analyzed in a direction transverse to **B**, say the x direction, we see that the absence of particles with small $v_\perp$ will cause the central region of the distribution function $f_0(v_x)$ to be depressed, relative to the normal Maxwellian (Gaussian) shape. This leads to double humping, as shown in Fig. 17. Thus the existence of a loss cone is sufficient to cause double humping in f_0 after it is integrated over the velocities parallel to **B**.

We may search for unstable electrostatic waves driven by loss-cone distributions by using our solution to the Vlasov equation, Eq. (122). In considering waves propagating in a dense plasma in a direction nearly perpendicular to the field $k_\parallel \ll k_\perp$, we are led to the dispersion relation[41]

$$\left(\frac{\omega_{pe}^2}{\Omega_e^2} + 1\right) = \frac{\omega_{pe}^2}{\omega^2}\frac{k_\parallel^2}{k^2} + \frac{\omega_{pi}^2}{k^2 \bar{v}_i^2} F_i(y) \qquad (235)$$

where

$$F_i(y) = \int_0^\infty \frac{\partial \psi(v_\perp)/\partial v_\perp}{(1 - v_\perp^2/y)^{1/2}} \, dv_\perp \qquad (236)$$

with $y = \omega/k\bar{v}_i$.

The quantity $\psi(v_\perp)$ is simply the normalized ion distribution function, $f(v_\perp^2, v_\parallel^2)$, after integration over $v_\parallel$. In a loss-cone distribution $\psi(0) = 0$. We see immediately that $F(y)$ must be complex, having real and imaginary parts of comparable magnitude.

If we now analyze the above dispersion equation for its stability properties we will find the disconcerting result that as long as $k_\parallel \neq 0$, no matter how small, unstable wave growth is always predicted for waves within that range of ω and k values for which Im $F(y)$ is negative. From the dispersion relation we can see that the most unstable waves correspond to those where $|\omega| \approx \omega_{pi}$ and $k_\parallel/k_\perp \approx (m/M)^{1/2}$. For such waves the electron and ion terms (first and second terms on the right side of Eq. (235)) are both comparable. Since there will always be some region in y, however small, for which Im $F(y)$ is negative as long as f_0 has a loss-cone character, this means that such plasmas will always be at least residually unstable against this mode. What saves the situation is that the loss-cone instability that we are here describing is found to be of the "convective" type, i.e., one which corresponds to a wave that grows only as it propagates. This is by contrast with "nonconvective" or "absolute" instabilities that may grow as bounded standing waves. In the case of the convective instability we are here considering, both the group and the phase velocity are directed nearly along the field lines, so that starting with some local initial perturbation one only expects to find growth as the wave propagates down the field lines. If the wave growth rate is slow enough, and if the wave is absorbed without strong reflection as it nears the end of the system, it cannot grow to sufficient amplitude within the available length to disrupt the plasma. One therefore expects to find a "critical length" for this instability, one that we may readily calculate from the dispersion relation.

Since $k_\parallel \ll k$ is required for the instability to occur at all (or else Landau damping by electron motion along the field lines would suppress it), we may readily solve (235) for the imaginary part of $k_\parallel$ in this limit, finding

$$\text{Im } k_\parallel \approx \tfrac{1}{2}(m/M)^{1/2}(\omega_{pi}/\bar{v}_i)(1 + \omega_{pe}^2/\Omega_e^2)^{-1/2} \cdot [y \text{ Im } F(y)] \qquad (237)$$

Remembering the form of our test wave solution we note that Im $k_\parallel < 0$ corresponds to unstable growth of the wave, a situation that implies from Eq. (237) the requirement that $[y \text{ Im } F(y)] < 0$. The growth length (for exponentiation of the wave), $L_0 = k_\parallel^{-1}$, can be written from Eq. (237) in the equivalent form

$$L_0 = 2\bar{a}_i(1 + b_e^2)^{1/2} \cdot [y \text{ Im } F(y)]^{-1} \quad (238)$$

where $b_e^2 = \Omega_e^2/\omega_{pe}^2 = B^2/(4\pi n m e^2)$ and $\bar{a}_i = \bar{v}_i \Omega_i$ is the mean ion orbit radius.

If the ion distribution function has a pronounced loss-cone nature, or is sharply peaked in energy, then $|y \text{ Im } F(y)|$ may be large (order 1) and we see that the growth length may become very short, i.e., of the order of a few ion orbit radii as measured along the field. In such cases one would expect the instability, even in a short plasma, to grow rapidly to disruptive amplitudes, causing rapid diffusion of the ions in velocity space and consequent end losses. On the other hand, in well-randomized plasmas confined between strong magnetic mirrors ($R_M \gg 1$), the loss-cone region of the ion distribution function becomes localized near $v = 0$ (see Fig. 5 for the case $R_M = 10$, for example), so that $|y \text{ Im } F(y)| \ll 1$ even at its maximum negative value. In such cases the growth length may become quite long so that the instability, though present, should not grow to appreciable amplitude within the plasma, provided its length is not too great and provided there is no strong reflection of the waves at the ends.[42] In this situation the problem of whether the instability grows to large amplitude or not is very much like the question of whether a laser setup will "lase" or not: Either an insufficiently inverted excited state population (in our case too small a loss-cone region in velocity space), too short a length, or too poor a reflection coefficient at the ends should quench the "instability" that laser action implies. In both cases, laser and loss-cone plasma, wave–particle coupling mechanisms are involved, and the existence of unstable waves depends sensitively on a competition between growth and damping effects.

When one specifically includes the effects of electron Landau damping (not covered by the simplified dispersion relation (235)) on the growth of the convective loss-cone mode, it is found that the growth lengths are markedly increased.[43] Similarly, relatively slight changes in the ion distribution function at low energies, near the loss cone, can theoretically result in strongly stabilizing effects. The upshot is that, while this mode could be strongly disruptive of confinement in poorly randomized distributions, it should not be an important loss mechanism in well-randomized plasmas contained between strong magnetic mirrors.

The convective loss-cone instability we have just discussed is but one of a number of electrostatic instabilities whose free energy reservoirs can be traced to departures from homogeneity of the distribution functions in velocity space. The first examples of such modes characteristically occurring at the ion cyclotron frequency or its harmonics were described by Harris[44]; all are sensitively dependent on the degree of peaking of the distribution functions in speed or in angle, being suppressed by a sufficient degree of randomization.

When one couples a density gradient with the existence of loss-cone-like distributions, new modes are predicted to appear that require both reservoirs for their existence.[45,46] One such instability occurs for frequencies near the ion cyclotron frequency. Here the carrier wave is an azimuthal drift wave moving perpendicular both to **B** ($k_\parallel = 0$) and to ∇n, involving **E × B** motion of the electrons and a Doppler-shifted resonance with the ion cyclotron motion. The dispersion equation for this instability may be determined from the electrostatic mode solution to the Vlasov equation, Eq. (122), that we have employed so often before. We need only insert an equilibrium f_0 appropriate to a plasma with a density gradient $\varepsilon = (1/n)\, dn/dx$, where $\varepsilon a_i \ll 1$ is assumed, and postulate a transversely moving test wave potential of the form

$$\phi = \hat\phi \exp[i(k_y y + \omega t)]$$

From these assumptions one obtains the general form of the dispersion relation for such waves as

$$\left(\frac{\omega_{pe}^2}{\Omega_e^2} + 1\right) = \frac{\omega_{pe}^2}{\omega}\frac{\varepsilon}{\Omega_e}\frac{k_\perp}{|k_\perp^2|} - \frac{8\pi^2 n_0 e^2 \omega}{k_\perp^2 M}$$
$$\times \int_0^\infty dv_\perp \int_{-\infty}^\infty dv_\parallel \frac{\partial f_0}{\partial v_\parallel} \sum_{n=-\infty}^\infty \frac{J_n^2(k_\perp v_\perp/\Omega_i)}{\omega + n\Omega_i} \quad (239)$$

Since it will turn out that this instability is of very short wavelength, $ka_i \gg 1$, the Bessel functions may be approximated asymptotically, i.e., $J_n^2(x) \approx 1/\pi x$, $x \gg 1$, so that the dispersion relation may be greatly simplified: The sum on n can be performed through the identity

$$\sum_{n=-\infty}^\infty [1/(x+n)] = \pi \cot \pi x \quad (240)$$

and defining $w = \pi(\omega/\Omega_i)$ the dispersion relation reduces to

$$w^2 \cot w + \eta w = \alpha \quad (241)$$

where

$$\eta = \pi(k\langle a_i\rangle)^3 \left(\frac{m}{M} + \frac{\Omega_i^2}{\omega_{pi}^2}\right) > 0 \quad \text{and} \quad \alpha = \pi^{2/3} k^2 \varepsilon \langle a_i \rangle^3 > 0$$

The weighted average orbit radius $\langle a_i \rangle$ is to be calculated from the relationship $\langle a_i \rangle = \Omega_i \langle v_i \rangle$, where

$$\langle v_i \rangle = \left[\int_{-\infty}^\infty dv_\parallel \int_0^\infty dv_\perp \frac{f_0}{v_\perp^3}\right]^{-1/3}$$

Note that $\langle v_i \rangle$ depends sensitively on the variation of f_0 near $v_\perp = 0$, that is, on the details of the loss-cone end of the ion distribution function.

As long as the roots of the dispersion equation are real, the drift wave will be stable. If the plasma parameters are such as to permit complex roots, however, one of the waves will be unstable. We may find the critical boundary between stability and instability for the present case

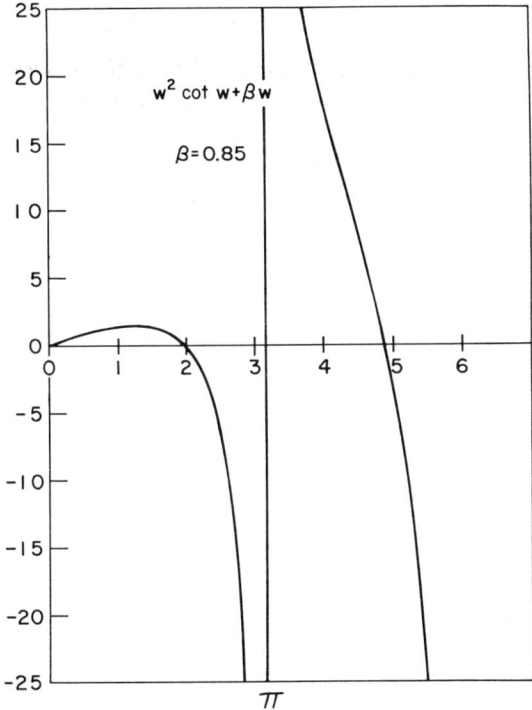

Fig. 18. Plot of portion of dispersion function for azimuthal plasma waves used in analyzing drift-cyclotron instabilities. Presence of extremum at $w = 1.24$ indicates possibility of unstable roots as explained in text.

merely by examining the properties of the left-hand side of the dispersion equation, $w^2 \cot w + \eta w$. This function is shown plotted over the range $0 < w < 2\pi$, for a particular value of η, in Fig. 18. Note that this function has a positive extremum in the interval $0 < w < \pi$, but none beyond. Returning now to the full equation,

$$w^2 \cot w + \eta w = \alpha \qquad (241)$$

note that if α is less than the height of this extremum the dispersion equation will have only real roots, two between 0 and π, and one for every interval of π thereafter, corresponding to the intersection of a horizontal line at height α with the plotted function: This situation is stable. But if α is greater than the height of the extremum, the two roots located between 0 and π will become complex, implying instability. The most unstable case, found by differentiation of (241), occurs at $w = 1.237$, $\eta = 0.85$ (plotted). Using this result we obtain a condition for stability, in terms of a critical density gradient

$$\varepsilon \langle a_i \rangle < (\varepsilon \langle a_i \rangle)_{\text{crit}} = 0.38(m/M)^{2/3}(1 + b_e^2)^{2/3} \quad \text{stable} \quad (242)$$

To each plasma density (each value of b_e^2), there will be a corresponding critical density gradient, measured in units of $\langle a_i \rangle$. If the actual density gradient is steeper than this critical value the plasma would be expected to be unstable against the drift cyclotron instability. To indicate the order of the parameter involved, consider an example for protons ($M/m = 1836$). In a magnetic field of 10 kG, $b_e^2 = 1$ at a density of 10^{13} ions/cm³. For this case $(\varepsilon \langle a_i \rangle)_{\text{crit}} \approx 0.004$, with the critical unstable frequency $\omega \approx 0.4 \Omega_i$. For any plasma with a poorly filled loss cone, such that $\langle a_i \rangle \approx \bar{a}_i$, this small a value of critical gradient implies that at high plasma densities the confined plasma column would have to have a very large radius $\approx \varepsilon^{-1} \approx 250 \bar{a}_i$ to avoid this instability. As the loss-cone region is progressively filled in, however, the critical radius drops—that is, the critical gradient increases—until it reaches a lower limit, that for a Maxwellian distribution. For this case the critical gradient is given by the relationship[47]

$$(\varepsilon \bar{a}_i)_{\text{crit}} = 2(m/M)^{1/2}(1 + b_e^2)^{1/2} \quad (243)$$

In this limit the free energy reservoir feeding the instability is simply the energy associated with expansion, manifesting itself through a relative drift of the ions and electrons. In this case the mode is of even shorter wavelength than the loss-cone drift cyclotron mode, and the frequency is very close to cyclotron resonance with the ions. This general type of instability—that is, one which needs only a density gradient to stimulate it—is sometimes gloomily called a "universal instability." Fortunately for the future of magnetic confinement, this particular instability can be suppressed by keeping the plasma density gradient below the critical value For the case $b_e^2 = 1$ and for protons, this value, from Eq. (243), is $\varepsilon a_i = 0.066$.

The electrostatic instabilities just considered were characterized by very small (or zero) values of $k_\parallel/k_\perp$, a requirement enforced by the effects of electron Landau damping, strong when $\lambda_\parallel/\lambda_\perp = k_\perp/k_\parallel$ is not $\gg 1$. In the case of the convective loss-cone instability we can readily see the origin of this restriction. Since instability occurs for $\omega \approx k_\perp \bar{a}_i \approx \omega_{pi}$ the phase velocity along the field lines will be $v_p \approx \bar{v}_i(k_\perp/k_\parallel)$. But to avoid electron damping this velocity must be large compared to $\bar{v}_e$. Since $\bar{v}_i \ll \bar{v}_e$ it follows that $k_\perp/k_\parallel \gg 1$ is required. A similar argument shows that $k_\parallel/k_\perp$ must be even smaller for the drift cyclotron modes, small enough to set it equal to zero in the analysis. However, the fact that in open-ended systems of reasonable length $k_\parallel$ is in fact finite means that we have idealized the situation in the analysis of the loss-cone drift cyclotron mode (Eq. (239)) by not including the end boundary effects *ab initio*. There is therefore reason to suspect the validity of the simple theory here presented. Since the mode in question has not as yet been detected experimentally, there is some doubt as to whether it will actually appear under ordinary circumstances.

It is, however, possible to find new unstable modes by considering waves propagating at a substantial angle to **B** so that $k_\parallel$ and $k_\perp$ are comparable. However, because of the strong influence of electron Landau damping for propagation at these angles, these new modes are highly resonant, it being only possible for them to grow at frequencies close to Ω_i or its harmonics where one of the resonant ion terms in the dispersion equation (see, for example, Eq. (239)) can overcome the damping contributions.[48] The requirement for unstable growth is therefore that $|\omega| - n\Omega_i \ll \Omega_i$. Recognizing that $\omega = \omega_r + i\omega_i$, it can be seen that the resonance requirement also implies that $|\omega_i| \ll \Omega_i$, that is, these instabilities are characterized by rather slow growth rates, especially where they try to appear in well-randomized plasmas, for which the relevant free-energy reservoirs are small. One expects, therefore, that whether or not the highly resonant ion cyclotron modes are important in a given situation will depend very much on the details of the plasma parameters. It will also depend on the shape and scale of the confining field, since spatial variations in **B** will lead to smearing out of the cyclotron resonances within the plasma.[49] The detailed theory of the effect of magnetic field inhomogeneity on resonant cyclotron modes is just now being developed, but it is already clear that important stabilizing effects can occur through this mechanism.

D. The Low Frequency "Universal" Instability

We have thus far discussed either low frequency instabilities of a non-resonant hydromagnetic type or high frequency electrostatic instabilities where wave–particle resonant effects play an essential role. We will now discuss a very important mode that bridges the gap, i.e., one, though a low frequency mode, requires wave-resonance exchange with longitudinal electron motions to provide the excitation. The mode is the so-called "low frequency universal instability."[50-52] Its importance lies in the fact that the only requirement for this mode to exist is that there exist a density gradient in the plasma; it can even occur in a uniform magnetic field. For this reason the mode is a direct threat to the basic concept of magnetic confinement. Again, however, armed with the dispersion equation for the mode it is possible to predict those plasma conditions which lead to its suppression.

We will not attempt to derive the dispersion equation for the low frequency universal mode although this can be readily accomplished by use of the same method of characteristics that we have used to solve the Vlasov equation for the other instabilities. The starting point for the analysis is to consider a near-flute mode ($k_\parallel \ll k_\perp$) in the presence of density gradients and to couple this mode with excitations arising from thermal motions of the electrons along the field lines. The fact that $k_\parallel \ll k_\perp$ means that even a low frequency drift wave can achieve a phase velocity, $\omega/k_\parallel$ sufficiently high to couple to free energy arising from the electrons as they expand across the field. The mode is found to occur over a wide range of plasma conditions, including ones with finite β. It persists, however, even in the limit $\beta \ll 1$, in fact being somewhat less easily stabilized in this limit.[53] We will here discuss the mode in its low β limit. As we shall see, the clue to stabilization of the mode lies in restricting the plasma length, just as it was in the case of the convective loss-cone mode.

We have already mentioned the basic waves of the universal instability. To repeat, the dispersion equation for these waves (Eq. (194)) in the absence of driving terms and of ion thermal effects is

$$\omega^2 - k_x\left(\frac{\kappa T_e c}{eB}\right)\varepsilon\omega - k_z^2\left(\frac{\kappa T_e}{M}\right) = 0 \qquad (244)$$

Depending on the plasma regime the drift wave branch (second term) or the drift wave and ion sound wave branch of this equation participate, as well as Alfvén wave contributions at high β. In the low β limit, $\beta < m/M$, where the drift mode is the dominant wave, the inclusion of Landau resonant effects yields the embellished dispersion equation in the form

$$1 + \frac{T_i}{T_e} + k_\perp^2\lambda_D^2 + I_0\left(\frac{\alpha^2}{2}\right)\exp\left[-\frac{\alpha^2}{2}\right]\left(1 - \frac{\omega^*}{\omega}\right)\sigma_i Z(\sigma_i)$$
$$+ \left(\frac{T_i}{T_e} + \frac{\omega^*}{\omega}\right)\sigma_e Z(\sigma_e) = 0 \quad (245)$$

where

$$\omega^* = \Omega_i(k_\perp \varepsilon a_i^2) \ll \Omega_i, \qquad a = k_\perp a_i, \qquad \sigma_{i,e} = \omega/k_\parallel \bar{v}_{i,e}$$

and

$$\lambda_{Di}^2 = \kappa T_i/4\pi n_0 e^2$$

In the limit $T_i \to 0$ this dispersion equation yields the drift wave terms of (244).

A dispersion equation as complicated as (245) would at first sight seem to be of little solace in trying to find stabilizing conditions. If we recognize, however, that marginal stability for this mode will be determined by the condition where the Landau damping effects of the ions just balance the Landau "antidamping" of the electron streaming motions along the field lines, the critical conditions can be more easily discerned. We first solve the dispersion equation for ω^*/ω and then for Im (ω), recognizing that Im $(\omega) < 0$ here corresponds to stability.

$$\frac{\omega^*}{\omega} = \frac{1 + T_i/T_e + k^2\lambda_{D_j} - I_0(\alpha^2/2)\exp[-\alpha^2/2]\sigma_i Z(\sigma_i) - (T_i/T_e)\sigma_e Z(\sigma_e)}{\sigma_e Z(\sigma_e) - I_0(\alpha^2/2)\exp[-\alpha^2/2]\sigma_i Z(\sigma_i)} \quad (246)$$

$$\text{Im}(\omega) = \frac{+\omega\{\sigma_i I_0(\alpha^2/2)\exp[-\alpha^2/2](1-\omega/\omega^*) \times \exp[-\sigma_i^2] - \sigma_e[1+(T_i/T_e)(\omega/\omega^*)]\exp[-\sigma_e^2]\}}{\sigma_e Z(\sigma_e) - I_0(\alpha^2/2)\exp[-\alpha^2/2]\sigma_i Z(\sigma_i)} \quad (247)$$

We note that since $\sigma_i \gg \sigma_e$, to keep Im (ω) from being exponentially small we must at least insist that $\sigma_e < 1$, i.e., $\omega < k_\parallel \bar{v}_e$. Similarly, the ion term will also be exponentially small unless $\sigma_i \approx 1$ at least. But if $\sigma_i \approx 1$, $\sigma_e \ll 1$, so that $\sigma_e Z(\sigma_e) \ll \sigma_i Z(\sigma_i)$, therefore $\omega^*/\omega < 1$ so that the ion term is seen to contribute damping. For instability this damping term must be less than the residual growth term from the electrons, that is, for growth one must require at the least that $\sigma_i I_0(\alpha^2/2)\exp[-\alpha^2/2] < \sigma_e$, i.e., that

$$I_0(\alpha^2/2)\exp[-(\alpha^2/2+\sigma_i^2)] < (m/M)^{1/2} \quad (248)$$

in this way putting lower limits on $k_\perp \bar{a}_i$ and $\omega/k_\parallel \bar{v}_i$. This requirement, in fact, gives a clue to a most effective way of stabilizing the universal mode—namely, by limiting the parallel wavelength to the point where the above requirement can no longer be satisfied. Inserting the definition of ω^*, Eq. (246) may be solved[54] for $\varepsilon/k_\parallel = (1/k_\parallel)(1/n)(dn/dr)$, in the appropriate limit $\sigma_e \ll 1$:

$$\frac{1}{k_\parallel}\left(\frac{1}{n}\frac{dn}{dr}\right) = \left|\sigma_i\left\{\frac{1+T_i/T_e+k^2\lambda_{D_i}^2+I_0(\alpha^2/2)\exp[-\alpha^2/2]\sigma_i Z(\sigma_i)}{I_0(\alpha^2/2)\exp[-\alpha^2/2]}\right\}\right| \quad (249)$$

The most unstable case corresponds to the least value of the right-hand side of the equation, found to occur at $\sigma_i \approx 1$. The magnitude of this minimum value decreases as a function of $a(k_\perp \bar{a}_i)$ down to $k_\perp a_i \approx 2$, thus determining an extremum. Since this extremum now defines the *least* value of $\varepsilon/k_\parallel$, i.e., the shortest wavelength that can become unstable for a given ε, it can be used to define a sufficient condition for stability against the universal mode, in terms of a critical length for the plasma. Setting this length equal to $L_u = \pi/k_\parallel = \lambda_\parallel/2$ (the longest wavelength that can satisfy the required boundary condition of zero amplitude at both ends of the system), we find

$$\frac{\pi}{k_\parallel}\left(\frac{1}{n}\frac{dn}{dr}\right) = \varepsilon L_u = 10\left(1 + \frac{T_i}{T_e} + \frac{4\lambda_{D_i}^2}{a_i^2}\right) \qquad (250)$$

For reasonable values of ε, this does not imply a very serious restriction on the plasma length. For example, if we approximate ε by R^{-1}, where R is the plasma radius, and set $T_i = T_e$, then the condition $L < 20R$ would be sufficient to maintain stability under all conditions. For $T_i/T_e > 1$, as is usual, the condition would be even less restrictive.

Stabilization by finite length effects is but one of the means for suppressing the universal instability. A closely related effect is that associated with sheared fields, or with field curvature. Thus, either sufficient shear or sufficiently favorable field curvature (as in a magnetic well) will also stabilize this mode.[54] We have dwelled at length here on the universal mode because it, like the hydromagnetic drift mode, being of low frequency would be expected to produce intolerably large plasma losses transverse to the field if allowed to grow. In addition, since it has the disconcerting property of requiring only a density gradient to stimulate it, it challenges the very concept of magnetic confinement. It is indeed fortunate that there appear to be effective ways to stabilize this mode.

E. Recapitulation

We have thus far discussed examples of some of the more important modes of unstable behavior in high temperature plasmas. Emphasis has been placed on determining those plasma regimes or external circumstances that promote stability against the modes in question, this being the goal of research aimed at achieving long-time plasma confinement. We have seen that for every theoretically predicted instability one can generally find an antidote—i.e., a combination of accessible parameters that will suppress it. Though the prescriptions vary, the stabilizing techniques can be categorized as follows: They either (*a*) consist in reducing or eliminating the particular free energy reservoirs, as in the weakening of density gradients below critical values or as in the use of magnetic wells, or (*b*) involve exploitation of natural damping or finite length effects. Up to the present it appears that none of the remedies are mutually exclusive—a fortunate circumstance indeed. Still this possibility cannot yet be ruled out, and it is certainly true that each added requirement narrows the range of possible systems and parameters that can be employed.

The instability modes that we have discussed have been ones that are either of special interest or in some sense fundamental, but many modes have been ignored. New sources and modes of instability may arise in theory if temperature gradients,[55] finite resistivity effects[56] (important only at relatively low plasma temperatures, where collision rates are appreciable), or static electric fields[26,57] are present. Under special circumstances an instability can theoretically be stimulated by a gradient of impurity ions in the plasma.[58]

A word of caution must again be introduced on the subject of the relevance of some of the calculations of instabilities. Since all theoretical calculations of instabilities involve idealizations of one kind or another there is always some question as to whether a theoretically predicted instability will in fact occur. In the case of the robust ones, or ones that have been unambiguously identified experimentally there is, of course, no doubt. But in the question of some of the esoteric ones (for example, some of the subtler loss-cone modes) that have not yet been detected, there is real question as to their relevance. These remarks do not imply, however, that the basic theoretical equations used in the analysis are in question, only that the idealizations used in solving them are sometimes too extreme.

F. Nonlinear Effects

All of the examples we have considered are picked from the general area of "linear" stability theory, that is, the theory of modes that are predicted to grow, starting from any small initial perturbations, such as those provided by the natural fluctuation levels in the plasma. But what of nonlinear instability phenomena in high temperature plasma? This important aspect of plasma physics has been the subject of intense study in the last few years, despite the great difficulty of applying analytical techniques to such problems.

Two aspects of nonlinear phenomena in plasmas have been explored. The first aspect has to do with the prediction of the time history and the macroscopic consequences of the growth of "linear" instabilities of the types we have been discussing. The second type of problem, far less well understood, is the prediction of nonlinearly unstable modes, i.e., modes that require a finite perturbation of the plasma before they can be initiated. Fortunately for fusion research enthusiasts, there do not seem to be too many of these latter beasts lying in wait, but the story is not yet completely told. As for the nonlinear consequences of ordinary instabilities, this is a topic of wide interest, not only to fusion buffs, but to anyone interested in the macroscopic behavior of plasmas, whether as observed in the laboratory or as seen in space. That this should be true is simply because the linear theory predicts that any unstable mode, however weakly driven and small in initial amplitude, will grow exponentially with time. This growth will perforce proceed unchecked until some nonlinear process steps in to limit its growth—or to quench the instability. The decision as to whether or not an instability is "important"—that is, whether or not it significantly affects the plasma's behavior—thus depends upon determining at what amplitude these nonlinear limiting effects step in. It therefore follows that plasma confinement is a relative thing—it is only "bad" when the level of instabilities is so intense as to seriously compromise the confinement, relative to that determined by "classical" collision-induced particle losses. It follows that a plasma need not be perfectly stable to permit adequate, even excellent, confinement: The level of unstable fluctuations needs only to be low enough in amplitude or high enough in frequency to limit the resulting transport to an acceptable level.

One way to estimate the maximum rate of plasma transport that can be caused by unstable plasma fluctuations is based on thermodynamic considerations.[59] One notes that the energy density associated with the fluctuations cannot exceed the free energy contained within a volume defined by the correlation length (wavelength of the instability) in the plasma. Using this datum and applying the arguments, discussed earlier, about the random walk "step lengths" caused by the unstable fluctuations, one can estimate the resulting transport.

One important result of this kind emerges from some simple considerations. We have already noted that the rate of diffusion of plasma across a magnetic field caused by instabilities can be estimated by assigning the following value to the diffusion coefficient:

$$D_\perp \approx \omega \lambda_\perp^2 \tag{251}$$

where ω is the frequency of the fluctuations and $\lambda_\perp$ their correlation length perpendicular to the field. If we now make the simple and physically reasonable assumption that the "frequency" ω in Eq. (251) is to be identified with the time for an element of the plasma to drift a distance $\lambda_\perp$ at the electric drift velocity $c(E/B)$, one obtains the result

$$D_\perp \approx (CE/B\lambda_\perp)\lambda_\perp^2 \tag{252}$$

To close the problem we now take $|E|$ to be of the order of the potential fluctuations across the distance $\lambda_\perp$, a value that in turn must be bounded by the electron temperature. Thus $E \approx \phi/\lambda_\perp$; $e\phi \approx \kappa T_e$.

Inserting these into (252) one finds that $\lambda_\perp$ drops out, leading to the result[60]

$$D_\perp \approx c(\kappa T_e/eB) \text{ cm}^2 \text{ sec}^{-1} \tag{253}$$

This expression is the now famous and yet equally mysterious "Bohm diffusion" coefficient, the exact origin of which, though clearly traceable to Bohm, is lost in the mists of antiquity. That this coefficient has a rugged and ubiquitous nature in the context of plasma research can be suspected from the fact that exactly the same coefficient emerges from a completely different set of assumptions! Consider the diffusion across a magnetic field to be expected in the presence of "collisions" that displace the guiding centers. Here

$$D_\perp \approx \frac{\bar{v}^2 \tau}{1 + (\Omega\tau)^2} \tag{254}$$

In the limit $\tau \gg \Omega^{-1}$ (infrequent collisions). $D_\perp \gg 0$; in the limit $\tau \ll \Omega^{-1}$, $D_\perp \approx \bar{v}^2 \tau$, the classical result for collisional diffusion in the absence of a magnetic field. Thus $D_\perp$ vanishes at both limits, $\tau = \infty$ and $\tau = 0$. Consider now the maximum value that $D_\perp$ may have, seen to occur when $\Omega\tau = 1$. At this point

$$D_\perp = (\bar{v}^2/2\Omega) = (\tfrac{1}{2} M \bar{v}^2/eB)c \tag{255}$$

If we now make the identification $\tfrac{1}{2} M \bar{v}^2 = \kappa T_e$, this again leads to the Bohm result.[61] Though these "derivations" are little more than dimensional arguments, they strongly suggest that the Bohm diffusion rate might define the characteristic confinement time in toroidal systems where well-developed instabilities are present. As we shall later discuss, this conjecture seems to be well supported by many experimental observations. Though the rate of plasma losses implied by Bohm diffusion may not be as rapid as that induced by gross instabilities, it is still far too rapid for comfort in the context of fusion research, especially when one considers the embarrassing fact that the Bohm relationship predicts more rapid diffusion with increasing temperature and only a linear decrease with increasing B. Contrast this result with that for classical diffusion (Eq. (40)), where $D_\perp$ varies inversely with $T^{1/2}$ and B^2. To illustrate the extreme disparity between these two predictions as applied to a high temperature plasma, note that at a density of 10^{14} ions/cm^3, an electron temperature of $10^{8\circ}$ K and in a magnetic field of 10^4 G, the classical diffusion velocity per unit logarithmic density gradient $[(1/n)(dn/dr) = 1]$ is, from Eq. (40), $v_c \approx 4$ cm/sec ($Z = 1$). For the same condition the Bohm formula predicts a diffusion velocity approaching 10^8 cm/sec! To be sure, in a realistic case the logarithmic gradients would perforce be much smaller than unity, but even so diffusion velocities approaching the Bohm value would be intolerably large at high plasma temperatures.

Fortunately it devolves that the constant appearance of Bohm diffusion as observed in low temperature plasma experiments in toroids does not necessarily imply that the same situation will persist at high temperatures. Not only does plasma resistivity seem to play a role in the present experiments, but it is clear that only relatively low frequency instability modes can induce transport rates approaching the Bohm value. Suppression of these modes, once they are clearly identified, should therefore in principle limit anomalous cross-field transport to acceptable values. Experimental evidence is now being accumulated that increasingly supports this contention.

In addition to analysis dealing with the consequences of low frequency instabilities, progress has also been made in predicting the nonlinear behavior of high frequency instabilities—for example, the two-stream instability. In those cases where the free energy reservoirs represent only a small fraction of the total plasma energy, perturbation methods can be used, thus rendering the problem analytically tractable. This situation, one in which fluctuations, though fully developed, carry only a small fraction of the total plasma energy, is called the "weak turbulence" limit. The classical example, first published[62,63] in 1962, concerned the growth and decay of a stream instability in a Maxwellian plasma through which a low density nonenergetic beam of electrons is passed, thus creating a "bump on the tail" distribution as indicated in Fig. 19a. The presence of this beam was shown to create a spectrum of growing waves that in turn caused the bump to flatten and finally to disappear, somewhat as indicated in Figs. 19b and 19c. This calculation, an example of what has come to be known as "quasi-linear theory,"[64] illustrates several important points: First, that wave–particle instabilities will give rise to a spectrum of turbulent waves more or less centered on the unstable waves of linear theory, and second that the situation is more or less self-healing in the sense that the interaction with these waves forces the distribution function toward a Maxwellian, through depletion of the free energy reservoirs that drive the unstable waves. The third point illuminated by quasi-linear theory is the importance of nonlinear wave–wave coupling processes in unstable plasmas.[64,65] As later calculations have shown, these wave–wave processes are predicted to give rise to entirely new phenomena. An example is "nonlinear Landau damping," a process in which one wave, of a phase velocity such as to be only weakly Landau damped, will mix with another wave or waves to produce a "beat wave" that is strongly Landau damped. The end result can therefore be a strong damping of the originally undamped wave. It also appears that the opposite effect can occur—a normally damped wave can be caused to grow by mixing with other waves so as to produce beat waves that satisfy resonant growth

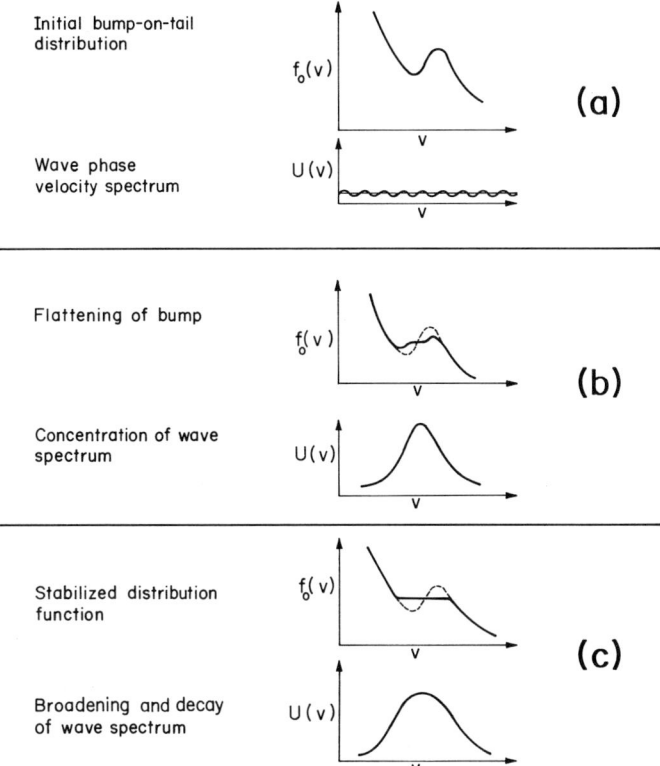

Fig. 19. Schematic illustrations of predictions of quasi-linear theory as to the smoothing of an unstable bump-on-tail particle distribution by interaction between the particles and the instability-stimulated wave spectrum.

conditions. The field of nonlinear plasma theory is still too much in its infancy to predict how many of these new effects will in fact turn out to be of practical significance, but there is little doubt that there is no area of greater significance for the understanding of how plasmas actually behave.

Before leaving the subject of nonlinear processes in plasmas we should note a situation where such processes can be, indeed have been, put to practical use. This is the area of so-called nonlinear wave–particle effects. Such effects offer a powerful means for the conversion of ordered energy (for example, a directed electrical current) in a plasma into thermal particle motion. This effect may occur in nature, as for example when currents streaming through a plasma in space exceed the critical value for instability, suddenly dumping their energy into random particle motion. In the laboratory, the phenomenon offers the possibility of heating a dense plasma to thermonuclear temperatures in microseconds, utilizing plasma instabilities to convert induced plasma currents into thermal motion.[66,67]

G. Summary of Instability Effects

To summarize our discussion of plasma instabilities, we have seen that their origin is rooted in the inevitable departures from thermodynamic equilibrium that will exist in almost every plasma situation of interest. Yet we have seen that the mere existence of a free energy reservoir is not sufficient to guarantee that instability will occur.

We dealt first with the most elementary free energy reservoir, that of plasma expansion. In a magnetically confined plasma this reservoir can lead to gross hydromagnetic instability or to so-called universal instabilities. We saw, however, that most such instabilities, those at low frequencies, can be overcome by the use of magnetic wells, average magnetic wells, or in some cases magnetic shear. Higher frequency modes fed from this reservoir but sensitive to the steepness of the density gradients at the boundaries of the confined plasma can in theory be controlled by reducing these gradients.

Other free energy reservoirs arise from streaming motions in the plasma, either directly associated with currents or arising indirectly from the existence of pressure anisotropy or loss-cone-like particle distribution functions. While the hydromagnetic instabilities always occur at relatively low frequencies ($\omega \ll \Omega_i$), the streaming instabilities and their cousins may occur over a wide range of frequencies, up to the order of the electron plasma frequency.

Analysis of each type of instability has thus far shown that it possesses an "Achilles' heel" in that it can be stabilized or avoided by discovering the critical conditions required for its onset or the environmental situations that prevent its occurrence. Even if an instability cannot be completely suppressed, the nonlinear consequences of instabilities vary so greatly that one cannot *a priori* assume that it is necessary to completely suppress all instabilities to preserve confinement; some instabilities are predicted to produce only relatively slow plasma transport.

That the existence of plasma instabilities enormously complicates the problem of confining a plasma, there is no doubt. But the intense study of the subject that the last few years has seen has shown that the problem is tractable, and that the methods of analysis that have been developed, based on the Boltzmann and Vlasov equations, are basically sound, failing only when the idealizations made in the theoretical analyses are too unrealistic.

Not only has the study of instabilities been essential for making progress toward controlled fusion, but it has also provided the rationale for the understanding of virtually all the complicated phenomena observed in plasmas in space, where plasma instabilities are involved in both the origin and the evolution of such plasmas.

V. The Present Status

Trying to describe what is known today about the behavior of high temperature plasmas is somewhat like trying to analyze the personality and the potential of a precocious, but almost totally uninhibited, youngster with an IQ of 200. One has little doubt but that in a few years some of his presently annoying and puzzling behavior will be understood in retrospect, and that great things are ahead for him, but today there is a considerable element of frustration and an uncomfortably large number of unanswered questions in the picture. This is by way of saying that despite the often brilliant conceptual, theoretical, and experimental advances that have been made in the field of high temperature plasma physics, some very basic questions have not yet been answered—either in the sense of providing ironclad theoretical arguments or introducing irrefutable experimental evidence. It is not that many of the physicists involved do not have strong personal opinions as to what the answers to these questions will be—it is just that these opinions are sometimes poles apart!

At the considerable risk of putting the cart before the horse, we will first discuss these yet-to-be-resolved problems in plasma physics before undertaking to discuss the present status of research in this field, particularly as it relates to fusion. Of course, we have throughout this article attempted to relate theoretical derivations to their experimental verification, but there has been too little opportunity to give coherence to the picture. We will attempt to do this by presenting a picture of the field as it might be seen by an experimentalist, and then by a theorist working actively in plasma physics. Following that, we shall relate what we have been discussing to the problem of fusion energy and to the "devices" that have been, and are being, used to carry out fusion plasma research, since it was the hope for fusion that launched the new era of modern plasma physics and that still provides its strongest single motivation.

To return to our first task of asking what is yet unknown, what are those remaining "key questions" in high temperature plasma physics? Even the answers to this query will vary greatly and will contain considerable elements of personal bias. But to this author they are as follows:

(*1*) *Does there exist any case of a totally quiescent high temperature, high density plasma confined by a magnetic field?* By "totally quiescent" state is meant one in which only statistical effects, such as those occurring in ordinary gas, contribute to the fluctuation level. By "high temperature, high density" we mean a density such that the plasma pressure represents an appreciable fraction of the magnetic pressure and yet the magnetic field and temperature are high enough to make collision frequencies negligible compared to cyclotron frequencies.

(2) If the answer to question (*1*) is "no" (as it is likely to be) the second question is: *What factors determine the degree to which a quiescent confined state can be approached*, i.e., *do there exist "practically stable" states in such plasmas?*

(3) *Do attainable high density magnetostatic equilibrium states exist for plasma confined in non-axially symmetric toroidal systems?* While theory and experiment agree as to the existence of equilibria in axially symmetric toroidal fields (for which interior conductors or directed currents are required) no such rigorous proof yet exists for the general toroidal system. An affirmative answer to this question would seem to be essential to the success of stellaratorlike devices.

(4) *What role is played by the static ambipolar electric fields that necessarily arise in achieving magnetostatic equilibrium in determining the stability and transport properties of a magnetically confined high density plasma?* Relatively little experimental or theoretical evidence exists on this point—important in understanding the nature of the confinement process itself.

(5) *What is the nature and structure of shock waves in a high temperature plasma?* A detailed answer to this question is important for the understanding of plasma heating effects in the laboratory, and for the interpretation of many phenomena in space plasma research and astrophysics.

To the list above one might add a whole series of more detailed questions as to transport processes, nonlinear instabilities, wave–wave interactions, and other topics. Also, more out of curiosity than any belief that it is a fundamental issue, one could add the question: "What is the explanation for the ubiquity of Bohm diffusion rates appearing in many toroidal confinement experiments, when such rapid rates (of cross-field transport) have never been observed in magnetic well systems?" The author has the personal opinion that the explanation of this paradox lies in the radically different plasma boundary conditions in the two cases. Toroidal systems, by necessity, impose boundary constraints on the plasma only outside the last closed magnetic surface on which plasma is confined. All interior flux surfaces are therefore free to take on electrostatic potentials that are little influenced by external conditions. By contrast, in open-ended systems, all lines on which plasma is confined leave the system, subsequently passing through the walls of the confinement chamber. Thus, as coupled through the end losses, the exterior regions can exert a powerful influence on the interior radial distribution of plasma potentials. In this way exterior conducting regions may be able effectively to "short out" evolving potential distributions that could lead to rapid cross-field transport. The existence of this "line-tying" effect is one of the oldest bits of experimental evidence in open-ended research, since it was early shown that this same effect could enforce gross stability on an otherwise hydromagnetically unstable plasma confined in a non-magnetic-well mirror field.[68]

To return to a more orderly development of remarks on "the present status," we will attempt to characterize the present knowledge of the high temperature plasma state by two capsule summaries—one as seen through the eyes of an experimentalist; the other as it might be depicted by a theorist.

A. High Temperature Plasma—The Experimentalist's View

In the early days of research in high temperature plasmas the key problems of this research were: How do you produce such a plasma for study? and How do you measure its properties and discern its behavior? By now, workable solutions to both of these problems have been devised. While it remains technically difficult, it is routine today to be able to produce plasmas at temperatures of tens or even hundreds of millions of degrees kinetic temperature at densities anywhere between the lowest useful densities of about 10^8 up to 10^{17} or 10^{18} ions per cubic centimeter. Perforce the highest densities have only been produced transiently, by rapid magnetic compression in intense magnetic fields.[69,70] Still, even the highest density plasmas live long enough (microseconds at 10^{17} cm^{-3}, for example) for useful studies to be made. At lower densities other techniques can be used, and observation times have in these cases been extended to fractions of a second, or even seconds, at the lowest densities.

While only a few years ago the cooling effect of impurity ions and atoms was a matter of serious concern for experiments in all density ranges, this problem has by now been virtually eliminated by developments in materials and vacuum technology within the plasma research effort. Similarly, the techniques for designing, fabricating, and energizing magnet coil systems capable of producing virtually any field configuration and intensity, up to hundreds of kilogauss, needed in the experiments are today well in hand. Rapidly coming into the picture are the exciting developments of the "hard" superconductors, which promise a revolution in some aspects of high temperature plasma research.

The always difficult problem of "plasma diagnostics" has evolved toward more and more sophisticated techniques, until it is by now possible in many experiments to deduce, by non-interfering methods, all of the critical parameters of a plasma—its density, gross motions, wave emissions, and the detailed distribution functions of its ions and electrons. Most recently, laser-scattering techniques have been applied to the problem with considerable success.

What has been established experimentally as a result of applying these many technological developments to the problem? First and foremost it has been learned that one must take seriously the prediction of hydromagnetic theory with respect to instability. Except for rather special circumstances, where finite orbit effects or "line tying" modify the picture, situations predicted to be hydromagnetically unstable (non-magnetic-well mirror fields, pinching currents,, etc.) are violently so, usually abruptly terminating the experiment through loss of plasma to the chamber walls in a matter of microseconds. With rare exceptions all early magnetic confinement experiments that were predictably hydromagnetically unstable were doomed to failure as far as their stated objectives—long-time confinement—were concerned.

The converse has been found to be often, but not always, true. The classic experiment in hydromagnetic stability, performed by Ioffe[71] in the USSR in 1961, showed that a magnetic well field, theoretically stable against hydromagnetic modes, does indeed suppress all such modes in an otherwise unstable plasma. This principle, of the magnetic well, has by now been incorporated into virtually every open-ended experiment, to their great benefit.

Less success has been had with applying other predictions of hydromagnetic theory with respect to stabilizing effects—magnetic shear stabilization of toroidal systems, for example. Here the results are mixed, only partial stabilization being achieved. But there is strong evidence that these failures result from trying to apply the theory to situations outside its purview. In particular, the assumption of "infinite conductivity" implicit in the simple hydromagnetic theory may be badly violated in experiments whose technological problems have prevented the achievement of sufficiently high plasma temperatures to reduce collisional effects to insignificance. Thus there is good reason to believe that the failures of hydromagnetic theory with respect to shear stabilization reflect failures of the experiments to achieve the conditions required, rather than fundamental deficiencies in the theory.

With respect to nonhydromagnetic phenomena, such as high frequency wave–particle effects, a great deal of information has been accumulated. Some of it can already be fitted neatly into the theoretical picture, while much is still too intertwined with competing effects to be sorted out. The classic fundamental experiment[29] in this field is the verification of Landau damping—the most basic of the wave–particle interactions. Beyond this the work has largely consisted of the step-by-step process of isolating and identifying the many wave–particle instability modes predicted by theory to occur. Some of these modes have now been identified beyond a reasonable doubt. These include: (*1*) the low frequency "universal" drift instability, explored by the use of cesium plasma columns,[72] (*2*) the "negative mass" instability (a precessional drift mode at ion cyclotron instabilities)[73,75] (*3*) the "two-stream" instability, in a variety of forms, investigated by use of electron beams in plasma,[76,77] and (*4*) the "double-humped" instability,[78] found in a confined plasma when either ions or electrons are characterized by distribution functions that show two energetically separated groups. Other authors might well add other examples, depending on their convictions as to the validity of the results that have been obtained.

In addition to the identification of some of the unstable wave–particle modes of high temperature plasma, a considerable amount of experimental work has been done in isolating and identifying the bewildering variety of stable wave modes of magnetically confined plasma. This work has been carried out both in plasma laboratories and in "nature's laboratory," through space research. Alfvén waves, "Landau waves," "whistlers," drift waves, ion sound waves, to name a few, have all been analyzed with a considerable degree of precision and detail. Since these waves are the carriers of the unstable modes, gaining an understanding of the stable waves is an essential element in our understanding of confinement and stability.

One must add to this picture another type of "wave" on which much work has been done, but with somewhat less complete success—the elusive "collisionless shock wave" in a high temperature plasma. Controversy still exists as to the exact nature of this beast, although it is clear from space research and from laboratory evidence that shock-like effects can exist in high temperature plasma, sustained by collective interactions, rather than by simple collisions, as is the case with ordinary shocks.

The subject of shocks introduces another facet of experimental plasma research—the investigation of nonlinear phenomena. This topic is only now coming into its own, but it is one of major importance. After all, whenever one observes an instability in the laboratory it has usually reached saturation level, where nonlinear effects are dominant. What the instability then "does" to the plasma has everything to do with nonlinear effects. Whether saturation sets in before the kinetic energy stored in the unstable waves is comparable to or much less than the total particle kinetic energy is very important to the theorist: If the latter is the case, "quasi-linear" or "weak turbulence" theory may be applicable, for which analytical techniques have been developed. If not the case, virtually no theoretical tools are presently available to handle such problems. Although experimental evidence on this point is recent and fragmentary, it appears that there may exist many important cases where "weak turbulence" theory applies. There is real hope, therefore, to be able to predict not only the onset, but the subsequent evolution and macroscopic consequences, of instabilities in confined plasmas. Some of the most marked successes in this area have been achieved in space research, where detailed predictions of particle loss processes in the earth's Van Allen radiation belts have been possible.[79]

In the experimental search for stable plasma confinement there still remain the two specific, crucial questions: (*1*) In toroidal systems, can we understand and eventually adequately control the Bohm diffusion process so as to approach the classical loss rates, or if not, can we at least reduce the leading coefficient of Bohm diffusion to a small fraction of its "normal" value? Progress toward this latter objective has been accelerating in the recent past and there is considerable hope that this objective, at least, will be reached. (*2*) For open-ended systems, will it be possible sufficiently to reduce the free-energy reservoirs and to control the boundary conditions of plasmas confined in open-ended systems so that their confinement time approaches that expected from collisional processes alone? Again, recent progress toward this objective has been encouraging.[80] Cases where quiescent or nearly quiescent long-time confinement has been achieved are multiplying. The use of magnetic wells has enormously simplified the problem, since their use excludes wide classes of instabilities, those with frequencies less than the ion cyclotron frequency.

In brief, we characterize the present situation of the earth-bound plasma experimentalist as follows: He finds it possible, with sufficient effort, to create high temperature plasma for study over a wide range of conditions. Though in earlier days empiricism was the rule, today the experimentalist finds his greatest successes in undertakings specifically designed to elucidate predictions of the theory, or to explore areas where theoretical predictions are presently inadequate. The real trick is to find situations and regimes where only one process—the one in question—plays the major role. In space, where we are only observers, but where we can really take an "inside" view through space probes, the problem is to sort out, aided by theory and laboratory experiment, just which ones of the possible plasma phenomena are operative at the point of observation.

B. The Theorists' View

No less than the plasma experimentalist's problem, the plasma theorist's task has been the development of workable tools and techniques to handle the complexities of plasma behavior. In this he has been remarkably successful. Still there are some that have felt that there remain some questionable points in the structure of the theory that were bypassed by the press to obtain specific results. To be specific on this point, though remarkable successes have been achieved through the use of the Boltzmann-Vlasov equations for solving plasma problems, there exist substantial questions as to how closely these equations model the rigorous N-body equations, particularly where transport processes are involved.[11] The situation perhaps resembles that obtaining in the early days of quantum theory, before wave mechanics. Though there were many successes to point to in the theory, there was probably a feeling that something was not quite right about the basic equations in use, something that limited their ability to give exactly the right answer, or even to predict some of the subtler effects. In plasma theory a major effort has been expended to put the basic equations in better shape, but as yet it appears that nothing as useful as the workhorse Vlasov equation has emerged.

The greatest volume of work, and it is really prodigious, has been done in the area of instability theory, using the linearized hydromagnetic and Vlasov equations. Today the feverish search for new unstable modes has largely subsided, and effort is turning increasingly toward more "realistic" calculations of the already known modes. In these calculations the difficult analytical problems associated with finite length effects, realistic distribution functions, magnetic gradients, and wave reflections are being tackled. These problems often tax classical analysis to its limits, and the going is slow. More often than not, after formulating and reducing the problem one is forced to finish it by numerical computation, with high speed computers, as pure analysis is not equal to the task.

At this point it appears that a virtually complete inventory of the linearly unstable modes of high temperature plasma has been assembled. Though the list is long it *is* finite. Furthermore, a large fraction of the instabilities in the list fall by the wayside as one begins to consider the stabilizing influences that may be brought into play—magnetic wells, shear, finite length effects, and so forth. Linear stability theory seems to be on a firm footing and to be a topic from which useful results will continue to emerge.

Though in the sequence of problems associated with magnetic confinement, magnetostatic equilibrium certainly antedates stability considerations, there still remain some difficulty problems concerned with equilibrium yet to be settled. We have already mentioned these among the "key questions." Here the difficulties of analysis arise for topological or geometrical reasons. It is of marginal utility to calculate equilibria in a torus by idealizing it as an infinite cylinder, but this is what often must be done to render the problem tractable. Even when the exact geometry can be used there remains a fundamental question: If one finds a solution to the equations of magnetostatic equilibrium, how is one sure that this solution is a unique one? Or, if it is not, may not the plasma find an excuse, thermodynamically speaking, to choose another, less well confined, solution? While the equilibria in magnetic wells appear to be so rugged that this problem does not arise, it is by no means clear that the same situation prevails for toruses.

Turning to nonlinear theory, it is here that some of the most significant developments in plasma theory are taking plase. Nonlinear theory is difficult at best, but already new effects have been predicted that could be of major importance. An example is the phenomenon of nonlinear Landau damping, to which we have already alluded. Though one might have hoped that nonlinear effects would be generally benign, i.e., they should limit, rather than exacerbate instability effects, there are now enough examples of possible new instability modes of nonlinear origin[81] to give one an uneasy feeling, at least until the matter is settled by experiment—or more theory.

Perhaps the future for plasma theory will not be as exciting as it has been in the recent past, but it will no doubt be more satisfying in that one will be able to predict in ever more detail what real plasmas do. If this comes about, theory will be able to extend and largely supplant the detailed exploration of effects that can now only be accomplished by direct experimentation. In this goal computers will no doubt play an increasingly important role. In the context of space and astrophyical research such developments will be a great boon, since we cannot meddle in cosmic plasma processes, but can only observe and interpret.

C. Fusion Research and "The Devices"

Though the author's interest and experience lie in the area of controlled fusion research, and though this discourse has everywhere consciously (or unconsciously) been slanted toward fusion-related problems, we have not as yet explicitly stated just what it is fusion research is trying to achieve. Nor will we here, except in briefest detail. The subject of fusion reaction cross sections, their energy release, radiation losses, fusion fuel cycles, the scaling laws for fusion reactors, and other thermonuclear matters could by itself form the subject material for another treatise. Enough is known already about these matters to be assured that, given solutions to the plasma questions, fusion power will without doubt become an important, probably the major, source of energy for the human race. In the short range, therefore, it is clear that the problems of fusion research are almost entirely scientific ones. Basically they are the ones we have been discussing: How to contain a plasma of fusion fuel at ultrahigh temperatures, without contact with physical walls, long enough for a net release of fusion to occur. The first idea proposed to solve this problem, magnetic confinement, still seems to offer the only workable solution.

To be more specific about fusion we must be quantitative: How hot is hot enough for fusion and how long a containment is long enough for net energy release? To answer these questions we must consider specific reactions. There are four that appear to be the most important, involving two heavy isotopes of hydrogen and one isotope of helium, as follows:

(1a) $D + D \to T + p + 3.25$ MeV

(1b) $D + D \to He_3 + n + 4.0$ MeV

(2) $T + D \to He_4 + n + 17.6$ MeV

(3) $He_3 + D \to He_4 + p + 18.1$ MeV

Reactions (1a) and (1b), the DD reaction (deuteron–deuteron), occur with roughly equal probability. Reaction (2), the triton–deuteron reaction, has the largest cross section (5×10^{-24} cm² at 80 keV) and a large energy release. Thus although the other reactions will no doubt be exploited in time, the TD reaction is the one usually considered. Since deuterium occurs naturally but tritium does not, a fusion reactor employing the TD reaction would have to provide for regeneration of tritium. This can be accomplished by using the 14 MeV neutron released by the reaction to induce $n,2n$ reactions in a "blanket" surrounding the reactor, and then capturing these neutrons in lithium, thereby producing tritium. The efficiency and economics of such a process appear to be entirely reasonable, and lithium is a relatively abundant element.

The rate of energy release from fusion reactions between two types of reacting nuclei (for example, T and D) at densities n_1 and n_2, respectively, is given by the expression

$$p_n = n_1 n_2 \langle \sigma v \rangle_{12} w_{12} \tag{256}$$

where $\langle \sigma v \rangle$ represents the average reaction rate parameter, reaction cross-section times relative velocity, averaged over the distribution of relative velocities, and w is the energy release per reaction. As seen from Fig. 20, the peak value of $\langle \sigma v \rangle$ for the TD reaction is about 10^{-15} cm³ sec⁻¹, as calculated for Maxwellian or near-Maxwellian velocity distributions. This peak occurs at about 50 keV kinetic temperature (about 500 million degrees Kelvin); below about 10 keV temperature $\langle \sigma v \rangle$ drops off sharply. This precipitous dropoff implies that there must be a practical "ignition

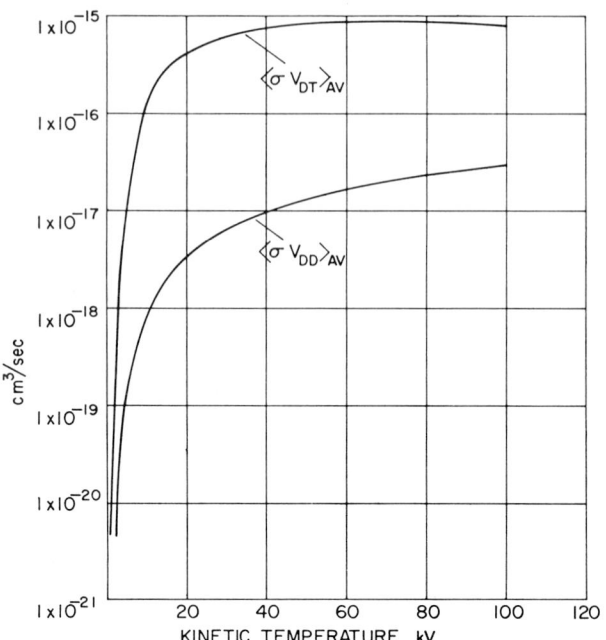

Fig. 20. Reaction rate parameters $\langle \sigma v \rangle$ for the deuteron–deuteron and deuteron–triton nuclear fusion reactions.

temperature" below which fusion reactions cannot sustain themselves. Such a temperature limit does in fact exist, and it is set by the rate of radiation losses from the plasma. Existing far from a state of equilibrium with its own radiation, high temperature plasma such as would be employed for fusion purposes emits radiation at a rate far lower than the Planck equilibrium value. (If this were not the case there would be no hope of ever producing a fusion reactor.) The actual rate of radiation losses in this case is set by the rate of bremsstrahlung or x-ray emission arising from near-collisions between the electrons and the ions in the plasma. An approximate expression for these radiation losses, in the case of a pure hydrogenic plasma, where $n_i = n_e$, is

$$p_b = 0.5 \times 10^{-30} n_e^2 T_e^{1/2} \text{ W/cm}^3 \tag{257}$$

For fixed plasma composition, both the fusion power release and the radiation losses vary as the square of the plasma particle density. In particular, for TD reactions in a 50:50 mixture of deuterons and tritons of total density n ions/cm³,

$$p_{TD} = 0.7 \times 10^{-12} n^2 \langle \sigma v \rangle \text{ W/cm}^3 \tag{258}$$

Comparing these two expressions one finds that fusion energy can only exceed radiation losses for ion temperatures above about 5 keV (taking $T_e = T_i$). Since bremsstrahlung radiation losses represent a lower limit to the plasma losses, any additional loss processes would only tend to increase the ignition temperature. For this reason it is usual to consider temperatures between about 10 and 100 keV to define the practical operating regimes.[82,83] The lower temperatures would perhaps be appropriate for toroidal systems if confinement times long compared to the Bohm values are achievable. The higher range would perhaps be appropriate for mirror machines, where high temperature helps in the competition between fusion reactions and collision-induced end losses.

We may readily derive one more important criterion for net energy release. One notes that the heating of a fusion plasma up to reactor temperatures involves an energy "investment" per unit volume of plasma not less than

$$U_H = \tfrac{3}{2} n \kappa T \tag{259}$$

Clearly, to sustain the reaction this heated fuel must be retained for a sufficient time, τ, to release nuclear energy in excess of U_H (ignoring inefficiencies in both heating and energy recovery). But

$$U_n = n^2 \langle \sigma v \rangle w \cdot \tau \tag{260}$$

We find upon comparison of Eqs. (259) and (260) that the ratio of nuclear energy released to heating energy is proportional to $n\tau$. The minimum value of $n\tau$ for balance (the "Lawson criterion") therefore gives the minimum required product of density and confinement time for self-sustaining reactions. For TD this product is about 10^{14} cm^{-3} sec. Presumably any confinement system that can exceed this limit could realize self-sustaining reactions.

The existence of the Lawson criterion suggests a natural division point between approaches to fusion power—steady state or pulsed. If one believes that plasma stability adequate to provide confinement times approaching 1 sec will be possible, then fusion reactors with densities of 10^{14}–10^{15} ions/cm^3 are indicated ($\sim 10^{-4}$ of the particle density of air at atmospheric pressures!) At these densities, as shown by Eq. (258), the fusion energy released will be of the order of 10–1000 W/cm^3 (or MW/m^3), appropriate to large steady-state central power plants. On the other hand, if one believes that confinement times will remain limited to a few milliseconds, one must contemplate densities of 10^{16}–10^{17} cm^{-3}. But such high densities are probably only possible for pulsed systems, since at these densities the fusion power rates, varying as n^2, are enormous, of order

10^6–10^8 W/cm³, and the plasma pressures will be very high, thousands of atmospheres, requiring extremely intense confining fields. Even higher density operation could in principle be contemplated, but in such cases one is beginning to approach explosive rather than controlled energy release and it is also not at all clear how to generate the enormous confining fields that would be required.

Fusion research therefore looks toward creating and confining plasmas over the density range of about 10^{14}–10^{17} ions/cm³, with kinetic temperatures between 10 and 100 keV. Whereas $n\tau$ products exceeding 10^{14} will be required, the highest values yet achieved in magnetically confined plasmas are about 10^{11}–10^{12}, obtained in pulsed magnetic compression experiments.

D. The "Devices"

Though it is not necessary to achieve thermonuclear plasma conditions in order to conduct significant experiments in magnetic confinement, it is obviously necessary to adopt and construct some type of confinement "device" in which to carry out these experiments. Unlike nuclear physics, where the particle accelerator is a "black box" used only as a tool to carry out the research, in fusion plasma research the device *is* the experiment. Furthermore, a systematic search for understanding of the general problem of stable magnetic confinement cannot be conducted with only one type of system, since stability is subtly dependent on many factors, only a few of which can be explored with a single device. For these reasons, and also because of the intense interest in solving the fusion confinement problem, fusion research has spawned a wide variety of experimental devices, employing many and diverse field configurations. We cannot here hope to list all of the types that have been studied or are now in use, but will instead select a few as representative, first discussing some of the older ideas.

1. Linear and Toroidal Pinches

These devices are by now virtually extinct, except for a few experiments. One of the earliest ideas in fusion confinement, the pinch was aimed at achieving both heating and magnetic confinement through self-generated magnetic fields associated with large electrical currents flowing axially through the plasma. Violently unstable hydromagnetically, pinches achieved only low temperatures and very short confinement times (typically microseconds). But they provided some of the earliest confirmations of hydromagnetic theory and served as test beds for the development of plasma diagnostic and pulsed high current technology.

2. Cusp Fields (Fig. 3d)

Several devices have been built to capitalize on the idea that the cusp-shaped field produced between circular coils carrying opposing currents should permit hydromagnetically stable confinement (which it does). Although this configuration predates and has similar properties to magnetic well mirror fields, it has a fundamental flaw. Because the field has a null value at its exact center and is therefore weak and sharply curved over much of the interior volume, it is not suitable for adiabatically confining the particles of a plasma. But since it is open-ended the result is that particle losses are extremely rapid, only possibly being reduced if plasma sheath effects, predicted but not yet observed, could step in and seal the cusps. Though little work is now being done with cusp fields, it is clear that they were the forerunners of the magnetic well devices.

3. Mirror Machines

One of the main avenues of research in plasma confinement has been the use of "mirror machines" in their various forms. Early work was carried out using simple mirror fields (Fig. 3b), which as we now know are potentially subject to the hydromagnetic flute instability. Still, it was found possible to create and confine hot plasmas in these experiments and much was learned from them about the nature of mirror confinement. The reasons for the observed lack of instability, achieved only under special conditions, were (1) the existence of "line tying" provided by low density plasma outside the mirrors and (2) the presence of finite-orbit effects. The complete elimination of the flute instability came with the introduction of the idea of magnetic well-type mirror fields, and its first test in the classic experiment of Ioffe.

Many methods have been developed for filling open-ended systems such as the mirror machines with hot plasma. One of the earliest, and still widely used, methods is plasma injection and magnetic compression. In this technique plasma guns are used to inject a burst of energetic plasma in through one of the mirrors while the magnetic field is relatively weak. The field is then pulsed on within a few hundred microseconds, trapping and compressing the plasma. The act of magnetic compression (explainable by the flux invariance of mirror-trapped particles (see Fig. 3b and Eq. (105)), also results in heating the plasma by virtue of the constancy of $\mu = \omega_\perp/B$. By this technique hot plasmas of near-thermonuclear density and temperature (5×10^{13} cm^{-3} density and 5 keV temperature) have been produced and held confined for periods of the order of 1 msec.

In another method, especially suitable for creating a hot electron plasma, high power microwaves tuned to resonate with the electron cyclotron frequency are directed at a trapped low temperature discharge plasma, raising its electron temperature to very high values. In such experiments, plasmas with β values up to about 0.4 have been produced and held stably confined by virtue of line tying and collisional damping effects associated with the presence of background plasma.

Steady state low density plasmas with high ion temperatures and long particle lifetimes (of order 1 sec) have been produced by injecting beams of energetic particles into the confinement chamber. One such method utilizes beams of energetic neutral atoms passed through the confinement chamber, which is maintained at ultravacuum conditions. Breakup of a portion of the injected neutral beam by "Lorentz ionization" (ionization occasioned by the electric field seen by fast particles traversing a magnetic field) or by collisional effects with already trapped plasma provides a continuous method for building the plasma.

Mirror machines have probably gone farther than any other system in overcoming low frequency instabilities and in establishing close contact between experiment and theory. The present task of mirror research is to eliminate the remaining unstable modes, ones observed to occur near the ion cyclotron frequency or its harmonics. The tools to accomplish this are ones we have mentioned: achieving more stable distribution functions and exploiting the stabilizing effects of electron Landau damping and of deep magnetic wells. There is a good chance that a virtually completely stable situation could in this way be achieved at high plasma density in a mirror system, making mirrors lively contenders for the solution of the fusion confinement problem.

E. "Theta Pinches"

The highest density hot plasmas of substantial volume attained in fusion research have been achieved in so-called "theta pinches." These devices are composed of long one-turn cylindrical magnet coils, sometimes capped with mirrors, surrounding a ceramic chamber filled with hydrogen gas at about $\frac{1}{100}$ atm pressure. By means of special high speed condenser banks the gas is first preionized and then shock heated and compressed by energizing the magnet coil and thereby applying a very high magnetic field (up to 100 kG) in a few microseconds. The resultant compressed plasma column is in some cases hot enough and dense enough to completely exclude the confining field ($\beta = 1$), yet can be held stably confined for many microseconds. Experiments of this kind not only provide a remarkable plasma for study but are the forerunners of experiments aimed at acheiving a "pulsed" reactor system, provided confinement times can be extended to the millisecond range.

F. The Astron

The Astron, invented by N. Christofilos, differs radically from all other plasma confinement systems. In this device it is proposed to create a magnetic well-type field, yet one with closed magnetic lines. This goal, one which cannot be achieved through the use of external conductors alone, is to be achieved in the Astron by "winding" a "coil," i.e., a tubular cylinder, of particles at relativistic energies—electrons in the present experiments—in turn held trapped by an externally generated field. If the circulating current in this trapped ring can be made intense enough the field inside the ring will be reversed in direction from that on the outside, creating an internal region of closed lines. In this way it is hoped that the advantages of the magnetic well can be combined with those of toroidal systems.

Thus far, work on the Astron has been confined to attempting to build up the trapped "E layer" and determine its stability properties. Fractional field reversals of about 0.06 have been achieved without the onset of limitations from instabilities. In the layer, two-stream instabilities are probably the major worry, but here relativistic effects come into play that may provide adequate stabilization up to the currents required for field reversal. If reversal can be achieved, the Astron field would appear to be very suitable for fusion purposes.

G. Closed Systems—Stellarators, Tokomaks, Levitrons, and Multipoles

The "closed" or toroidal systems form the largest and most widely studied class of devices in fusion research. The first proposed of these, the stellarator, was invented by L. Spitzer. We have earlier discussed the basic idea of the stellarator: a toroidal confining field possessing rotational transform. This concept is fundamental to all devices in this class except the multipole. In the stellarator both rotational transform and magnetic shear are provided by means of special helical multifilar windings fitted to the chamber, inside the windings producing the main solenoidal confining field. In the Tokomak devices, studied intensively in the USSR, shear and rotational transform are simultaneously provided by passing a heavy current through the plasma, the current being induced by transformer action. As in the stellarator, the confining fields are produced by external windings. The operating regime is then optimized by adjusting the ratio of internally produced field to that applied externally. Tokomaks are seen to be inherently transient in operation, while the stellarator is aimed at steady state confinement. In the Levitron type of device the interior field distribution is very much like that in the Tokomaks, except that the circulating interior current is carried by a levitated metallic ring, immersed directly in the plasma, but insulated from it by its magnetic field. Such devices as the Levitron are not visualized as leading directly to fusion confinement, but rather as tools for the systematic study of confinement in toroidal systems. The multipole systems, in which the confining fields are produced entirely by current circulating in ring conductors immersed in the plasma, we have already described.

Many methods have been used to create hot plasma within closed systems. One of the most widely used in the past has been to fill the confinement chamber with cold gas, partially ionize this gas by means of a radiofrequency discharge, and then render it totally ionized and heat it to operating temperatures by passing a heavy electrical current through it, induced by transformer action. This "ohmic heating" method becomes relatively ineffective, because of falling plasma resistivity, above temperatures of about 200 eV, but is one of the simplest methods in use. Other heating methods used include ion cyclotron resonant heating and electron heating by high power microwaves. Another method of creating the plasma, coming into greater use in closed systems, is to inject plasma directly into the chamber, across the confining fields, from plasma guns. This method has been employed in multipole and other devices. The reason that one can utilize guns at all to inject into closed systems is an interesting bit of plasma lore. By directing the gun at a region of the field where the field lines are convex toward the plasma confinement region, one launches the plasma blob toward a region of "bad" curvature as looked at from the outside. In this situation local fluting instability will develop that allows the plasma to penetrate the field. Once inside, however, the plasma stream encounters the good curvature regions on the opposite side of the chamber. These it cannot penetrate, but is instead forced to flow azimuthally around the chamber, filling it.

Research on closed systems is today in a state of ferment and intense activity. For many years stellarators and related experiments continued to show confinement times only as long as those allowed by the Bohm diffusion rates. This unhappy situation was encountered under a wide variety of plasma conditions, over a temperature range from a fraction of a volt to hundreds of volts. There was strong suspicion that finite resistivity effects were involved in this state of affairs, but no direct proof. The log jam was broken by the multipole experiments. These were the first to demonstrate strong suppression of unstable fluctuation within the plasma, in agreement with theory, and evidence for confinement times longer than that allowed by Bohm diffusion. But because the first multipole experiments perforce required the use of supports to hold the interior conductors in place, particle losses to these supports clouded the issue and prevented an unambiguous measurement of the actual confinement times achieved. Nevertheless, encouraged by the multipole results, stellarator experimenters explored new, more nearly collisionless, regimes finding improvements by a factor of 10 or more over Bohm losses. At the same time a new generation of multipole-like devices was designed, many including floating rings (superconducting, in some cases) to eliminate the support problem. There is good reason to hope that the "Bohm barrier" will be thoroughly breached in these new experiments, and that they may point the way toward improved stellarators not requiring internal conductors.

From the fusion standpoint the great advantage of closed systems is the fact that a large enough torus could provide confinement adequate for thermonuclear purposes even in the presence of appreciable levels of instability. The classical rate for diffusion across a strong magnetic field (Eq. (40)) is so slow that even if residual instabilities were to enhance this rate by several orders of magnitude, confinement would still be adequate in a large system. Looked at from the other end of the scale, an improvement in confinement time by about 10^3 over the Bohm rate would put large closed systems into a regime adequate for fusion containment purposes. The key question is, of course, whether the present encouraging leads provided by the multipole experiments can be followed up, extended, and reproduced in stellarator-like systems.

Conclusion

Where will it lead us? High temperature plasma research occupies today a relatively minor position, as far as level of research support is concerned, in the world's physics research activities. Yet it is addressing itself to some exceedingly important questions: What are the characteristics of nature's most elusive form of matter, the plasma state? And, can we manipulate and control matter in this state in a practical way? Understanding the plasma state has everything to do with understanding what goes on in the far reaches of space, and on earth, with solving the problem of fusion power. Fusion power would not only provide mankind with an inexhaustible source of energy for his use on earth, but also is most likely to be the source of power that would someday propel him to the stars. We are today at best only glimpsing a portion of what might be in the future with respect to plasma and its uses. But there is no doubt in this author's mind that these and other wonders will indeed come about.

References

1. L. Spitzer, *Physics of Fully Ionized Gases*, 2nd ed., Interscience, New York, 1962.
2. C. L. Longmire, *Elementary Plasma Physics*, Interscience, New York, 1963.
3. N. C. Christofilos, *J. Geophys. Res.*, **64,** 869 (1959).
4. R. F. Post, in *Plasma Physics and Thermonuclear Research*, C. Longmire, J. L. Tuck, and W. B. Thompson, Eds., Vol. I, Pergamon Press, New York, 1959.
5. L. Spitzer, *Proc. U.N. Intern. Conf. Peaceful Uses At. Energy*, 2nd, **32,** 181 (1958).
6. J. B. Taylor, *Phys. Fluids*, **6,** 1529 (1963).
7. J. B. Taylor and R. J. Hastie, *Phys. Fluids*, **8,** 323 (1965).
8. D. J. BenDaniel, *J. Nucl. Energy*, Pt. C, **3,** 235 (1961).
9. R. F. Post, *Phys. Fluids*, **4,** 902 (1961).
10. E. E. Yushmanov, *Soviet Phys.—JETP*, **22,** 409 (1966).
11. D. C. Montgomery and D. A. Tidman, *Plasma Kinetic Theory*, McGraw-Hill, New York, 1964.
12. S. Chapman and T. G. Cowling, *The Mathematical Theory of Non-Uniform Gases*, Cambridge University Press, Cambridge, England, 1953.
13. M. N. Rosenbluth, D. MacDonald, and W. Chuck, *Phys. Rev.*, **107,** 2 (1957).
14. I. P. Shkarofsky, T. W. Johnston, and M. P. Bachynski, *The Particle Kinetics of Plasmas*, Addison-Wesley, Reading, Mass., 1966.
15. J. A. Roberts and M. L. Carr, Univ. Calif. Lawrence Radiation Lab. Rept. UCRL-5651-T, 1960.
16. D. J. BenDaniel and W. P. Allis, *J. Nucl. Energy*, Pt. C, **4,** 31 (1962).
17. S. Chandrasekhar, *Principles of Stellar Dynamics*, University of Chicago Press, Chicago, 1942.
18. R. F. Post, *Ann. Rev. Nucl. Sci.*, E. Segre and L. I. Schiff, Eds., **9,** 367 (1959).
19. E. Fermi, *Phys. Rev.*, **75,** 1169 (1949).
20. H. Alfvén, *Cosmical Electrodynamics*, Oxford University Press, Oxford, England, 1950.
21. W. B. Thompson, *An Introduction to Plasma Physics*, Addison-Wesley, Reading, Mass., 1962.
22. T. G. Northrop, *The Adiabatic Motion of Charged Particles*, Interscience, New York, 1963.
23. T. G. Northrop and E. Teller, *Phys. Rev.*, **117,** 215 (1960).
24. W. Gibson, W. Jordan, E. Lauer, *Phys. Rev. Letters* **5,** 141 (1960).
25. V. M. Balebonov and N. N. Semasko, *Nucl. Fusion*, **7,** 207 (1967).
26. M. N. Rosenbluth, N. A. Krall, and N. Rostoker, *Nucl. Fusion Suppl.*, **1,** 143 (1962).
27. L. Landau, *J. Phys. (USSR)*, **10,** 25 (1946).
28. B. D. Fried and S. D. Conte, *The Plasma Dispersion Function*, Academic Press, New York, 1961.
29. J. H. Malmberg, and C. B. Wharton, *Phys. Rev. Letters*, **17,** 175 (1966).
30. T. H. Stix, *The Theory of Plasma Waves*, McGraw-Hill, New York, 1962.
31. G. Schmidt, *Physics of High Temperature Plasmas*, Academic Press, New York, 1966.
32. C. B. Wharton, R. H. Huddlestone, and S. L. Leonard, in *Plasma Diagnostic Techniques*, Academic Press, New York, 1965.
33. M. N. Rosenbluth and C. L. Longmire, *Ann. Phys.* **1,** 120 (1957).
34. A. B. Mikhailovskii, *Soviet Phys.—JETP*, **16,** 364 (1963).

35. N. A. Krall and M. N. Rosenbluth, *Phys. Fluids*, **6**, 254 (1963).
36. I. Bernstein, E. Frieman, M. Kruskal, and R. Kulsrud, *Proc. Roy. Soc. (London)*, **A244**, 17 (1958).
37. T. Okhawa and D. W. Kerst, *Phys. Rev. Letters*, **7**, 41 (1961).
38. H. P. Furth and M. N. Rosenbluth, *Phys. Fluids*, **7**, 764 (1964).
39. S. Chandrasekhar, A. N. Kaufman, and K. M. Watson, *Proc. Roy. Soc. (London)*, **A245**, 435 (1958).
40. J. D. Jackson, *J. Nucl. Energy, Pt. C*, **1**, 171 (1960).
41. M. N. Rosenbluth and R. F. Post, *Phys. Fluids*, **8**, 547 (1960).
42. R. E. Aamodt and D. L. Book, *Phys. Fluids*, **9**, 143 (1966).
43. C. W. Horton, Jr., Ph.D. Dissertation, University of California at San Diego, 1967.
44. E. G. Harris, *J. Nucl. Energy, Pt. C*, **2**, 138 (1961).
45. T. K. Fowler and Y. Shima, *Phys. Fluids*, **8**, 2245 (1965).
46. R. F. Post and M. N. Rosenbluth, *Phys. Fluids*, **9**, 730 (1966).
47. A. B. Mikhailovskii and A. V. Timofeev, *Soviet Phys.—JETP*, **17**, 626 (1963).
48. G. E. Guest and R. A. Dory, *Phys. Fluids*, **8**, 1853 (1965).
49. M. N. Rosenbluth, H. Berk, and L. D. Pearlstein, *Bull. Am. Phys. Soc.*, **13**, 297 (1968).
50. A. B. Mikhailovskii and L. I. Rudakov, *Soviet Phys.—JETP*, **17**, 621 (1963).
51. B. B. Kadomstev and A. V. Timofeev, *Soviet Phys.—Doklady*, **7**, 826 (1963).
52. N. A. Krall and M. N. Rosenbluth, *Phys. Fluids*, **8**, 1488 (1965).
53. N. A. Krall and L. D. Pearlstein, in *Plasma Physics and Controlled Nuclear Fusion Research*, Vol. I, International Atomic Energy Agency, Vienna, 1966, p. 735.
54. N. A. Krall, in *Advances in Plasma Physics*, Vol. 1, A. Simon and W. B. Thompson, Eds., Interscience, New York, p. 153.
55. L. J. Rudakov and R. Z. Sagdeev, *Soviet Phys.—JETP*, **10**, 952 (1960).
56. H. P. Furth, J. Killeen, and M. N. Rosenluth, *Phys. Fluids*, **6**, 459 (1963).
57. B. D. Fried, M. Gell-Mann, J. D. Jackson, and H. W. Weld, *J. Nucl. Energy, Pt. C*, **1**, 190 (1960).
58. B. Coppi, H. P. Furth, M. N. Rosenbluth, and R. Z. Sagdeev, Report 1C/66/79, International Centre for Theoretical Physics, Trieste, 1966.
59. T. K. Fowler, *Phys. Fluids*, **8**, 459 (1965).
60. R. Z. Sagdeev, in *Plasma Physics*, International Atomic Energy Agency, Vienna, 1965, p. 555.
61. M. N. Rosenbluth, *Nucl. Fusion Suppl. Pt. 1*, 21 (1962).
62. W. F. Drummond and D. Pines, *Nucl. Fusion Suppl. Pt. 3*, 1049 (1962).
63. A. A. Vedenov, E. P. Velikov, and R. Z. Sagdeev, *Nucl. Fusion Suppl. Pt. 2*, 465 (1962).
64. W. F. Drummond, in *Plasma Physics*, International Atomic Energy Agency, Vienna, 1965, p. 527.
65. B. B. Kadomtsev, in *Plasma Physics*, International Atomic Energy, Vienna, 1965, p. 543.
66. M. V. Babykin, P. P. Gavrino, E. K. Zavoiskii, S. L. Nedoseev, L. I. Rudakov, and V. A. Skoryupin, in *Plasma Physics and Controlled Nuclear Fusion*, Vol. II, International Atomic Energy Agency, Vienna, 1966, p. 851.
67. J. Alexeff, W. D. Jones, R, V, Neidigh, W. F. Peed, and W. L. Stirling, in *Plasma physics and Controlled Nuclear Fusion Research*, Vol. II, International Atomic Energy Agency, Vienna, 1966, p. 781.
68. R. F. Post, R. E. Ellis, F. C. Ford, and M. N. Rosenbluth, *Phys. Rev. Letters*, **4**, 166 (1960).

69. W. F. Quinn, E. M. Little, F. L. Ribe, and G. A. Sawyer in *Plasma Physics and Controlled Nuclear Fusion Research*, Vol. I, International Atomic Energy, Vienna, 1966, p. 237.
70. A. C. Kolb, W. H. Lupton, R. C. Elton, E. A. McLean, M. Swartz, and M. P. Young, in *Plasma Physics and Controlled Nuclear Fusion Research*, Vol. I, IAEA, Vienna, 1966, p. 261.
71. Y. T. Baiborodov, M. S. Joffe, V. M. Petrov, and R. I. Solbolev, *Soviet J. At. Energy*, **14,** 459 (1964).
72. H. Lashinsky in *Plasma Physics and Controlled Nuclear Fusion Research*, Vol. I, IAEA, Vienna, 1966, p. 499.
73. C. E. Neilsen, A. M. Sessler, and K. R. Symon in *Proc. Intern. Conf. High Energy Accelerators and Instrumentation, Geneva, 1959*, p. 239.
74. H. Postma, J. L. Dunlap, R. A. Dory, G. R. Haste, and R. A. Young, *Phys. Rev. Letters*, **16,** 265 (1966).
75. L. G. Kuo-Petravik, E. G. Murphy, M. Petravic, R. M. Sinclair, D. R. Sweetman, and E. Thompson, *Bull. Am. Phys. Soc.*, **12,** 817 (1967).
76. I. F. Kharchenko, Y. B. Fainberg, R. M. Nikolaev, E. A. Kornilov, E. A. Lutsenko, and N. S. Pedenko, *Soviet Phys.—JETP*, **38,** 493 (1960).
77. L. O. Smullin and W. O. Getty, *J. Appl. Phys.*, **34,** 3421 (1963).
78. W. A. Perkins and W. L. Barr in *Plasma Physics and Controlled Nuclear Fusion Research*, Vol. II, IAEA, Vienna, p. 115.
79. C. F. Kennel and H. E. Petschek, *J. Geophys. Res.*, **71,** 1 (1966).
80. K. G. Moses, C. C. Damm, J. H. Foote, A. H. Futch, Jr., A. L. Gardner, J. E. Osher, and R. F. Post, in *Proc. Colloq. Interaction Stationary Progressive Electromagnetic Fields Within a Plasma*, Saclay, France, Jan. 1968.
81. T. M. O'Neil, *Phys. Fluids*, **10,** 1027 (1967).
82. R. F. Post, *Rev. Mod. Phys.* **28,** 338 (1956).
83. T. H. Jensen, O. Kofoed-Hansen, and C. F. Wandel, *Proc. Intern. Conf. Peaceful Uses At. Energy 2nd, Geneva, 1958*, Vol. 32, p. 431.

Numerical Simulation of Turbulent Flow*

C. E. LEITH

Lawrence Radiation Laboratory, University of California, Livermore, California

During the fifteen years since the establishment of the Livermore branch of the Lawrence Radiation Laboratory, Edward Teller has been an enthusiastic believer in the value of numerical simulation for the analysis of complex physical systems. Such a belief, quite common now, had originally to be sustained against the recognition of such difficulties as imperfect error control, lack of human understanding, and loss of intuition. There have been by now many successful applications of the massive numerical approach to physical problems. The use of computers for numerical weather prediction represents one of the more successful of these. But there remain nagging problems, and this seems a good time to review one of them.

In any finite numerical model of a highly turbulent hydrodynamic flow one must be content to compute explicitly the larger scales of motion and to ignore or at most treat statistically the remaining smaller scales. This is true because of the necessarily finite nature of the representation no matter whether it be in terms of values on a space mesh or in terms of Fourier coefficients of the hydrodynamic fields. Unfortunately, the hydrodynamic equations are essentially nonlinear and there can be no natural division into independent large and small scales; thus the statistically treated small scales must produce a random perturbation to the motions of the large scales, rendering these also increasingly unpredictable with the passage of time. To reduce these qualitative arguments to a quantitative theory of nonlinear error, one must use the somewhat imperfect tools provided by turbulence theory. The notion of scale can be made more precise by considering the Fourier transform or wavevector representation.

Let us assume that we can treat explicitly scales characterized by wavevectors $\mathbf{k}$ with $|\mathbf{k}| \leq k_*$. We can call k_* the truncation wavenumber and generally will indicate truncation-dependent quantities by an asterisk. It is in the nature of the nonlinear equations of hydrodynamics that two

* This work was performed under the auspices of the United States Atomic Energy Commission.

wavevectors interact in such a way as to transfer energy to wavevectors which are sums or differences of the original two. In wavevector representations this energy transfer across the truncation boundary is normally cut off, leaving too much energy inside. In space mesh representations the finite difference approximations to the nonlinear terms usually do something even more damaging. On a regular mesh, wavenumbers k differing by integer multiples of $2k_* = 2\pi/\Delta x$ lead to the same function values on the mesh. Such an indistinguishable wavenumber with $k_n = k_0 + 2nk_*$ is called an alias of the wavenumber k_0. That k_n with smallest absolute value is called the principal alias k_p. Since $|k_p| \leq k_*$ the principal alias lies within what we have called the explicit range. Any energy transfer to a wavenumber outside the explicit range must appear then as a transfer to its principal alias and can destroy the significance of the low wavenumber amplitudes.

This difficulty, called nonlinear instability, was clearly described by Phillips,[1] whose suggested cure was the removal at regular time intervals of all energy residing in wavevectors with $k_*/2 \leq |\mathbf{k}| \leq k_*$. This can be done very simply in wavevector space; in configuration space one could transform to wavevector space, make the necessary deletion, and transform back to configuration space. It is, however, far simpler to recognize this as a smoothing and energy-dissipating process and to add an appropriate artificial viscosity or diffusion term to the equations being integrated.

Although artificial viscosities have been introduced into hydrodynamic calculations since the proposal of von Neuman and Richtmyer[2] to handle shock problems in this way, it was in the work of Smagorinsky[3] that the connection between artificial viscosity and nonlinear energy cascading was clearly recognized. This artificial viscosity is introduced then in order to remove from the explicitly described scales of motion that energy which in the full turbulent system would cascade through higher wavenumbers to be finally dissipated by true molecular viscosity at a scale much smaller than can be explicitly described. The artificial viscosity may also be viewed as an eddy viscosity describing the diffusive effects of the smaller scales (or eddies) on the larger.

The simplest estimate from turbulence theory of an energy cascading rate is that for stationary isotropic turbulence.[4] For very large Reynolds numbers the viscous dissipation wavenumber range can be high enough to be separated from the energy-containing range by an inertial range through which energy cascades at a rate independent of wavenumber. According to the dimensional arguments of Kolmogorov transformed into wavenumber space by Obukhov, the only physically significant quantity in this range in addition to wavenumber k of dimension $[L^{-1}]$ is this constant energy cascading rate $\varepsilon[L^2 T^{-3}]$. From this follows the $-\frac{5}{3}$ power law for the energy distribution $E[L^3T^{-2}]$

$$E = \alpha \varepsilon^{2/3} k^{-5/3} \tag{1}$$

with α a dimensionless coefficient observed to be about 1.5.

We assume similarly that if k_* is in the inertial range the artificial viscosity $\nu_*[L^2T^{-1}]$ depends only on $\varepsilon[L^2T^{-3}]$ and $k_*[L^{-1}]$ and must then on dimensional grounds have the form

$$\nu_* = \beta \varepsilon^{1/3} k_*^{-4/3} \tag{2}$$

with β a dimensionless coefficient.

In a space mesh representation we have $k_* = \pi/l_*$, where $l_* = \Delta x$ is the mesh interval. If we know an average value of ε for the flow considered, then Eq. (2) with β about 1.0 provides an estimate for the artificial viscosity coefficient ν_*. In practice such a viscosity coefficient is chosen to be as small as possible without leading to erratic behavior identified subjectively as indication of nonlinear instability. The resulting artificial viscosity is linear; that is, the coefficient is independent of the nature of the flow itself.

A more satisfactory procedure is to deduce ε from the calculation itself and indeed in a space mesh representation to attempt to compute a local value of ε appropriate to each mesh interval. This leads in turn to a local value of ν_* and a nonlinear artificial viscosity.

Whatever the local value of ν_* the rate of energy dissipation will be computed as

$$\varepsilon = \nu_* |D_*|^2 \tag{3}$$

where $|D_*|^2$ is the squared rate of strain in finite difference approximation. Substituting into Eq. (2) we see that

$$\begin{aligned}\nu_* &= \beta^{3/2} |D_*| k_*^{-2} \\ &= \gamma |D_*| l_*^2\end{aligned} \tag{4}$$

is the local estimate of ν_* based on the computed $|D_*|$ and known l_* rather than on a prior estimate of ε. Here $\gamma = \beta^{3/2}/\pi^2$ is also dimensionless.

A test of the validity of this technique would be the independence of ε, a flow-dependent quantity, and l_*, a mesh-dependent quantity. That is, we might hope that as the mesh was refined and l_* became smaller the average value of ε would remain unchanged. From Eqs. (3) and (4) we have

$$\varepsilon = \gamma |D_*|^3 l_*^2 \tag{5}$$

and the required independence of ε and l_* would arise only because with smaller l_* the finite difference estimate $|D_*|$ will increase as $l_*^{-2/3}$ due to finer scales of motion being explicitly computed. (We have used an asterisk throughout to indicate mesh-dependent quantities.)

These rather crude arguments reflect the inadequacy of turbulence theory to give more precise estimates. It is unlikely that in any real calculation k_* lies within an inertial subrange and that Eq. (5) holds. Such artificial viscosity terms have been used nonetheless with some success in convection calculations by Lilly,[5] who has also examined more sophisticated extensions of this general approach.

For numerical models of atmosphere or ocean circulations on a global scale, the horizontal mesh interval is large compared to the fluid thickness, and the flow is more nearly two- than three-dimensional. We turn then to two-dimensional turbulence theory for a guide to proper smoothing.

In two-dimensional incompressible flow there are many new constraints arising from the conservation of vorticity. In particular the enstrophy (defined as one-half the squared vorticity) is conserved in inviscid flow. Associated with this integral there is an inertial range through which enstrophy is cascaded at a constant rate η to be removed at some sufficiently high wavenumber by dissipative processes.

The dimensional arguments now change to depend on $\eta [T^{-3}]$, and the energy spectrum becomes

$$E = \alpha' \eta^{2/3} k^{-3} \tag{6}$$

This -3 power law has been observed for planetary scales of atmospheric motion.[6] For it the energy cascade rate ε is zero, according to detailed analysis[7] of nonlinear interactions as well as the use of a diffusion approximation[8] to these. We have now a spectrum with no energy transfer, but with enstrophy being cascaded to higher wavenumbers where it will be lost.

The fact that the energy spectrum drops more rapidly and that the energy is locked into the lower wavenumbers makes the smoothing problem less serious for two-dimensional problems and probably accounts for such calculations having worked as well as they have. Still the improper treatment of enstrophy cascading can lead to nonlinear instability and some smoothing is required.

Now the artificial viscosity by dimensional argument becomes

$$\nu_* = \beta' \eta^{1/3} k_*^{-2} \tag{7}$$

The vorticity equation is

$$\frac{d\omega}{dt} = \nabla(\nu_* \nabla \omega)$$

Let the enstrophy be $G = \tfrac{1}{2}\omega^2$. The enstrophy equation is

$$\frac{dG}{dt} = \nabla(\nu_* \nabla G) - \nu_* |\nabla \omega|^2$$

We may estimate the local enstrophy dissipation rate as

$$\eta = \nu_* |\nabla_* \omega|^2 \tag{8}$$

and thus the artificial viscosity as

$$\nu_* = \gamma' |\nabla_* \omega| l_*^3 \tag{9}$$

where $|\nabla_* \omega|$ is the finite difference approximation to the magnitude of the gradient of the vorticity.

Again here we would hope that

$$\eta = \gamma' |\nabla_* \omega|^3 l_*^3 \tag{10}$$

would in fact be independent of l_*. In other words, the average value of $|\Delta_* \omega|$ from one mesh point to another should be independent of the mesh interval, an interesting aspect of the -3 power law.

The artificial viscosity, Eq. (9), has been used satisfactorily in some ocean circulation calculations,[9] and will soon be tried in a numerical model of the atmosphere.[10]

References

1. N. A. Phillips, *An Example of Nonlinear Computational Instability* (Rossby Memorial volume), The Rockefeller Institute Press, New York, 1959.
2. J. von Neumann and R. D. Richtmyer, *J. Appl. Phys.*, **21**, 232 (1950).
3. J. Smagorinsky, *Monthly Weather Rev.*, **91**, 99 (1963).
4. G. K. Batchelor, *The Theory of Homogeneous Turbulence*, Cambridge University Press, Cambridge, 1953.
5. D. K. Lilly, "The Representation of Small-Scale Turbulence in Numerical Simulation Experiments," in *Proceedings IBM Scientific Computing Symposium on Environmental Sciences*, IBM Data Processing Division, White Plains, N.Y., 1967, pp. 195–210.
6. A. Wiin-Nielsen, *Tellus*, **19**, 540 (1967).
7. R. H. Kraichnan, *Phys. Fluids*, **10**, 1417 (1967).
8. C. E. Leith, *Phys. Fluids*, **10**, 1409 (1967).
9. W. P. Crowley, private communication.
10. C. E. Leith, "Numerical Simulation of the Earth's Atmosphere," in *Methods in Computational Physics*, Vol. 4, Alder, Fernbach, and Rotenberg, Eds., Academic Press, New York, 1965, pp. 1–27.

A Statistical Mechanical Treatment of Macroscopic Change with Time

JOSEPH E. MAYER

*Department of Chemistry, University of California,
San Diego, California*

1. Introduction

The general task of statistical mechanics is to make the connection between the microscopic view of matter as consisting of atoms and molecules obeying the laws of mechanics and the macroscopic behavior of bulk matter. The general macroscopic laws are to be derived as a consequence of those of mechanics, and, in addition, the values of the functions and parameters entering those laws are to be evaluated in terms of the mechanical characteristics of the particular species of molecules or other units composing the macroscopic system. The task is a broad one.

Ever since the time of Gibbs the general procedures for systems in thermodynamic equilibrium have been known; the Gibbs outline was so comprehensive and concise that the transition from the axiomatic assumptions of classical mechanics to those of quantum mechanics was almost trivial. The treatment of systems changing toward equilibrium is far less advanced, and, for the most part, the approaches have been tailored separately for special problems, without much of a generalized procedure being in evidence.

The reason for this difference is not hard to find. There exists a concise and rigorous set of macroscopic laws for the behavior of any equilibrium system, namely the laws of thermodynamics. The task of equilibrium statistical mechanics is then straightforward: to derive the laws of thermodynamics from those of mechanics; to present a general algorithm for computing the appropriate thermodynamic function for any thermodynamic single phase whose macroscopic state is defined; and to devise methods by which, with appropriate approximations, numerical results can be obtained, at least for simple systems. No concise set of macroscopic laws covers the behavior of all systems out of equilibrium, and each case seems to present a special problem.

However, in the last three decades a general, reasonably concise, and rigorous set of laws has been formulated for the behavior of a very considerable class of systems. These laws are based on the reciprocity theorem of Onsager[1] and are known under vague titles such as "non-equilibrium thermodynamics." They refer to the behavior of systems in which the displacement from equilibrium is not too drastic and a local equilibrium is defined in terms of a set of macroscopic variables at each position in three-dimensional space, or at least for each of several classes of degrees of freedom at each position in space. If the displacements from a single equilibrium state are not too great, the rate of approach to equilibrium is linear in the displacements, except in rather esoteric low temperature quantum cases. In these cases of superflow it would be fair to say that even the dynamic flowing system is at equilibrium.

The essential feature, for our purposes, of this discipline is that fluxes, representing the flow of extensive thermodynamic variables toward equilibrium, can be defined rather arbitrarily, but once defined the corresponding displacements conjugate to these fluxes are determined by a definite rule. If the displacements are correctly chosen, the real positive definite matrix of the coefficients, $k_{\alpha\beta}$, relating the flux, J_α, to the displacements, Δx_β, is then symmetric,

$$J_\alpha = \sum_\beta k_{\alpha\beta} \Delta x_\beta, \qquad k_{\alpha\beta} = k_{\beta\alpha} \tag{1.1}$$

The rule for determining the conjugate displacement is that the product, $J_\alpha \Delta x_\alpha$, be the entropy production per unit volume and unit time.

With this limitation to small displacements with local equilibrium, and the added condition that the flows are linear, one may hope that a concise and general formulation of the same order of rigor and generality as the Gibbs statistics could be redacted. Progress in this direction has been made, and we propose to outline the characteristics that appear to be emerging.

Our treatment is of the properties of matter under the most usual conditions that occur for bulk material in the human environment, that of near, but less than perfect, equilibrium. The excuse for its inclusion is that the author knows little else, and that Edward Teller's catholic interest includes the usual as well as the unusual. Probably he will find the treatment unusual.

2. *The Formalism of Equilibrium Statistical Mechanics*

In order to discuss the character of the method it is desirable to review the general formulation for the equilibrium case. In conventional thermodynamic practice the state of the system may be determined by specifying at least three, and usually more, extensive variables, proportional to the system size, such as the volume V; the mass; the number of moles, or, better for our purposes, the number of molecules, N_a, of each species, a, of molecules present; the entropy, S; and other possible variables such as total electric or magnetic moment, or the strain tensor components. For this specified state the internal energy, E, is minimum at equilibrium. The intensive thermodynamic force, f_a, conjugate to the extensive variable X_α is defined by

$$f_\alpha = (\partial E/\partial X_\alpha)_{X_\beta}, \ldots \qquad (2.1)$$

so that the negative pressure, $-P$, is conjugate to V, the chemical potential μ_a to N_a, and the temperature T to S.

A less conventional formulation is suitable for statistical mechanics. Replace S as variable by the internal energy, E, so that now the dimensionless quantity,

$$\sigma(V, N_a, \ldots, E, X_\alpha \cdots) = -S/k \qquad (2.2)$$

with k Boltzmann's constant, is the function which is minimum at equilibrium. The intensive variable x_α conjugate to X_α is now defined as

$$x_\alpha = -(\partial \sigma/\partial X_\alpha)_{X_\beta}, \ldots \qquad (2.3)$$

so that

$$\left. \begin{array}{l} \phi = +P/kT \\ \nu_a = -\mu_a/kT \\ \beta = +1/kT \\ x_\alpha = -f_\alpha/kT \end{array} \right\} \qquad (2.4)$$

are now the intensive variables conjugate to $V, N_a, \ldots, E$, and X_α, respectively.

The thermodynamic function to be minimized at equilibrium if one or more of the intensive variables x_α is substituted in the state specification for its conjugate X_α is obtained by adding $x_\alpha X_\alpha$ to σ, for each substitution used (subtracting $f_\alpha X_\alpha$ from E in the usual formulation). Thus

$$\sigma + \beta E = (1/kT)[-TS + E] = A/kT \qquad (2.5)$$

with A the Helmholtz free energy is the natural function of $V, N_a, \ldots, \beta, X_\alpha, \ldots$. Of particular interest is the function

$$\theta(V, \nu_a, \ldots, \beta, x_\alpha, \ldots) = \sigma + \sum_{\alpha \text{ other than } V}' x_\alpha X_\alpha \qquad (2.6)$$

$$= -PV/kT$$

which takes minimum value at equilibrium when the state variables are the extensive volume V, and otherwise only the intensive set, $x_\alpha, x_\beta, \ldots$. (That θ is $-PV/kT = \phi V$ follows from the fact that σ is linear in the X_α's and hence $\sigma = -\sum_{\text{all }\alpha} x_\alpha X_\alpha$.)

Now except for V, the extensive variables X_α are all values, or average values, of some function, or operator, $\chi_\alpha(\mathbf{p}^{(N)}, \mathbf{q}^{(N)})$ of the momenta and coordinates of the molecules of the system,

$$X_\alpha = \langle \chi_\alpha(\mathbf{p}^{(N)}, \mathbf{q}^{(N)}) \rangle \tag{2.7}$$

thus the function corresponding to N_a is the trivial function unity for each molecule of type a, E is the average of the Hamiltonian, and, for instance, the total electric moment is the sum of that of the neutral molecules, plus the product of vector position times charge of any charged entities present. Indeed, in otherwise doubtful cases the fact that X_α is the average value of a function $\chi_\alpha(\mathbf{p}^{(N)}, \mathbf{q}^{(N)})$ distinguishes which is to be regarded as the "extensive" function X_α with x_α "intensive" rather than Nx_α extensive with $X_\alpha/N = \xi_\alpha$ intensive. The volume is unique in being a boundary condition in the statistical mechanical viewpoint.

Now the complete Gibbs equilibrium statistical mechanics can be formulated in a concise statement. Imagine an ensemble consisting of an infinite number of single equilibrium systems, prepared by a macroscopic prescription giving a sufficient number of thermodynamic variables, of which at least one is extensive. The probability, $W_\mathbf{K}$, that a single system selected at random be found in a quantum state of the whole system, $\mathbf{K}$, is always given by

$$W_\mathbf{K} = \exp\left\{\sigma + \sum_\alpha x_\alpha[X_\alpha - \langle \lambda_\alpha(\mathbf{q}^{(N)}\mathbf{p}^{(N)}) \rangle_\mathbf{K}]\right\} \tag{2.8}$$

for all states $\mathbf{K}$ for which $W_\mathbf{K}$ is nonzero. In the classical case, the probability density in the 2Γ-dimensional gamma space, $\mathbf{p}^{(N)}, \mathbf{q}^{(N)}$, is

$$W(\mathbf{p}^{(N)}, \mathbf{q}^{(N)}) = \exp\left\{\sigma + \sum_\alpha x_\alpha[X_\alpha - \chi_\alpha(\mathbf{q}^{(N)}\mathbf{p}^{(N)})]\right\} \tag{2.8'}$$

Let $\mathscr{I}$ be a summation integration operator, defined in the quantum case to be

$$\mathscr{I} \equiv \sum_{N_a \geq 0} \sum_{N_b \geq 0} \cdots \sum_{\mathbf{K}(N_a, N_b, \ldots)} \tag{2.9}$$

or in the classical limit

$$\mathscr{I} = \sum_{N_a \geq 0} \sum_{N_b \geq 0} \int\!\!\int_V \cdots \int \frac{d\mathbf{p}^{(N)} d\mathbf{q}^{(N)}}{N_a! N_b! \cdots h^\Gamma} \tag{2.9'}$$

The normalization of the probability of a quantum state, or of the probability density, is

$$\mathscr{I} W = 1 \qquad (2.10)$$

and this condition determines the value of the thermodynamic function which is a minimum for the specified variables. If the state specification is by the complete set X_α rather than any x_α, then only states for which $X_\alpha - \langle \chi_\alpha(\mathbf{p}^{(N)} \mathbf{q}^{(N)}) \rangle_K$ is zero have nonzero values of W_K and the product $x_\alpha(X_\alpha - \chi_\alpha)$ can be omitted. If instead of any X_α the corresponding conjugate x_α is used, then $x_\alpha X_\alpha$ can be added to σ to give the corresponding minimal function, and the summation operator $\mathscr{I}$ must be extended over states of all values of $\chi_\alpha(\mathbf{p}^{(N)}, \mathbf{q}^{(N)})$.

The two extreme cases are that the state is that of the microcanonical ensemble specified solely by the extensive set $V, X_\alpha, X_\beta, \ldots$, in which case $W = e^\sigma$ for all states consistent with the given variables, and zero otherwise. The other extreme is that of the grand canonical ensemble of variables $V, x_\alpha, x_\beta, \ldots$ specifying the state and

$$W = \exp\left[\theta - \sum_{\alpha \neq V}' x_\alpha \chi_\alpha(\mathbf{p}^{(N)}, \mathbf{q}^{(N)})\right] \qquad (2.11)$$

where the summation-integration includes all numbers of molecules and all quantum states or phase space. Various in-between ensembles follow the rule of Eq. (2.8).

3. Orthonormal Macroscopic Displacement Variables

It is convenient to replace the extensive variables, X_α, by pseudo-intensive variables, ξ_α, of the same dimensions giving (nearly) the value of X_α per molecule. Actually we define $\xi_\alpha(r)$ for a system displaced from a given equilibrium state by a small amount as the local density of X_α per unit volume divided by the *equilibrium* number density, $\rho_0 = \bar{N}/V$,

$$\xi_\alpha(\mathbf{r}) = \lim_{v \to 0} \frac{1}{v\rho_0} [X_\alpha \text{ in } v \text{ at } \mathbf{r}] \qquad (3.1)$$

and for the equilibrium system,

$$\xi_\alpha^{(0)} = X_\alpha / \bar{N} \qquad (3.1')$$

where $\bar{N}$ is the total average number of molecules in V.

The state of a nonequilibrium system displaced from a definite equilibrium state characterized by a volume, V, and values $\xi_\alpha(\mathbf{r})$ or $x_\alpha(\mathbf{r})$ can be specified by giving the equilibrium values, $\xi_\alpha^{(0)}$ or $x_\alpha^{(0)}$, and the (small) displacements,

$$\Delta\xi_\alpha(\mathbf{r}) = \xi_\alpha(\mathbf{r}) - \xi_\alpha^{(0)} \qquad (3.2)$$

or of

$$\Delta x_\alpha(\mathbf{r}) = x_\alpha(\mathbf{r}) - x_\alpha^{(0)} \qquad (3.2')$$

for each α, as functions of position $\mathbf{r}$. The statement that there exists local equilibrium means that at each $\mathbf{r}$ value the relations of $\xi_\alpha(\mathbf{r})$ and $x_\alpha(\mathbf{r})$ are those of equilibrium, or at least of a conceivable metastable state.* In order that $\xi_\alpha(\mathbf{r})$ be meaningfully defined by (3.1) it is necessary that the gradients be small enough that $\xi_\alpha(\mathbf{r})$ is effectively constant in volumes v large enough to contain a statistically significant number of molecules.

In equilibrium statistical mechanics we use the macroscopic laws of thermodynamics, once these are established as following from mechanics, for all systems containing any complicated array of single thermodynamic phases, and content ourselves with the already sufficiently difficult task of evaluating the thermodynamic functions for single phases. So in this case we can limit ourselves to displacements in space which are single Fourier components with a given wavevector $\mathbf{k}$, which is necessarily small in units of the third root of the equilibrium number density. A complete analysis of the approach to equilibrium with arbitrary displacement amplitudes, $\Delta x_\alpha(\mathbf{k})$ for any one $\mathbf{k}$ value suffices to determine the coefficients, $k_{\alpha\beta}$, of the linear Onsager matrix.

However, instead of using the more conventional thermodynamic functions ξ_α, x_α, we propose, for a reason which will become apparent later, to use a set of dimensionless linear combinations of these. The transformation starts with an amusing and harmless, but also apparently pointless, observation. Consider the two matrices $\boldsymbol{\sigma}$ and $\boldsymbol{\theta}$ whose elements are the second derivatives, $\sigma_{\alpha\beta}$ and $\theta_{\alpha\beta}$,

$$\begin{aligned} \sigma_{\alpha\beta} &= N(\partial^2\sigma/\partial X_\alpha \partial X_\beta) & X_\alpha, X_\beta \neq V \\ \theta_{\alpha\beta} &= -N^{-1}(\partial^2\theta/\partial x_\alpha \partial x_\beta) & x_\alpha, x_\beta \neq \phi \end{aligned} \qquad (3.3)$$

where the variables do *not* include the V, P/kT pair, and where σ is $-S/k$ and $\theta = -PV/kT$. The factors of N are chosen so that both $\sigma_{\alpha\beta}$ and $\theta_{\alpha\beta}$ are independent of the system size. Since $x_\alpha = -\partial\sigma/\partial X_\alpha$ and $X_\alpha = \partial\theta/\partial x_\alpha$ we have that

$$dx_\alpha = -\sum_\beta \sigma_{\alpha\beta}\, d\xi_\beta \qquad (3.4)$$
$$d\xi_\alpha = -\sum_\alpha \theta_{\alpha\beta}\, dx_\beta$$

and

$$\sum_\gamma \sigma_{\alpha\gamma}\theta_{\gamma\beta} = \sum_\gamma \theta_{\alpha\gamma}\sigma_{\gamma\beta} = \delta(\alpha - \beta) \qquad (3.5)$$

* One may wish to include such displacements as a small difference in temperature between, say, translational and internal degrees of freedom, which would be considered esoteric in conventional thermodynamics.

In short, $\boldsymbol{\sigma}$ and $\boldsymbol{\theta}$ are reciprocal matrices. Now since the negative entropy σ is minimum at equilibrium it is easy to prove that

$$\sigma_{\alpha\alpha} \geq 0 \tag{3.6}$$

and that this relation holds for any linear combination of variables. The equality is excluded for any single-phase system away from a critical point. The matrices are positive definite and

$$\sigma_{\alpha\beta} \leq (\sigma_{\alpha\alpha}\sigma_{\beta\beta})^{1/2} \tag{3.7}$$

In a space of dimension of the number of thermodynamic variables less one, represent the displacement of a thermodynamic state from some standard state given by the $\xi_\alpha^{(0)}$, $x_\alpha^{(0)}$, values along vectors $\mathbf{i}_\alpha$, with scalar products

$$\mathbf{i}_\alpha \cdot \mathbf{i}_\beta = \sigma_{\alpha\beta} \tag{3.8}$$

which gives the length of vector $\mathbf{i}_\alpha$ the reciprocal dimension of X_α which is the dimension of x_α. Vectors $\mathbf{i}_\alpha^*$,

$$\mathbf{i}_\alpha^* = \sum_\beta \theta_{\alpha\beta}\mathbf{i}_\beta, \qquad \mathbf{i}_\alpha = \sum_\beta \sigma_{\alpha\beta}\mathbf{i}_\beta^*$$

$$\mathbf{i}_\alpha^* \cdot \mathbf{i}_\alpha^* = \sum_\gamma \sum_\delta \theta_{\alpha\gamma}\sigma_{\gamma\delta}\theta_{\delta\beta} = \theta_{\alpha\beta} \tag{3.9}$$

are related to the $\mathbf{i}_\alpha$'s as a reciprocal set, or contravariant basis set, namely,

$$\mathbf{i}^* \cdot \mathbf{i}_\beta = \sum_\gamma \theta_{\alpha\gamma}\mathbf{i}_\gamma \cdot \mathbf{i}_\beta = \sum_\gamma \theta_{\alpha\gamma}\sigma_{\gamma\beta} = \delta(\alpha - \beta) \tag{3.10}$$

For a given displacement by magnitudes $\Delta\xi_\alpha(\mathbf{r})$ define a dimensionless vector function $\mathbf{t}(\mathbf{r})$ in this space to represent the displacement by

$$\mathbf{t}(\mathbf{r}) = \sum_\alpha \Delta\xi_\alpha(\mathbf{r})\mathbf{i}_\alpha$$

$$= \sum_\beta \left(\sum_\alpha \Delta\xi_\alpha \sigma_{\alpha\beta}\right)\mathbf{i}_\beta^* = -\sum \Delta x_\beta(\mathbf{r})\mathbf{i}_\beta^* \tag{3.11}$$

where the last relation follows from (3.4). We have, from (3.8) and (3.9) that

$$\mathbf{t}(\mathbf{r}) \cdot \mathbf{t}(\mathbf{r}) = \sum_\alpha \sum_\beta \Delta\xi_\alpha(\mathbf{r})\sigma_{\alpha\beta}\Delta\xi_\beta(\mathbf{r})$$

$$= \sum_\alpha \sum_\beta \Delta x_\alpha(\mathbf{r})\theta_{\alpha\beta}\Delta x_\beta(\mathbf{r}) \tag{3.12}$$

Now the negative entropy function σ is given to second order by an integral over $\mathbf{r}$,

$$\Delta\sigma = \rho_0 \int d\mathbf{r}\left[-\sum_\alpha x_\alpha^{(0)}\Delta\xi_\alpha(\mathbf{r}) + \tfrac{1}{2}\sum_\alpha\sum_\beta \Delta\xi_\alpha(\mathbf{r})\sigma_{\alpha\beta}\Delta\xi_\beta(\mathbf{r})\right] \tag{3.13}$$

For conservative variables such as $N_a, \ldots, E$, the displacement from equilibrium is necessarily such that $\langle \Delta \xi_\alpha(\mathbf{r}) \rangle$ is zero. For nonconservative variables the condition that σ be a minimum at equilibrium is that $x_\alpha^{(0)}$ be zero. We have

$$\Delta \sigma = \tfrac{1}{2} N \sum_\alpha \sum_\beta \langle \Delta \xi_\alpha(\mathbf{r}) \sigma_{\alpha\beta} \Delta \xi_\beta(\mathbf{r}) \rangle$$

$$= \tfrac{1}{2} N \langle \mathbf{t}(\mathbf{r}) \cdot \mathbf{t}(\mathbf{r}) \rangle \tag{3.14}$$

from (3.12).

The analogy of the representations of displacements by the set $\xi_\alpha(\mathbf{r})$ or by the set $x_\alpha(\mathbf{r})$ to a vector that can be defined in a basis set or its contravariant basis set immediately suggests a transformation to an orthonormal, or cartesian, basis, and, as we shall find later, such a transformation is indeed useful for statistical mechanical reasons. Transform the extensive variable set X_α to dimensionless extensive variables,

$$Y_\nu = \sum_\alpha r_{\nu\alpha} X_\alpha \tag{3.15}$$

using coefficients $r_{\nu\alpha}$ of dimensions reciprocal to X_α, for which the reciprocal equation is

$$X_\alpha = \sum_\nu s_{\alpha\nu} Y_\nu \tag{3.16}$$

$$\sum_\nu s_{\alpha\nu} r_{\nu\beta} = \delta(\alpha - \beta), \quad \sum_\alpha r_{\nu\alpha} s_{\alpha\mu} = \delta(\nu - \mu) \tag{3.17}$$

It is quite straightforward to show that the variables y_ν conjugate to Y_ν, $y_\nu = -(\partial \delta/\partial Y_\nu)_{Y_\mu}, \ldots$ are given by

$$y_\nu = \sum_\alpha x_\alpha s_{\alpha\nu}, \quad x_\alpha = \sum_\nu y_\nu r_{\nu\alpha} \tag{3.18}$$

If now the transformation coefficients are chosen so that with $s_{\nu\alpha}^{(t)} = s_{\alpha\nu}$, etc.,

$$\sum_\alpha \sum_\beta s_{\nu\alpha}^{(t)} \sigma_{\alpha\beta} s_{\beta\mu} = \delta(\nu - \mu) \tag{3.19}$$

then, from (3.5) that σ and θ are reciprocal, one has

$$\sum \sum r_{\nu\alpha} \theta_{\alpha\beta} r_{\beta\mu}^{(t)} = \delta(\nu - \mu) \tag{3.19'}$$

One finds that with

$$\left. \begin{array}{l} \Delta \eta_\nu = \sum_\alpha r_{\nu\alpha} \Delta \xi_\alpha, \quad \Delta \xi_\alpha = \sum_\nu s_{\alpha\nu} \Delta \eta_\nu \\ \Delta y_\nu = \sum_\alpha s_{\nu\alpha}^{(t)} \Delta x_\alpha, \quad \Delta x_\alpha = \sum_\nu r_{\alpha\nu}^{(t)} \Delta y_\nu \end{array} \right\} \tag{3.20}$$

and (3.4) relating Δx_α to the $\Delta \xi_\beta$'s,

$$\Delta y_\nu = -\sum_\mu (\sum_\alpha \sum_\beta s_{\nu\alpha}^{(t)} \sigma_{\alpha\beta} s_{\beta\mu}) \Delta \eta_\mu = -\delta(\nu-\mu) \Delta \eta_\nu$$

For these, then, the pseudointensive displacements $\Delta \eta_\nu$ and the displacements in the variable conjugate to Y_ν are merely negative of each other,

$$\Delta y_\nu = -\Delta \eta_\nu \tag{3.21}$$

Correspondingly the off-diagonal second derivatives of σ and θ become zero, and the diagonal elements unity,

$$\sigma_{\nu\mu} = N^{-1}(\partial^2 \sigma/\partial \eta_\nu\, \partial \eta_\mu)_{\nu,\eta\lambda}$$
$$= -(\partial y_\nu/\partial \eta_\mu) = -(\partial y_\mu/\partial \eta_\nu) = \delta(\nu - \theta) \tag{3.22}$$
$$\theta_{\nu\mu} = N^{-1}(\partial^2 \theta/\partial y_\nu \partial y_\mu) = \delta(\nu - \mu) \tag{3.22'}$$

The dimensionless vector set

$$\boldsymbol{\lambda}_\nu = \sum_\alpha s^{(t)}_{\nu\alpha} \mathbf{i}_\alpha \tag{3.23}$$

$$\mathbf{i}_\alpha = \sum_\nu r^{(t)}_{\alpha\nu} \boldsymbol{\lambda}_\nu \tag{3.23'}$$

contains orthonormal vectors,

$$\boldsymbol{\lambda}_\nu \cdot \boldsymbol{\lambda}_\mu = \delta(\nu - \mu) \tag{3.24}$$

and the vector **t** of Eq. (3.11) is now

$$\mathbf{t}(\mathbf{r}) = \sum_\alpha \Delta \xi_\alpha(\mathbf{r}) \mathbf{i}_\alpha = \sum_\nu \Delta \eta_\nu(\mathbf{r}) \boldsymbol{\lambda}_\nu = -\sum_\nu \Delta y_\nu(\mathbf{r}) \boldsymbol{\lambda}_\nu \tag{3.25}$$

so that from (3.14)

$$\Delta \sigma = \tfrac{1}{2} N \sum_\alpha \langle(\Delta \eta_\nu(\mathbf{r}))^2\rangle = \tfrac{1}{2} N \sum_\nu \langle(\Delta y_\nu(\mathbf{r}))^2\rangle \tag{3.26}$$

The example of a simple fluid system with only one chemical component shows that of the two variables, number density of molecules and temperature, one combination of orthogonal displacements is along axes of constant entropy and of constant pressure, respectively. These are displacements which, as functions of position in the system, correspond to the initial states for observing the decay of a standing sound wave or of thermal diffusion, and are those that diagonalize the Onsager matrix in this simple case.

One might also add that the transformation matrices **R** and **S** can be more readily computed from tabulated quantities by diagonalizing the second derivatives of the internal energy.

Since nonconservative quantities X_α have zero values of their conjugates, $x_\alpha \equiv 0$, at equilibrium, the matrix $\boldsymbol{\sigma}$ has only zero off-diagonal elements $\sigma_{\alpha\beta}$ when one of the variables, X_β, is conservative and the other, X_α, is nonconservative. The transformation matrices **R** and **S** of elements $r_{\nu\alpha}$, $s_{\alpha\nu}$ can always be chosen so that the Y_ν are linear combinations of solely conservative or solely nonconservative variables.

4. Probability Density for Nonequilibrium Ensembles

The language becomes simpler if we restrict our discussion to the classical case, which we shall do. The functions $\chi_\alpha(\mathbf{p}^{(N)}, \mathbf{q}^{(N)})$ in the phase space whose average value is X_α can always be written as a sum over the single molecules, i, of the system, $1 \leq i \leq N$, of functions $\zeta_{\alpha i}$ that depend on the coordinates and momenta of the molecule i, and perhaps also on coordinates and momenta of other molecules, but if so only as sums of functions that depend on a limited number of other molecules which are spatially close to i. The functions are then sums of functions each depending on coordinates and momenta of relatively small subset of the 10^{24} molecules comprising the system. We write them as

$$\chi_\alpha(\mathbf{p}^{(N)}, \mathbf{q}^{(N)}) = \sum_{i=1}^{i=N} \zeta_{\alpha i}(\mathbf{p}_i, \mathbf{q}_i; \mathbf{p}^{(N-i)}, \mathbf{q}^{(N-i)}) \qquad (4.1)$$

As an example, in the case of a potential which is a sum of pair terms only, the Hamiltonian of an ensemble of monatomic molecules is

$$H(\mathbf{p}^{(N)}, \mathbf{q}^{(N)}) = \sum_{i=1}^{i=N} \left[\frac{\mathbf{p}_i \mathbf{p}_i}{2m_i} + \frac{1}{2} \sum_{j \neq i} u_{ij}(r_{ij}) \right] \qquad (4.2)$$

if half of the mutual pair potential, u_{ij}, is assigned to each molecule. This arbitrary assignment is immaterial since the pair potential is short range and either each term is zero or both contributions occur at the same position on a macroscopic scale. (The case of an r^{-1} potential requires a little more care.)

It is then intuitively suggestive to write the probability density $W(\mathbf{p}^{(N)}, \mathbf{q}^{(N)})$ for the initial $t = 0$ state of an ensemble of laboratory systems prepared according to a macroscopic prescription giving $x_\alpha(\mathbf{r})$ for each α, in a volume V, from Eq. (2.11) as

$$W(t = 0, \mathbf{p}^{(N)}, \mathbf{q}^{(N)}) = \left[\exp \theta - \sum_\alpha \sum_{i=1}^{i=N} x_\alpha(\mathbf{r}_i) \zeta_{\alpha i} \right] \qquad (4.3)$$

That this simple guess is correct can be confirmed by showing that it obeys two criteria. These are, first, that the quantity $\xi_\alpha(\mathbf{r})$ which is

$$\rho_0 \xi_\alpha(\mathbf{r}) = \mathscr{I} W \sum_{i=1}^{i=N} \delta(\mathbf{r}_i - \mathbf{r}) \zeta_{\alpha i} \qquad (4.4)$$

does, to terms of order of the square of the gradients in the $x_\alpha(\mathbf{r})$'s, reproduce the equilibrium relations at $\mathbf{r}$ between the set of ξ_α's and x_α's, and, second, that (4.3) gives the probability density of maximum entropy satisfying these relations.

The quantity θ in (4.3) is to be determined from the condition $\mathscr{I}W = 1$, and can be shown to be $\bar{P}V/kT$ with $\bar{P}$ the space average of the pressure.

The specified set of $x_\alpha(\mathbf{r})$'s determines a given average number, $\bar{N}_a, \bar{N}_b, \ldots,$ of each species, $a, b, \ldots,$ of molecules and an average energy $\bar{E}$, and for these numbers of molecules and energy we write the equilibrium probability density, $W^{(0)}(\mathbf{p}^{(N)}, \mathbf{q}^{(N)})$ from (2.11), characterized by a set $x_\alpha^{(0)}$. With

$$\left.\begin{array}{r}\Delta x_\alpha(\mathbf{r}) = x_\alpha(\mathbf{r}) - x_\alpha^{(0)} \\ \Delta \Phi(t=0) = \sum\limits_\alpha \sum\limits_{i=1}^{i=N} \Delta x_\alpha(\mathbf{r}_i)\zeta_{\alpha i} \\ W(t=0) = W^{(0)} \exp[-\Delta\Phi^{(0)}]\end{array}\right\} \quad (4.5)$$

The future evolution of W in an isolated system is given by the imaginary Hermitian Liouville operator,

$$\mathscr{L} = i\sum_{\nu=1}^{\nu=\Gamma}(\partial H/\partial p_\nu)(\partial/\partial q_\nu) - (\partial H/\partial q_\nu)(\partial/\partial p_\nu) \quad (4.6)$$

$$W(t) = W^{(0)} \exp[-\Delta\Phi(t)] = e^{it\mathscr{L}} W(t=0) \quad (4.7)$$

and, since $\mathscr{L}W^{(0)} = 0$ and $\mathscr{L}$ contains only first derivatives,

$$\Delta\Phi(t) = e^{it\mathscr{L}}\Delta\Phi(t=0) \quad (4.8)$$

With $W(t)$ determined, Eq. (4.4) gives the time development of the $\xi_\alpha(\mathbf{r})$'s.
Certain general conditions apply to integrals of W, namely,

$$\mathscr{I}W = \mathscr{I}W^{(0)} = 1; \qquad \mathscr{I}W^{(0)}(e^{-\Delta\Phi} - 1) = 0 \quad (4.9\text{a})$$

$$\mathscr{I}N_a W = \mathscr{I}N_a W^{(0)} = \bar{N}_a; \qquad \mathscr{I}W^{(0)}N_a(e^{-\Delta\Phi} - 1) = 0 \quad (4.9\text{b})$$

$$\mathscr{I}HW = \mathscr{I}HW^{(0)} = \bar{E}; \qquad \mathscr{I}W^{(0)}H(e^{-\Delta\Phi} - 1) = 0 \quad (4.9\text{c})$$

If one assumes that any other quantity X_α is conserved, in short if within the times to be considered its value does not change, then

$$\mathscr{I}W\chi_\alpha(e^{-\Delta\Phi} - 1) = 0 \quad (X_\alpha \text{ conservative}) \quad (4.9\text{d})$$

The general statistical mechanical formula for $\sigma = -S/k$ is that

$$\sigma = \mathscr{I}W \ln W = \mathscr{I}W[\ln W^{(0)} - \Delta\Phi] \quad (4.10)$$

Since $\ln W^{(0)}$ contains additively only constant terms, the numbers $N_a, N_b, \ldots$ and the Hamiltonian, H, plus possible other functions χ_α regarded as "conservative," we find from (4.9a) to (4.9d) that

$$\mathscr{I}W \ln W^{(0)} = \mathscr{I}W^{(0)} \ln W^{(0)} = \sigma^{(0)} \quad (4.11)$$

the equilibrium value, $\sigma^{(0)} = -S^{(0)}/k$, of the negative entropy, so that

$$\Delta\sigma = \sigma - \sigma^{(0)} = -\mathscr{I}W^{(0)}\Delta\Phi e^{-\Delta\Phi} \quad (4.12)$$

Subtract Eq. (4.9.1) from (4.12) and develop the exponentials to find

$$-\Delta\Phi e^{-\Delta\Phi} + 1 - e^{-\Delta\Phi} = \tfrac{1}{2}(\Delta\Phi)^2 + O(\Delta\Phi)^3$$

We then have

$$\Delta\sigma \simeq \tfrac{1}{2}\mathscr{I}W^{(0)}(\Delta\Phi)^2 \tag{4.13}$$

up to terms quadratic in $\Delta\Phi$.

5. Orthonormal Displacement Functions in Phase Space

Having a limited ambition to treat only first-order effects in the displacement from equilibrium, which should give us the linear results contained in the Onsager relations, we assume that, except for the negative entropy increment of Eq. (4.13), which necessarily appears only quadratically, we can develop the exponential exp $(-\Delta\Phi)$ to first order only, and ignore the higher terms,

$$e^{-\Delta\Phi} - 1 \simeq -\Delta\Phi \tag{5.1}$$

It now seems natural to attempt to express $\Delta\Phi(t)$ in terms of some orthonormal set of functions. The unit value defining normalization of a function, $F(\mathbf{p}^{(N)}, \mathbf{q}^{(N)})$, is chosen so as to be independent of the system size, namely, if

$$F \cdot F \equiv \frac{1}{N} \mathscr{I}W^{(0)}[F(\mathbf{p}^{(N)}, \mathbf{q}^{(N)})]^2 = 1 \tag{5.2}$$

F is said to be normalized. Orthogonality is defined by

$$F \cdot G = \frac{1}{N} \mathscr{I}W^{(0)}F(\mathbf{p}^{(N)}, \mathbf{q}^{(N)})G(\mathbf{p}^{(N)}, \mathbf{q}^{(N)}) = 0 \tag{5.2'}$$

It is here that the usefulness of our thermodynamic play in Section 3 becomes apparent. The dimensionless extensive variables, Y_ν, of Eq. (3.15), $Y_\nu = \sum_\alpha r_{\nu\alpha} X_\alpha$, are the average values of functions $\sum_\alpha r_{\nu\alpha} \chi_\alpha(\mathbf{p}^{(N)}, \mathbf{q}^{(N)})$. For any given matrix $\mathbf{R}$ of elements $r_{\nu\alpha}$ obeying (3.19') that they transform $\boldsymbol{\theta}$ of elements $\theta_{\alpha\beta}$ to the unit matrix $\theta_{\nu\alpha} = \delta(\nu - \mu)$ we define

$$\psi_{\nu i} = \sum_\alpha r_{\nu\alpha} \chi_{\alpha i} \tag{5.3}$$

$$\Psi_\nu = \sum_{i=1}^{i=N} \psi_{\nu i} \tag{5.3'}$$

so that, at equilibrium

$$Y_\nu^{(0)} = \mathscr{I}W^{(0)}\Psi_\nu \tag{5.4}$$

To the set of $x_\alpha(\mathbf{r})$'s the corresponding displacements are, from (3.18),

$$\Delta y_\nu(\mathbf{r}) = \sum_\alpha \Delta x_\alpha(\mathbf{r}) s_{\alpha\nu} \tag{5.5}$$

and the function $\Delta\Phi(t=0)$ is now

$$\Delta\Phi(t=0) = \sum_\nu \sum_{i=1}^{i=N} \Delta y_\nu(t=0, \mathbf{r}_i) \psi_{\nu i} \tag{5.6}$$

we make a Fourier integral analysis of the displacements $\Delta y_\nu(\mathbf{r})$ and define functions

$$\Delta\Psi_{\mathbf{k},\nu} = \sqrt{2} \sum_{i=1}^{i=N} \sqrt{2} \, (\cos 2\pi \mathbf{k}\cdot\mathbf{r}_i) \psi_{\nu i} \tag{5.7}$$

so that

$$\Delta\Phi(t=0) = \sum_\nu \int d\mathbf{k} \, \Delta y_\nu(\mathbf{k}) \, \Delta\Psi_{\mathbf{k}\cdot\nu} \tag{5.8}$$

One can then show that the functions $\Delta\Psi_{k,\nu}$ form an orthonormal set,

$$\Delta\Psi_{\mathbf{k},\nu} \cdot \Delta\Psi_{\mathbf{k}',\mu} = \Delta(\mathbf{k} - \mathbf{k}') \, \delta(\nu - \mu) \tag{5.9}$$

where $\Delta(\mathbf{k} - \mathbf{k}')$ is the Dirac delta function, and $\delta(\nu - \mu)$ the Kronecker delta. The proof is direct but involves a certain amount of algebraic manipulation* and, as usual, a neglect of terms proportional to the squares of the gradients. Using dimensionless coordinates, $\mathbf{r}$, such that the equilibrium number density is unity, displacements with nonzero amplitudes $\Delta y_\nu(\mathbf{k})$ are permitted only for very small $|\mathbf{k}|$ values, $|\mathbf{k}| \ll 1$, and the terms neglected are of order $|\mathbf{k}|^2$.

Conditions (4.9a) to (4.9d) on the integrals of $W^{(0)}[\exp(-\Delta\Phi) - 1]$ require that the amplitudes $\Delta y_\nu(\mathbf{k} = 0)$ of conservative quantities Y_ν be zero. For nonconservative variables nonzero amplitudes at $\mathbf{k} = 0$ may occur.

The expression (4.13) for the negative entropy increment, and the fact that the functions $\Delta\Psi_{\mathbf{k},\nu}$ are orthonormal in the sense of Eqs. (5.2) and (5.3) give us, for $t = 0$, that

$$\Delta\sigma(t=0) = \tfrac{1}{2}N \int d\mathbf{k} \sum_\nu [\Delta y_\nu(\mathbf{k})]^2$$

$$= \tfrac{1}{2}N \sum_\nu \langle (\Delta y_\nu(\mathbf{r}))^2 \rangle \tag{5.10}$$

in agreement with the thermodynamic conclusion of Eq. (3.26), a not completely trivial, but satisfying, result.

* Algebra of this, and of much of the manipulation, is given in more detail in an article by Bauer et al.[2]

Actually the fact that this is general is the justification for using $-k\mathscr{I}W\ln W$ as the statistical mechanical definition of entropy. The thermodynamic value of the entropy is only defined if the local values of the macroscopic values are meaningful, and in all cases, then, $\mathscr{I}W\ln W$ does give σ in agreement with the macroscopic definition.

6. Empirical Considerations

We have so far carried out a considerable amount of preparatory mathematical manipulation, with little regard to or discussion of the physical behavior. The value of this manipulation lies largely in its usefulness as a language tool for the discussion of what should be, and what cannot be, proven.

In the first place we incline to make an assumption which later requires more examination, namely that we know initially all of the macroscopically measurable quantities, X_α, in the type of physical system that we discuss, and that some experimental method exists for measuring the local concentrations $\xi_\alpha(\mathbf{r})$ or $x_\alpha(\mathbf{r})$. We also assume that the equilibrium properties of the systems have been investigated, and the elements

$$\sigma_{\alpha\beta} = -Nk^{-1}(\partial^2 S/\partial X_\alpha \partial X_\beta)$$

are known. We can then find a transformation matrix $\mathbf{S}$, of elements $s_{\nu\alpha}$ (3.19) which makes $\boldsymbol{\sigma}$ the unit matrix, $\sigma_{\nu\mu} = \delta(\nu - \mu)$.

We further know the macroscopic laws of transport based on the Onsager reciprocity relations. Fluxes account for the time change in $\eta_\nu(\mathbf{r})$. These fluxes may be vector flow, $\mathbf{J}_\nu(\mathbf{r})$ of the quantity η_ν per unit time across unit area normal to the vector, and only fluxes of this nature occur for conservative variables. In the case of nonconservative quantities, η_μ, such as the displacement from equilibrium of a chemical reaction, or the local electric or magnetic moment density, direct fluxes, $\dot\eta_\mu$, may also occur. The corresponding "forces" are the gradients, $\boldsymbol{\nabla} y_\nu(\mathbf{r})$, of the conjugate variables y_ν in the case of the vector fluxes, and the values $\Delta y_\mu(\mathbf{r}) = y_\mu(\mathbf{r})$ in the case of $\dot\eta_\mu(\mathbf{r})$. We know from the proof of Onsager that the matrix $\mathbf{K}$ of elements $k_{\nu\lambda}$,

$$\mathbf{J}_\nu(\mathbf{r}) = \sum_\lambda k_{\nu\lambda} \boldsymbol{\nabla} y_\lambda(\mathbf{r})$$

$$\eta_\mu(\mathbf{r}) = \sum_\varkappa k_{\mu\varkappa} y_\varkappa(\mathbf{r}) \tag{6.1}$$

is real, symmetric, positive definite, and has zero elements connecting terms of differing tensorial order. (Fluxes of higher tensorial order than vectors may also exist.)

A real unitary transformation matrix exists which diagonalizes **K**, and such a transformation leaves σ the unit matrix. It follows that at least one, or possibly a degenerate continuum of choices, exists for the matrices **S** and **R** such that the Onsager **K** matrix is diagonal

$$\mathbf{J}_\nu(\mathbf{r}) = k_{\nu\nu}\nabla y_\nu(\mathbf{r})$$
$$\eta_\mu(\mathbf{r}) = k_{\mu\mu}y_\mu(\mathbf{r}) \tag{6.2}$$

Furthermore we know that the macroscopic equations (6.2) have independent Fourier solutions, the amplitudes $\Delta y_\nu(t, \mathbf{k}) = -\Delta \eta_\nu(t, \mathbf{k})$ decreasing exponentially with time. (The case of sound waves in which the solution is oscillatory with decreasing amplitude is due to complications introduced into the simple equations above by macroscopic motions in the system.)

In terms of our discussion of the statistical mechanical probability density functions in the 10^{24}-dimensional gamma space, $\mathbf{p}^{(N)}$, $\mathbf{q}^{(N)}$, we can now draw conclusions as to what to expect and what to look for to be proven. We have *assumed* that only a finite small number of macroscopic variables $\xi_\alpha(\mathbf{r})$, say n of them, $1 \leq \alpha \leq n$ can be measured. These, in turn, can be expressed in terms of the amplitudes of a set of orthonormal displacement functions, $\Delta\psi_{\mathbf{k},\nu}$, of which, for each wavevector, $\mathbf{k}$, there are only n, $1 \leq \nu \leq n$. We know empirically that these functions can be so chosen that the amplitudes of each decrease exponentially in time (or oscillate with decreasing amplitude), independently of the amplitudes of the others. Namely, to a certain degree of approximation at least, we can find functions $\Delta\Psi'_{\mathbf{k},\nu}$ such that

$$\Delta\Psi'_{\mathbf{k},\nu} \cdot e^{-it\mathscr{L}} \Delta\Psi'_{\mathbf{k}',\mu} \simeq \Delta(\mathbf{k} - \mathbf{k}')\,\delta(\nu - \mu)e^{-t/t(\mathbf{k},\nu)} \tag{6.3}$$

The relaxation times $t(\mathbf{k}, \nu)$ can then be readily related to the coefficients $k_{\nu\nu}$ of (6.1).

7. Difficulties and Paradoxes

The general field of statistical mechanics, and particularly the treatment of time evolution by the methods of non-equilibrium statistical mechanics, has long been plagued by apparent paradoxes and consequent difficulties. A contemplation of the actual experimental situation in even relatively simple classically behaved systems convinces one shortly that any general "mathematical proof" of the statements made about Eq. (6.3) must be hedged with many qualifications.

Examine first our assumption that the possible measurable macroscopic variables X_α are known, and limited to n, $1 \leq \alpha \leq n$. The situation has an exact analog in the case of the ergodic theorem, considered generally to be a necessary assumption in the Gibbs formulation of equilibrium statistics. The minimum number of thermodynamic variables necessary to define the state of a thermodynamic equilibrium system is, in addition to the volume V, the number, c, of independent chemical compounds plus one, and c is the total number of chemical components less the number of independent chemical reactions that are in equilibrium. This definition already leaves ambiguous the question of whether a given chemical reaction proceeds to equilibrium, or is so slow that within the time of experimentation it can be disregarded. A general purely theoretical statistical mechanical prediction of the rate is usually impractical. One resorts to experimental observation to determine the choice of the number of variables to introduce in a given case. But even the minimum number of variables so defined is frequently insufficient, since transitions between different thermodynamic phases, such as graphite and diamond, often take geologic time, and the phase must often be specified. All this merely shows that a mathematical "proof" of the ergodic theorem must be limited to very special cases, and extremely simple ones at that. Although of great mathematical interest, such proofs are of very limited practical value to general applications. This appears to leave the situation in an unsatisfactory condition for practical calculations.

The difficulty is obviated if we examine the possible behaviors of a non-ergodic system. Two extremes and an intermediary evolve. The first extreme is that the systems, although truly non-ergodic in the sense that no transitions occur between certain classes of microscopically defined states, are always prepared from a truly ergodic condition, and the ensemble of such systems is indeed random. Its properties are properly computed by summing over all states, just as if it were ergodic. This case occurs, for instance, in the random up–down configuration of CO molecules in the CO crystal at very low temperatures, and in many other examples of supercooled systems. The other extreme is that transitions do not occur between classes of microscopic states differing in some macroscopic property, measurable by an as yet undefined variable X_α. The macroscopic states of the systems must be redefined by identifying and measuring the value of this variable. One would like to misquote the historical fact, and identify the variable giving the fraction of H_2 in even and odd rotational states as an example of this. The intermediate case is really that in which the significant variable, or variables, X_α, have never been identified, namely one in which the systems are non-ergodic and nonrandom and nonreproducible, depending on unidentified differences in their preparation. Any laboratory chemist knows examples, and abjures them! The systems which are experimentally and macroscopically intractable are those in which the ergodic hypothesis fails.

This discussion of the difficulty in equilibrium statistics is chiefly in order to emphasize that the almost identical difficulty in the time-dependent treatment is not new in character. For sufficiently simple systems, those n variables, $\xi_\alpha(\mathbf{r})$, that can be measured and displaced from equilibrium, and that when displaced with a sufficiently small wavevector $\mathbf{k}$ relax to equilibrium with long relaxation times, are reasonably well known. As long as observations are limited to longer time scales the behavior can be expected to conform to the behavior implied by Eq. (6.3) with a set of variables, $n_\nu(\mathbf{r})$, containing only linear combinations of these n variables, $\xi_\alpha(\mathbf{r})$. But the bane of the theoretician is the clever experimenter, and techniques such as those developed by Manfred Eigen's group in Göttingen have pushed the time scale of measurement down into the microsecond region and shorter. Not surprisingly, one finds that more and more significant variables ξ_α emerge even in comparatively simple systems as the time scale is shortened.

The best that one can expect to "prove" in any mathematical sense then becomes somewhat lame. First introduce dimensionless length, mass, and time such that the equilibrium number density is unity, the average molecular mass is unity, and the energy kT is unity. The unit time is then approximately a time between molecular collision of the order, in dense media, of 10^{-12} sec at room temperature. If then we restrict ourselves to n variables $\eta_\nu = -y_\nu$ at wavenumbers $\mathbf{k} \ll 1$ for which all relaxation times $t(\mathbf{k}, \nu) \gg 1$ *and* if we assume that all *other* variables η_μ have $t(\mathbf{k}, \mu) \ll t(\mathbf{k}, \nu)$, then only can we expect the relations implied by (6.3) to hold for an appropriate choice of the η_ν's.

Before turning to a more detailed discussion of mathematics let us examine some of the older apparent paradoxes in the time behavior of macroscopic systems.

Even with our hedges on the assumption that we know, *a priori*, for a given system all the n significant variables ξ_α, the total number n must remain comparatively small. The n orthonormal functions $\Delta\Psi_{\mathbf{k},\nu}$ for each wavenumber $\mathbf{k}$ that give the Fourier components of $\Delta\eta_\nu(\mathbf{r})$ then span a very small subspace of the complete Hilbert space in the 10^{24}-dimensional $\mathbf{p}^{(N)}\mathbf{q}^{(N)}$ phase space. Our assumption is then that *all* macroscopically measurable quantities are determined by the amplitudes of these functions in $\Delta\Phi$. All functions orthogonal to these are "noise," or rather undetectable, and hence silent, noise!

One method of constructing a complete set of orthonormal functions in the phase space would be according to a procedure as follows. First construct a complete orthonormal set that are sums over all molecules of functions of the coordinates and momenta of the single molecule, for instance sin–cosine functions of the center-of-mass coordinates, times normalized Hermite polynomials of the center-of-mass momenta divided by $\sqrt{2mkT}$, multiplied by members of a complete set in any internal coordinates and momenta. Add, then, to this limited subset of the Hilbert space a complete set of functions which are sums over all molecular pairs, of functions each involving the coordinates and momenta of this pair of molecules only. Out of these project all sums of functions of singlets, so that all members of the sums of pairs are orthogonal to each in the sum of singlets. Proceed step by step to add sums of triplet functions, of quadruplets, and so on up to those involving functions of N, each set orthogonal to all members of the preceding sets.

The macroscopically significant functions, $\Delta\Psi_{k,\nu}$ will be linear combinations of a limited number from the singlet and doublet set only, in most cases. For instance the zeroth and second Hermite polynomials of the singlet functions contribute to the number density and kinetic energy, respectively, and the mutual potential energy terms are generally assumed to be sums of pair functions only, in all practical models.

The Liouville operator contains additively the operators $(\mathbf{p}_i/m_i) \cdot \mathbf{\nabla}_{ri}$ and $-\sum_j \mathbf{\nabla}_{ri} u(r_{ij}) \cdot \mathbf{\nabla}_{pi}$ so that in every two operations on a function involving only n molecules a nonzero amplitude of a function necessarily involving $n + 1$ molecules is produced. No function expressible as sums of functions of limited subsets of the molecules can be an eigenfunction of the operator $\mathscr{L}$.

Although $\Delta\Phi(t = 0)$ given in Eq. (5.6) contains nonzero amplitudes only of functions of the macroscopically significant Hilbert subspace of the $\Delta\Psi_{k,\nu}$'s the function $\Delta\Phi(t)$ of Eq. (4.8) obtained by operating with $\exp[it\mathscr{L}]$ on $\Delta\Phi(t = 0)$ necessarily, as t increases, contains nonzero amplitudes of functions expressing correlations between larger and larger subsets of the molecules.

One of the persistent difficulties in the treatment of time dependence in statistical mechanics has been the well-known fact that the only satisfactory mechanical expression for the entropy

$$S = -k\mathscr{I}W \ln W \qquad (7.1)$$

is time independent. The operator $\mathscr{L}$ leaves the magnitude of $\Delta\Phi \cdot \Delta\Phi$ invariant,

$$(d/dt)\,\Delta\Phi(t) \cdot \Delta\Phi(t) = 0 \qquad (7.2)$$

so that the negative increment of entropy, $\Delta\sigma$ of Eq. (4.12'), which is

$$\Delta\sigma(t) = \tfrac{1}{2}N(\Delta\Phi \cdot \Delta\Phi) \qquad (7.3)$$

does not decrease. However, since the sum of the squared amplitudes of *all* orthonormal constituents of $\Delta\Phi(t)$ remains constant, the sum of those in the macroscopically significant Hilbert subspace does decrease. It is the sum of these that corresponds to the entropy that would be inferred from a knowledge of the local values of all $\xi_\alpha(\mathbf{r})$'s or $x_\alpha(\mathbf{r})$'s and the assumption of local thermodynamic equilibrium.

The picture that emerges is that of a constant flow of the statistically defined negative entropy up the ladder of increasingly complicated correlations between increasingly large numbers of molecules; correlations that, because of the orthogonality of the lower states, do *not* influence the average correlations between smaller subsets of molecules.

The majority of the paradoxes that have plagued the understanding of the role of a reversible mechanics in producing macroscopic irreversibility have originated in the insistence on the existence of an hypothesized perfectly isolated system with a completely time-independent Hamiltonian, even at the walls. Such an idealization obviously never exists in nature, except for the universe as a whole, if it is finite. The isolated bulk system is one in which the time development of the macroscopic variables, the $\xi_\alpha(t, \mathbf{r})$'s, is independent of the surroundings. The Hamiltonian at the walls is always subject to random temporal fluctuations due to impinging and emitted radiation, if nothing else. These stochastic processes in turn remove negative entropy from the system itself into nonzero amplitudes of functions involving correlations between the molecules of the system and of the surroundings, or with the radiation field outside the system. In this view the statistically defined entropy is not an additive function for two systems, $S_{A+B} \leq S_A + S_B$, even of bulk systems, but the negative entropy of the universe is continually flowing into terms involving molecules or other entities separated by enormous distances.

Whether or not a considerable fraction of the initial negative entropy is still stored in correlations between the molecules within one system, or has escaped into correlations with the surroundings in a particular macroscopically isolated bulk system which has progressed a considerable fraction of the way from an initial low entropy state toward macroscopic equilibrium, must depend on details of the system and its extent of isolation. As long as no experimental method of recovery is available, the answer would be of no practical importance.

However, in one case, that of the Hahn spin echo, the recovery of what might appear to be macroscopically lost negative entropy has been achieved. As pointed out by John Blatt[3] the process of recovery is strongly reminiscent of the famous paradox of Loschmidt, namely that an exact reversal of all the momenta in a classically completely isolated system would cause it to return to its initial low entropy state. In the Hahn spin echo experiment, a once oriented nuclear spin system, precessing in a constant magnetic field, appears to become more randomly oriented as evidenced by the macroscopic magnetic moment magnitude. It then is returned to nearly its initial oriented configuration. It is perhaps worth pointing out that the initial loss of correlation in spin direction is, in this case, not due to interactions between the spins, but due to inhomogeneities in the static magnetic field, and the constancy of these inhomogeneities is utilized in the recovery. The negative macroscopic entropy had, in this case, not been distributed into multiple correlations between the spins. The fact that complete recovery is not achieved is due partly to such processes, but also largely to its loss to the surroundings, which in this case are the lattice degrees of freedom.

A purely classical thermodynamic examination of the experimental situation shows that no violation of the second law is involved, since a removal of the system from its surroundings after the magnetic moment magnitude has decreased, and a subsequent evaluation by a thermal cycle to measure its entropy will be *irreversible*. When later returned to the echo apparatus it is incapable of demonstrating the echo. A *reversible* path from the low magnitude of magnetic moment is possible by inducing the echo, using a thermal measurement at the peak, and then permitting the system to "run down" to the low magnetic moment magnitude. The classical macroscopic entropy measured by this classical reversible thermal procedure is then found to follow the monotonically increasing values deduced from the peak values of the echoes, and does not follow the oscillations inferred from the magnitude of the macroscopic magnetic moment.

In terms of our previous discussion there is a macroscopic variable, X_α, measured by the values of the magnetic moment at the peaks, and the value of this variable determines the entropy. In microscopic interpretation this variable is the amplitude of a rather complicated function correlating the spins in a manner determined by the spatial inhomogeneities of the particular static field. The phase variable of this motion can be reversed to obtain the echo, and it is this phase variable that determines the total magnetic moment magnitude. The behavior has some superficial resemblance to a damped standing sound wave.

The spin echo example is amusing but has no present analog at room temperature. The general answer as to how much of the lost thermodynamic entropy is still stored within the system, and how much has been lost to the surroundings, would be given by the answer to a hypothetical experimental realization of the Loschmidt paradox. In a classical system *exactly* reverse all momenta and follow how far the system echoes its past state. I personally would guess that in a room temperature experiment involving, say, heat conduction, the best attainable isolation would show "echo" for less than a million collisions of single molecules before some stochastic event at the walls, propagated with sound velocity through the system, would destroy the reversal. If so, the negative entropy stored would be only that which had been lost from the thermodynamic reservoir in approximately the last microsecond.

In a macroscopically isolated system subject to small but random stochastic processes at the walls, the bogey of a predictable, even if enormous, Poincaré time of return to the initial state disappears. In its place is a fluctuation probability of $\exp[-\Delta\sigma]$ that a randomly timed observation will show the initial state of entropy less than equilibrium, $\Delta S = -k\Delta\sigma$.

There remains of course the one nonremovable paradox that for an initial state $\Delta\Phi(t = 0, \mathbf{p}^{(N)}, \mathbf{q}^{(N)})$ wholly within the Hilbert subspace of the thermodynamically significant function set of $\Delta\Psi_{\mathbf{k},\nu}$'s, which are necessarily even in all momenta components, the predicted amplitudes $y_\nu(\mathbf{k}, t)$ at time t will be even in t, namely that the equations predict the same macroscopic state for future and past. The difficulty is essentially one of language connected intimately with our biologically determined direction of times arrow. We remember the past, and try to predict the future. We never, of course, predict the future behavior of any system unequivocally, but only its behavior *if* it remains isolated. So, correspondingly we find that, lacking any other knowledge, if a system has remained isolated for all past time but is found in a state of negative entropy displacement $\Delta\sigma = -\Delta S/k$ from equilibrium, the most likely event is that it has attained this state by fluctuation from equilibrium, even if this probability has the fantastically low value of $\exp[-\Delta\sigma]$. But our knowledge of history tells us that a far more improbable situation is the case, namely that all our surroundings are enormously displaced from equilibrium by a far greater negative entropy amount, and that human intervention has produced the system we find standing in the laboratory in a non-equilibrium state.

8. *Nature of the Formal Proofs and Equations*

The time-dependent matrix

$$D(t) = \exp it\mathscr{L} \tag{8.1}$$

having general elements, $D_{\gamma,\delta}(t)$ in the representation of a complete orthonormal set of functions $F_\gamma(\mathbf{p}^{(N)}, \mathbf{q}^{(N)})$ in the phase space

$$D_{\gamma,\delta}(t) = F_\gamma^* \cdot e^{it\mathscr{L}} F_\delta \equiv \mathscr{I} W_0 F_\gamma^* e^{it\mathscr{L}} F_\delta \tag{8.2}$$

determines uniquely the time evolution of the displacement function $\Delta\Phi(t, \mathbf{p}^{(N)}, \mathbf{q}^{(N)})$. Since the thermodynamically significant portion of the Hilbert space is that spanned by the set $\Delta\Psi^c_{\mathbf{k},\nu}$, the thermodynamic development is completely determined by the limited submatrix of elements $D_{\mathbf{k}\nu;\,\mathbf{k}'\mu}(t)$. It is comparatively easy to show that to order $|\mathbf{k}|^2$ (less than those between equal $\mathbf{k}$ vectors), transitions between differing $\mathbf{k}$ vectors are negligible. More strictly stated, in nonisotropic media the directional vectors can be so chosen that this is so. We suppress the wavevector notation and discuss, for a fixed wavevector $\mathbf{k} = \mathbf{k}'$ the elements $D_{\nu,\mu}(t)$ for which

$$\Delta y_\nu(t, \mathbf{k}) = \sum_\mu D_{\nu,\mu}(t)\,\Delta y_\mu(t = 0, \mathbf{k}) \tag{8.3}$$

In the preceding discussion we have made an omission for the sake of simplicity. Macroscopic motion of the system at any position $\mathbf{r}$ is a macroscopic observable. In standing sound waves, for instance, such motion occurs, and the macroscopic analysis in these cases shows that the variables that diagonalize the Onsager matrix are sums of the real purely thermodynamic variable (the amplitude of the dimensionless constant-entropy pressure wave) and an imaginary velocity variable. Since the bulk velocity has associated kinetic energy, it contributes a term to the negative entropy at constant total energy. Corresponding to this we enlarge the macroscopically significant Hilbert subspace to include certain vector functions of odd momenta parity, and the mathematical treatment is greatly simplified if these are introduced with imaginary amplitudes. The eventual normalized function that diagonalizes the Onsager matrix is then $2^{-1/2}$ times the sum of the real thermodynamic displacement function plus the imaginary normalized vector function. These contributions occur in conjugate complex pairs, and the elements $D_{\nu\nu}$ have equal damping factors and oscillatory frequencies.

As discussed, then, in Section 6, our knowledge of the behavior of macroscopic systems leads us to believe that in some representation of this Hilbert subspace we have

$$\begin{aligned} D_{\nu,\nu}(t) &= \exp{(-\gamma_\nu + i\omega_\nu)t} \\ D_{\nu,\mu}(t) &= 0 \end{aligned} \tag{8.4}$$

where γ_ν is the reciprocal relaxation time and ω_ν will be zero in many cases. However, as discussed in Section 7, a mathematical proof of (8.4) will necessarily involve a circuitous argument that may require the addition of new variables if the conditions are not satisfied.

We require, therefore, that the set of macroscopic variables be complete, and define completeness as follows. Start with any set $\Delta \eta_\nu(t = 0, \mathbf{k})$

of initial displacements and observe both

$$\Delta\eta'_\nu = \Delta\eta_\nu(t = t_1, \mathbf{k}) \quad \text{(A)}$$

and

$$\Delta\eta''_\nu = \Delta\eta_\nu(t = t_1 + t_2, \mathbf{k}) \quad \text{(B)}$$

Now in another observation, start with the displacements at zero time

$$\Delta\eta'_\nu(t = 0, \mathbf{k}) = \Delta\eta'_\nu \quad \text{(C)}$$

and observe $\Delta\eta'_\nu(t = t_2, \mathbf{k})$. *If the set of variables is complete*, then we should observe

$$\Delta\eta'_\nu(t = t_2, \mathbf{k}) = \Delta\eta''_\nu \quad \text{(D)}$$

Thus, for example, in any case showing behavior analogous to the spin echo, we would not be content with the variable corresponding to the magnitude of magnetic moment as sole macroscopic variable, but seek a new variable which, with the moment magnitude, completely determines the behavior at future times.

Only the most general properties of the operator $\mathscr{L}$—namely that it contains first derivatives only, is odd in the momenta, is purely imaginary, and is Hermitian—are available to us for formal proofs in a general case. These, however, are sufficient to show, by relatively simple steps, that the requirements of A, B, C, and D suffice to demand Eq. (8.4) and conversely that (8.4) requires A to D. But now, for a system with nonsingular forces (no hard spheres, please) one can also prove that (8.4) cannot be rigorously correct. For instance, this follows from the case of initial real functions of even momenta parity. In this case $D_{\nu,\nu}(t)$ is even in t and analytic at the origin. One can, of course, construct analytic functions even in t that simulate $\exp[-\gamma t]$ closely, for instance $(\cosh t)^{-\gamma}$ which differs from $\exp[-\gamma t]$ only in terms of order γ. Since our unit time is of order 10^{-12} sec, even very rapidly decaying variables with microsecond relaxation times obeying $(\cosh t)^{-\gamma}$ would be experimentally indistinguishable from $\exp[-\gamma t]$. One could, in general, anticipate that times of the order of a collision time or greater would be necessary before the asymptotic exponential decay will be reached.

In any case Eq. (8.4) can be expected to hold only to order γ_ν, and more generally only to order $\gamma_\nu/\gamma_\varkappa$, where $\varkappa$ is any hidden variable not included in the macroscopic set $\nu, \mu, \ldots$. Similarly the requirements A to D will be only asymptotically satisfied. We can expect proofs to be valid only to such orders. In the case that ω_ν in (7.4) is not zero we also find that the requirement $\gamma_\mu \ll \omega_\mu \ll 1$ must be made.

The Liouville operator, being Hermitian, generates a set of eigenfunctions, the eigenvalues, z, of which must be continuous. The form of the spectral functions, $|a_\nu(z)|^2$, where $a_\nu(z)$ is the amplitude of eigenfunctions of eigenvalue z in $\Delta\Psi_\nu$, can be deduced from the asymptotically valid relations of Eq. (8.4) and the assumption that $\gamma \ll 1$. From this, in turn, the order of the elements $M_{\mu\nu}^{(n)}$,

$$M_{\mu\nu}^{(n)} = \Delta\Psi_\mu^* \cdot \mathscr{L}^n \Delta\Psi_\nu \tag{8.5}$$

in the matrix of nth power of the operator $\mathscr{L}$ can be deduced. The conclusions arrived at are that

$$\begin{aligned} M_{\mu\nu}^{(0)} &= \delta(\mu - \nu) + 0(\gamma) \\ M_{\mu\nu}^{(1)} &= \omega_\nu \, \delta(\nu - \mu) + 0(\gamma\omega) \\ M_{\mu\nu}^{(2n)} &= 0(\gamma) \quad n \geq 1 \\ M_{\mu\nu}^{(2n+1)} &= 0(\gamma\omega) \quad n \geq 1 \end{aligned} \tag{8.6}$$

and are in agreement, at least, with examination of the terms obtained by successive operations with $\mathscr{L}$ on simple displacement functions of temperature and composition in fluids, as well as with the standing sound wave displacement. The conclusion is that for a properly constructed set of orthonormal displacement functions we must have for all ν, μ,

$$M_{\nu\mu}^{(2)} = \Delta\psi_\mu \cdot \mathscr{L}^2 \Delta\psi_\nu \ll 1 \tag{8.7}$$

in order that all relaxation times be small.

In the simple case of a one-component fluid, the two thermodynamic displacements—one being that of the standing sound wave with displacement at constant entropy and the other that of the constant-pressure sinusoidal temperature displacement—are orthonormal. The condition of diagonalizing the two-by-two submatrix of $\mathscr{L}$, namely $M_{\mu\nu}^{(1)}$, is sufficient to select uniquely these two displacements as those that diagonalize the Onsager matrix. The known expression for the sound frequency is indeed found. In all more complicated cases the matrix $\mathbf{M}^{(1)}$ of elements $M_{\nu\mu}^{(1)}$ is degenerate with most diagonal elements zero, $\omega_\nu = 0$.

Finally, the formal equation for determining the correct representation of functions to diagonalize the relaxation time matrix and evaluate the times themselves uses elements $\gamma_{\nu\mu}$ of a time-independent matrix, $\boldsymbol{\Gamma}$,

$$(\gamma_{\nu\mu}/\lambda) = \lim_{k \to 0} \lim_{\lambda \to 0} \frac{1}{2} \int_{-\infty}^{+\infty} dt\, e^{-ikt} |\lambda^{-1}\mathscr{L}\, \Delta\psi_\nu^* \cdot e^{it\mathscr{L}} \Delta\Psi_\mu| \tag{8.8}$$

where the limits are taken in the order given, first $\lambda \to 0$ and then $k \to 0$. This matrix can then be diagonalized, and the diagonal elements are the reciprocal relaxation times. The double limiting process has been used before by Zwanzig[4] and by Luttinger.[5] The general formalism for obtaining the relaxation times is essentially that of Kubo,[6] but the method of derivation given in Ref. 2 is definitely different in character. In particular the emphasis there has been mainly on the selection of the independently decaying variables.

9. Possible Numerical Methods

It is, of course, disappointing, but also completely to be expected, that a general method of treating the matrix of relaxation times results in as abstract and generally unwieldy an equation as (8.8). One cannot expect an equation pretending to complete generality to be numerically tractable without further extensive simplification. The possible simplifications can be expected to be limited to rather simple systems and to depend in nature on the particular cases. Most of the successful numerical work has been limited to cases of weak interactions between single molecules, as in the near-perfect gases, or between normal coordinates as in crystals. Methods that are both completely satisfactory from the standpoint of one interested in at least near-mathematical rigor, and that also lead to satisfactory numbers, are hardly available for dense fluid systems such as normal liquids. Probably the most nearly satisfactory method is that due to Stuart Rice[7] in which an ingenious and plausible separation of the effects due to close hard-sphere type collisions and those due to the longer range attractive field is introduced.

Based on ideas of Bogolubov[8] one might expect to develop a density power series expansion for the treatment of imperfect gases, an expansion vaguely reminiscent of the cluster expansion in the equilibrium case. Unfortunately, recent work indicates that logarithmic divergences occur early.

One possible simplification in the selection of the independently decaying functions can be conjectured, but a satisfactory mathematical derivation has so far evaded us. Equation (8.6) demands that, to order $\gamma\omega$, the elements $M^{(1)}_{\mu\nu}$ of the first power of $\mathscr{L}$ have only diagonal terms, $\omega_\nu \delta(\nu - \mu)$. In the one simple case of the one-component fluid, this uniquely selects the correct function, one being the standing sound wave of frequency ω, and the other the sinusoidal temperature displacement at constant pressure which has zero frequency. Unfortunately, in any more complicated system the number of zero-frequency displacements becomes large, and since the submatrix is now degenerate its diagonalization is not unique. The time-dependent elements, $D_{\nu\mu}(t)$ of the matrix $\exp[it\mathscr{L}]$ obeying (8.4) would all be zero for $\nu \neq \mu$ if the relations were exact, and must be asymptotically small of order γ_ν. This would be satisfied if all off-diagonal elements $M^{(n)}_{\nu,\mu}$ were zero. One might conjecture that diagonalization of $M^{(2)}_{\nu,\mu}$ would select the sought-for representation. This requirement would pose a reasonably tractable numerical problem in many cases.

The apparently complete failure of one approach has puzzled the writer. With any Hermitian operator, such as $\mathscr{L}$, one can project a sequence of orthonormal functions by repeated operation, starting with a single normalized function F_0. Namely, from $\mathscr{L} F_0$ project out F_0 and normalize the remainder to find F_1, and in general define F_{n+1} as the normalized part of $\mathscr{L} F_n$ with F_{n-1} and F_n projected out. The matrix of the operator in the representation of this function has nonzero elements on the diagonal and at one unit off the diagonal only. For the operator $\mathscr{L}$, if F_0 has even momenta parity, the diagonal elements are also zero. Since the set of functions spans the Hilbert space of all powers, $\mathscr{L}^n F_0$, one can develop $\exp[it\mathscr{L}]F_0$ by time-dependent coefficients multiplied by these functions,

$$\exp[it\mathscr{L}]F_0 = \sum_{n \geq 0} a_n(t) F_n = \Delta\Phi(t) \qquad (9.1)$$

and, for short times, only a finite number of amplitudes will be nonzero.

In terms of the elements $M^{(1)}_{n,m}$ of $\mathscr{L}$ in this representation it is relatively simple to write the equation for the values of any finite number of the coefficients, $a_n(t)$, $0 \leq n \leq m$ for which

$$da_n(t)/dt = -\gamma a_n(t) \qquad (9.2)$$

and to do this for *any* γ value, positive or negative. If the first $m + 1$ coefficients obey the derived relation, then up to the m' the derivative $a^m a_0/dt^m = (-\gamma)^m a_0$. In short, a_0 will change with time like $a_0(t=0) \times \exp[-\gamma t]$ for times of the order of m times a collision time.

This accomplishment is a sort of abstraction of a generalized Loschmidt paradox—namely, with any positive or negative relaxation time it is possible to define a displacement in phase space, $\Delta\Phi(t)$, for which, if $F_0 = \Delta\Psi_\nu$, the macroscopic displacement variable $\Delta\eta_\nu$ changes its amplitude in a pseudoexponential manner for considerable times.

That this can be done is not completely amazing if one thinks of the dilute gas case. The transport coefficients there depend on the mean free path. By sufficiently clever selection of an initial state with correlations in position and momenta of all subsets of m molecules, the mean free path can be made abnormally large or small even up to m collisions. However, one would tend to expect that such a *tour de force* would be recognizable in some abnormal complexity of the required distribution. No obviously significant feature of the equations that satisfy (9.2) seems to depend in any critical way on the value of the arbitrarily chosen reciprocal time constant, γ.

The explanation of this appears to be the following. The normal time development of a macroscopically displaced isolated system produces many-body correlations. These have the character that an exact reversal of all momenta will cause the temporal evolution to reverse. Such a time reversal occurs, in first order, if the sign of the amplitudes of the odd-parity momentum functions is changed in $\Delta\Phi(t)$. In addition, by altering the amplitudes of the components in this correlation function in a prescribed, but not drastic, manner, the rate of time evolution can be altered to any prescribed value for a finite time interval.

One approach appears to offer some hope of success in leading to not impossibly difficult equations for numerical evolution. Reduced probability density functions, $W_n^{(r)}(\mathbf{p}^{(n)}, \mathbf{q}^{(n)})$, in the phase space $\mathbf{q}^{(n)}$, $\mathbf{p}^{(n)}$ of small subsets, $n = 1, 2, 3, \ldots$ of n molecules can be defined by integration over all other $N - n$ molecules of the complete probability density. In a classical system a slightly altered Liouville operator $\mathscr{L}_n^{(r)}$ determines the time evolution of these functions in terms of an average force field acting on the molecules. If the potential energy is a sum of pair terms, only these forces acting on the members of the set of n molecules are due to the direct forces and to the effect of the others, which, in turn, is given by an integral over the reduced probability density, $W_{n+1}^{(r)}$, for n plus one molecules.

Considering a small volume element in a system undergoing steady state transport—for instance, a two-component fluid of constant temperature and pressure with a gradient of composition and a diffusion flux of molecules related by an unknown diffusion constant D. The singlet probability density, $W_i(\mathbf{r}_i, \mathbf{p}_i)$ for a molecule i (of type i), can be written for this case. Require that it be stationary in time. This requirement suffices to determine the average singlet force on the molecule i as a function of position and momentum. This known force, in turn, is due to nonequilibrium displacements in the reduced pair function, $W_{ij}^{(r)}$. Demand that this be stationary in time, and this requires average forces $_{ij}\mathbf{f}_i$ acting on molecule i when the positions and momenta of both i and a close molecule j are given. These forces are now fixed by integrals over $W_{ijk}^{(r)}$.

A closure is necessary. Make the closure analogous to the Kirkwood superposition closure in the equilibrium case: assume that the *nonequilibrium* correlation in the logarithm of $W_{ijk}^{(r)}$ is the sum of those in the logarithms of the three singlet and three pair functions.

Carrying out the procedure to second order does give an expression for the diffusion constant. The expression is in terms of integrals over functions which, in turn, are solutions of an integral equation.* The equations are cumbersome but not impossibly difficult.

No justification other than intuition can yet be given for the closure used. It is an extension of that found valid in the dilute gas case, utilized in the Boltzmann equation which assumes that the pair distribution function is a product of the two singlet distributions without significant two-body correlations in the nonequilibrium past. The method may be rather widely applicable.

* To be given in an as yet unpublished dissertation by Harold Raveché, University of California, San Diego (1967).

References

1. L. Onsager, *Phys. Rev.*, **37,** 405, 2265 (1931).
2. M. Bauer, J. R. Jordan, P. C. Jordan, and J. E. Mayer, *Ann. Phys.*, **35,** 96 (1965).
3. J. M. Blatt, *Progr. Theoret. Phys.*, **22,** 745 (1950).
4. R. Zwanzig, *J. Chem. Phys.*, **40,** 2527 (1964).
5. J. M. Luttinger, *Phys. Rev.*, **135,** A1505 (1964).
6. R. Kubo, *J. Phys. Soc. Japan*, **12,** 570 (1957); R. Kubo, M. Yokota, and S. Nakajima, *ibid.* **12,** 1203 (1957).
7. S. A. Rice and P. Gray, *The Statistical Mechanics of Simple Liquids*, Interscience, New York, 1965, particularly Chapter 5.
8. N. N. Bogolubov, *Studies in Statistical Mechanics*, Vol. 1, J. de Boer and G. E. Uhlenbeck, Eds., Interscience, New York, 1962.

Approximate Symmetries in Atomic and Elementary Particle Physics

H. P. DÜRR

*Max-Planck-Institut für Physik und Astrophysik,
Munich, Germany*

Elementary particle physics in its phenomenological appearance has many features in common with atomic physics. This similarity is rarely emphasized. There is, in fact, a tremendous contrast in their theoretical description, arising essentially from the relativistic features of elementary particle physics, which have prevented the construction of solvable models and introduced difficulties where one is still uncertain whether they are of a principal or only of a technical nature. On the other hand, atomic physics is without these difficulties. We are well acquainted with the H atom—its energy eigenvalues can be calculated by any physics graduate student—and perhaps a little bit with the He atom or the H_2^+ ion, or even the H_2 molecule. But we are rarely confronted with the problem of calculating the energy spectrum of a B atom, or an Fe atom, or even a complicated molecule, as it is attempted in quantum chemistry. These calculations are, indeed, very complicated and require very refined approximation methods, e.g., variational methods, for which in most cases no mathematical proof of convergence exists. Still, this does not cause any uneasiness on our part, because from the extremely good agreement between theory and experiment in the exactly calculable cases, we derive our conviction that our dynamical formulation is, without doubt, correct (up to a certain, well-defined accuracy), and that all the higher atoms and molecules known in nature will also exist as stationary solutions in our mathematical description if it is appropriately carried through. Our long and intense struggle with the fundamentally unclear and intricate dynamics of elementary particle physics on the one hand, and our relatively short occupation with the simple H atom on the other hand, has affected our relation to these problems by, to our mind, unduly exaggerating this contrast. Many people have come to believe that the solution of the dynamical problem in elementary particle physics requires some completely new features, let us say, in the same way as quantization was necessary for atomic physics.

In the present article we try to emphasize the common features of atomic and elementary particle physics, in particular with respect to their symmetry properties, or better, broken symmetry properties. These analogies have played, at least, an intuitive role in the formulation of Heisenberg's nonlinear spinor theory,[1] in which the present author has been involved for many years.

In the first section we will reflect on the similarities between atomic and elementary particle physics and try to evaluate their relevance. In the second we shortly discuss the approximate symmetries of elementary particle physics. As an example, for higher approximate symmetries in atomic physics the energy spectrum of the B atom is investigated in Section 3. In Section 4 we indicate another mechanism of symmetry violation which is caused by an asymmetry of the ground state of very large systems. In the last section we will try to apply the mechanisms of symmetry violation to elementary particle physics.

1. *Comparison of Elementary Particle Physics with Atomic Physics*

High energy experiments have taught us that elementary particles are not indestructible, that they also are not the "atoms" of the Greek, the indivisible units of matter, a role which was assigned to them after the experiments of Rutherford had clearly disqualified the atoms for it. The number of known elementary particles has increased tremendously during the last years. It was observed that these particles generally interact strongly with each other and that they transform into each other in a number of complicated ways. There seem to be no indestructible smaller particle units, the number of which remains unchanged in these transmutations, like the number of electrons and the nucleus stays intact in all atomic reactions. Only certain properties are still conserved, such as electric charge, baryonic number, or mechanical properties like energy, momentum, and angular momentum. From Noether's theorem[2] we know that these conservation laws are related to certain symmetry properties of the dynamics. In particular, the mechanical conservation laws are related to the invariance of the dynamics under the relativity group (Poincaré group). The formulation of causality in a relativistic invariant theory leads to the concept of an antiparticle, and the possibility of creating particle–antiparticle pairs from energy and of annihilating them into energy.

It is this property, linked to the causal (local) and relativistic structure of the dynamics of matter, which offers an exciting possibility of establishing a principle limit to a further dissolution of matter: Elementary particles may, indeed, qualify to be the ultimate building blocks of matter—not in the sense of being indestructible units, but in the sense that in their destruction again only elementary particles of the same or of some other sort are produced, with the energy applied for their destruction being partly used to restore the masses. From this point of view, elementary particles will differ distinctly in their description from atoms, or from molecules, or from higher composite systems such as a macroscopic crystal.

Technically, the complication in elementary particle physics arises exactly from the necessity of employing a relativistic dynamics. As a consequence, elementary particles appear as systems with an infinite number of internal degrees of freedom, in contrast to an atom, where the number of internal degrees of freedom is finite and related to the number of electrons, and also in contrast to a crystal or a general state of a solid, where the number of degrees of freedom can be made arbitrarily large, but only by sufficiently increasing the volume (the degrees of freedom per volume element is still finite). Intuitively this enrichment of the internal structure in elementary particles may be expressed in terms of many-particle states, into which these particles can virtually decompose. A π meson, for instance, can virtually decompose into $\pi\rho$, $\pi\pi\omega$, $\bar{N}N$ etc., but also into a $\pi x \bar{x}$ system, where x may be a very heavy object, e.g., a protein molecule. However, we suspect that the probability of creating virtually such heavy objects inside the π meson is very small, provided the interaction is not too singular at small distances, because they differ too much in energy from the original π, and hence can only appear in regions where, as a result of the interaction, a sufficiently high potential energy is present for compensation. If the interaction is very singular at small distances, as for example in nonrenormalizable theories, we would expect the elementary particles to also have important projections on these higher configurations—a practically hopeless situation—if the picture is not drastically changed by renormalization or regularization effects, in such a way that the singularity is essentially screened. In any case, it appears that in order to make a dynamical theory of elementary particles at all interpretable (technically speaking: to avoid the divergence difficulties), some kind of a limitation of configurations should be expected, in the sense that every elementary particle should be dominated by a reasonably small number of virtual configurations, which roughly may consist of the many-particle configurations closest in energy (nearest singularities in an analytical description).

In this sense there may be particles or resonances which have large projections on only a few channels, others may need for their description a large number of different configurations, at least if we choose an inappropriate basis. For example, the N_{33}^* resonance seems to be rather well represented by a πN-system in a p state, the ρ meson by a $\pi\pi$ system in a p state. As a consequence the overlap of the wavefunctions between the πN scattering state and the N_{33}^*, or the $\pi\pi$ scattering state and the ρ, is quite large and therefore the transition probabilities quite big. The N_{33}^* and the ρ hence are rather broad resonances. Broad resonances, therefore, should be taken as an indication that the system may be well approximated by resonance states of the particles which appear in the decay channel.

There are many examples of resonances which are surprisingly sharp[3] although they are way above threshold. This would indicate that their overlap with the configurations of the open channels is small, i.e., due to the strong interaction in their interior the structure is only effectively expressible in a high dimensional space, and the probability will be small for the system to fluctuate into that subspace where it can actually decay. (A description of this type was used by N. Bohr in his "sandbag" model of the atomic nucleus.)

It is rather obvious that for the description of the local structure the different physical channels do not represent an appropriate basis, because only in the asymptotical region are they orthogonal to each other. If, for instance, we consider virtually the exchange of a $N\bar{N}$ pair and also of a π meson, we may have counted certain parts twice, because a π meson off the mass shell can transform into a $N\bar{N}$ pair. The important point here is, of course, that elementary particles do not preserve any individuality if they are off their mass shells, which could be used for a characterization of a local system. Depending on how far they are off their mass shell and how close their competitors (particle or scattering states with the same set of quantum numbers) are, they will submerge, sooner or later, into this soup of strongly interacting matter. We would, however, expect that in the description of our matter configuration, into which configurations of low-lying elementary particles of a certain limited energy will "dissolve" in the interaction region, only a limited number of degrees of freedom will play an essential role. In this case, these particles may be obtained as stationary, or approximately stationary, solutions of dynamical equations which involve only a finite set of functions. The main difficulty, of course, will be to find the most appropriate finite set for this purpose. Every approximation procedure implies a particular selection. Hence the choice of an adequate approximation procedure will be of decisive importance, if we wish to obtain high accuracy.

If a limitation of the effective number of degrees of freedom, as described above, really occurs in a relativistic theory, then the difference between atomic physics and elementary particle physics is greatly reduced. In the atomic case these degrees of freedom are exactly given by the composing particles, i.e., the electrons and the nucleus. The dynamical problem can be solved exactly if the interaction between these particles is given.

In the elementary particle case these degrees of freedom are, at best, only approximately limited, and will be related to rather abstract objects. One may hope, however, that these abstract objects are connected to matrix elements of finite products of a fundamental field operator, for which a dynamical law can be formulated. Then the analogy to atomic physics would be rather close.

2. *Approximate Symmetries in Elementary Particle Physics*

One of the biggest puzzles in elementary particle physics is the existence of certain conservation laws which are only approximately valid, like the conservation law for isotopic spin, hypercharge, parity, etc. These approximate conservation laws have drastic consequences in nature. They are the origin of this curious hierarchy of interactions, i.e., the fact that phenomenologically elementary particles participate in a number of different interactions of decreasing strength which are commonly classified as superstrong, strong, electromagnetic, weak, PC-violating, and gravitational interactions. Aside from their rather different strength, these interactions are distinguished by the different selection rules they imply.

Conventionally the approximate conservation laws are interpreted as the consequence of a corresponding approximate symmetry of the underlying dynamical law. The interaction of elementary particles, in fact, is written as a sum of terms of decreasing strength and symmetry, and involving field operators of some, if not all, elementary particles. These symmetries are also reflected in the mass spectrum of the one-particle states, which transform according to an irreducible representation of the dynamical symmetry group. If these symmetries are slightly invalidated the one-particle states split up into characteristic mass multiplets. These multiplets are actually used to determine the representations of the particles, e.g., to determine their isospin or their $SU(3)$ quantum numbers.

This more phenomenological approach in the dynamical formulation was quite successful, in particular with respect to electrodynamics, but from a principal point of view, it is completely unsatisfactory, because it does not answer the main question—why elementary particles are governed by this peculiar hierarchy of interactions. The various interactions in this formulation seem to exist accidentally side by side. On the other hand, there is mounting evidence for a rather close relation between all these interactions.

It therefore may be more adequate to think about these interactions as different aspects of one universal interaction of high symmetry which, by some reason or the other, happens to be slightly distorted. The different phenomenological interactions merely reflect a sequence of approximations which, step by step, takes into account these various distortions. We may formulate such a universal interaction in terms of a dynamical equation for a fundamental field (e.g., a quark field), and derive the elementary particles from it, similarly to the way in which we deduce the spectra of atoms from a Schrödinger equation for electron wavefunctions. But to express the symmetry distortions, several interaction (or mass) terms of various reduced symmetries have to be introduced. They will be characterized by certain dimensionless constants which are a direct measure of these distortions, such as the Sommerfeld fine structure constant, which characterizes the isospin distortion. These parameters seem to be completely arbitrary from a theoretical point of view. Since, however, even quantum electrodynamics is far from being a mathematically rigorous theory [and we would not expect, nor even regret, this, since otherwise its observed linkage to the strong interactions (pair creation of hadrons) would appear as an unnecessary complication] it is argued that a more rigorous treatment of an interacting quantum field theory will limit or even fix the values of these constants in some way. On the other hand, it appears difficult for us to understand how by this reasoning, the value zero of the constants could be excluded, which would restore the full symmetry of the theory. Hence the principal question remains, why the symmetry is incomplete, and—after this is understood—why it is violated by exactly this amount.

Incomplete symmetries or incomplete conservation laws are not peculiar to elementary particle physics. In fact, we encounter them in various ways in atomic, solid state, and nuclear physics. There, of course, their origin is directly apparent. Our main interest, therefore, will be to investigate these cases with respect to their possible applicability in elementary particle physics.

3. Approximate Symmetries in Atomic Physics

The appearance of approximate symmetries is quite frequent in any dynamical system where more than one particle is involved. As an example we may take any atom with more than one electron, e.g., the B atom.

In a first approximation, two of the five electrons of the B atom occupy the first shell with the principal quantum number $n = 1$, and three electrons the second shell with $n = 2$. For the following discussion of the energy spectrum we can disregard the two $1s$ electrons, because their shell is closed. In a very coarse approximation we then may imagine the three outer electrons to move independently of each other in the same, somewhat screened Coulomb potential. This will give rise to a hydrogen-like spectrum, where the energy depends solely on the principal quantum number n. Group-theoretically speaking, n is connected with the irreducible representations of the four-dimensional rotation group $O(4)$.[4] Since the electrons are endowed with spin $\frac{1}{2}$, they themselves will transform according to an irreducible representation of the two-dimensional unitary group $SU(2)$. Due to the spin independence of the interaction and the dynamical independence of the electrons in this approximation, the energy eigenstates correspond to a representation of the huge group

$$G_0 = O(4) \otimes O(4) \otimes O(4) \otimes SU(2) \otimes SU(2) \otimes SU(2)$$
$$= [O(4)]^3 \otimes [SU(2)]^3$$

The lowest eigenvalues are characterized by all electrons occupying the $n = 2$ shell (the $n = 1$ shell is inaccessible, because it is filled). The $n = 2$ shell contains $2n^2 = 8$ different eigenstates [4-dimensional irreducible representation of $O(4) \otimes$ 2-dimensional irreducible representation of $SU(2)$]. Taking into account the Pauli principle (antisymmetrization of the total wavefunction) of the $8^3 = 512$ different states of the combined 3-electron system, only $\binom{8}{3} = 56$ are physically acceptable. Hence, in this rough approximation the ground state of the B atom appears as a state which transforms according to a 56-dimensional representation of the symmetry group G_0; it is a degenerate 56-plet (see Table I).

As a first correction we have now to take into account the different screening effects on the wavefunctions with different numbers of radial nodes, which correspond to the spherically symmetric part of the electron–electron interaction. The "elliptical" orbits (s states) will see a higher positive charge than the "circular" orbits (p states), and hence will get comparatively lower energy. These screening effects will "break" the individual $O(4)$ symmetry group, and will leave intact only the individual three-dimensional rotation group $O(3)$, which is a subgroup of $O(4)$. As a consequence the levels split up according to their $O(3)$ content into three distinct levels: $(2s)^2(2p)$, $(2s)(2p)^2$, and $(2p)^3$, which contain $\binom{6}{1} = 6$, $\binom{2}{1}\binom{6}{2} = 30$, and $\binom{6}{3} = 20$ levels, respectively, and can be distinguished by the p-level ($l = 1$) occupation number $n_1 = 1, 2, 3$. They are representations of the subgroup G_1 of G_0, with $G_1 = [O(3)]^3 \otimes [SU(2)]^3$.

If one now considers the nonspherically symmetric part of the electron–electron interaction, the levels will be split up further, according to their total intrinsic spin S, and their total angular momentum L. We will do this in several steps, in order to exhibit the group-theoretical structure.

TABLE I

n	n_l	$S_l(S)$	$v_l(v)$	L	J (lowest order)
$(1)^2(2)^3$ [56]	$(2p)^3$ [20]	$(2p)^3$, $S_1 = \frac{1}{2}$ [16]	$(2p)^3$, $S_1 = \frac{1}{2}$, $v_1 = 1$ [6]	^2P [6]	$^2P_{3/2}$ [4]; $^2P_{1/2}$ [2]
			$(2p)^3$, $S_1 = \frac{1}{2}$, $v_1 = 3$ [10]	^2D [40]	$^2D_{5/2}$ [6]; $^2D_{3/2}$ [4]
		$(2p)^3$, $S_1 = \frac{3}{2}$ [4]	$(2p)^3$, $S_1 = \frac{3}{2}$, $v_1 = 3$ [4]	^4S [4]	$^4S_{3/2}$ [4]
	$(2s)(2p)^2$ [30]	$(2s)(2p)^2$ $S_1 = 0$ [12]	$(2s)(2p)^2$, $S_1 = 0$, $v_1 = 0$ [2]	^2S [2]	$^2S_{1/2}$ [2]
			$(2s)(2p)^2$, $S_1 = 0$, $v_1 = 2$ [10]	^2D [40]	$^2D_{5/2}$, $^2D_{3/2}$ [6 + 4]
		$(2s)(2p)^2$ $S_1 = 1$ [18]	$(2s)(2p)^2$, $S_1 = 1$, $v_1 = 2$ [18]	^4P ^2P [12 + 6]	$^4P_{5/2}$, $^2P_{5/2}$ [6 + 4]; $^4P_{3/2}$, $^2P_{3/2}$ [4 + 2]; $^4P_{1/2}$ [2]
	$(2s)^2(2p)$ [6]	$(2s)^2(2p)$ $S_1 = \frac{1}{2}$ [6]	$(2s)^2(2p)$, $S_1 = \frac{1}{2}$, $v_1 = 1$ [6]	^2P [6]	$^2P_{3/2}$ [4]; $^2P_{1/2}$ [2]
$[O(4)]^3$ $\otimes [SU(2)]^3$	$[O(3)]^3$ $\otimes [SU(2)]^3$	$\prod_l U(2l+1) \otimes SU(2)$	$\prod_l O(2l+1) \otimes SU(2)$	$O(3) \otimes SU(2)$	$SU(2)$

If there are more than one particle in the same level, e.g., the p electrons in the levels $n_1 = 2, 3$ of our example, we may use the following group-theoretical descriptions.[5] A state with angular momentum l consists of $(2l + 1)$ linearly independent states (not counting the spin degrees of freedom). A state with n_l particles in this l state hence transforms like a component of a tensor of rank n_l in a $(2l + 1)$-dimensional space. These tensors are not irreducible under the general unitary (or even linear) transformations $U(2l + 1)$ of the wavefunctions. A decomposition with respect to the irreducible representations of $U(2l + 1)$ is identical to a decomposition with respect to the permutation group (symmetrical group). In order for the total wavefunctions to be fully antisymmetrical, these orbital wavefunctions have now to be appropriately combined with a spin wavefunction of the adjoint symmetry type. Since the symmetry type of the spin wavefunctions is uniquely characterized by the total intrinsic spin S_l of the particles which occupy this l level, the different irreducible representations of the group $U(2l + 1)$ can be simply labeled by S_l. Since the repulsive electron–electron interaction will depend on the orbital symmetry character of the wavefunctions, the levels will split up according to the S_l quantum number, favoring energetically antisymmetrical orbital configurations (large S_l). The symmetry group G_1 is now reduced to a subgroup

$$G_2 = \prod_l [U(2l + 1) \otimes SU(2)]$$

where in the direct product only partially filled l levels have to be included.

In the level $n_1 = 1$ we have two s electrons and one p electron. Hence there is no splitting. The level can be labeled by the spin quantum numbers $S_0 = 0$, $S_1 = \frac{1}{2}$, or even by the total intrinsic spin $S = \frac{1}{2}$. Group theoretically we have a $3 \times 2 = 6$-dimensional representation of the group $G_2 = U(3) \otimes SU(2)$. In the case of the $n_1 = 2$ level the $3^2 = 9$ components of the tensor split up into the antisymmetrical part with three components ($U(3)$ triplet) and the symmetrical part with six components ($U(3)$ sextuplet). The triplet will be multiplied by the symmetrical spin wavefunction $S_1 = 1$, hence incorporating totally nine levels; the sextuplet will be multiplied by the $S_1 = 0$ wavefunction, hence incorporating totally six levels. The number of levels are further multiplied by 2 due to the double spin orientation of the $2s$ electron. The effective symmetry group is $G_2 = U(1) \otimes U(3) \otimes [SU(2)]^2$. The upper level $S_0 = \frac{1}{2}$, $S_1 = 0$ corresponds also to $S = \frac{1}{2}$. The lower level $S_0 = \frac{1}{2}$, $S_1 = 1$ leads to levels with $S = \frac{1}{2}$ and $S = \frac{3}{2}$. The $S = \frac{1}{2}$ and $S = \frac{3}{2}$ levels, however, will only have different energies if real spin–spin interactions are included. The

$n_1 = 3$ level, finally, will split up into a four-dimensional ($U(3)$ singlet $\otimes$ $SU(2)$ quadruplet) and a 16-dimensional ($U(3)$ octuplet $\otimes$ $SU(2)$ doublet) representation of $G_2 = U(3) \otimes SU(2)$.

As a next step we may classify our levels in each l orbit according to $O(2l + 1) \subset U(2l + 1)$. The irreducible representations of $O(2l + 1)$ are characterized by the seniority quantum number v_l. Intuitively, the seniority quantum number is given by the number of electrons in a particular orbital l, the spin of which is not saturated (i.e., number of electrons minus $S = 0$, $L = 0$ pairs). For repulsive interaction saturated pairs cost energy (symmetrical orbital wavefunction) and hence are pushed up. Our levels therefore split according to their content with respect to the group*

$$G_3 = \prod [O(2l + 1) \otimes SU(2)]$$

In our case the level $n_1 = 2$, $S_1 = 0$ will split into levels with $v_1 = 0$ and $v_1 = 2$; the levels $n_1 = 3$, $S_1 = \frac{1}{2}$ will split into levels with $v_1 = 1$ and $v_1 = 3$, as can be easily seen in Table I. In the first case the $U(3)$ sextuplet is decomposed into an $O(3)$ singlet and quintuplet, in the second case the $U(3)$ octuplet is decomposed into an $O(3)$ triplet and quintuplet.

Next we consider the classification according to the total orbital angular momentum L, i.e., according to the subgroup $G_4 = O(3) \otimes SU(2)$. In our simple case the L quantum number, or better the seniority, is a redundant label, and we do not encounter any additional splitting. In more complicated cases, however, this will be different. The different L values are indicated in Table I. The multiplicity of the levels is simply $(2L + 1)(2S + 1)$.

We finally take into account the relativistic effects, which lead to a weak spin–orbit coupling. In lowest approximation the energy levels will split according to their total angular momentum J, preserving their configuration. In higher approximations, also different configurations (of the same J) will be mixed (configuration mixing). The total angular momentum characterizes the irreducible representations of the only symmetry group left, the group $G_5 = SU(2)$, which reflects the rotational symmetry of the Schrödinger equation.

The group $G_5 = SU(2)$ is the fundamental group of our dynamics. All the other, higher, groups are only approximate and are related to the structure of our particular configuration. We therefore may call them "structural" symmetries. The apparent fine structure of the energy levels is given by the best of the approximate symmetry groups, here the group G_4 which carries S (better S_l) and L as labels. States with half-integral S,

---

* To be more precise, the electron–electron interaction would merely lead to a seniority splitting, if the interaction had zero range (δ-function pair forces), the finite, or even here the infinite range, produces also a splitting with respect to S and L (see below).

as in our case, will therefore normally give rise to doublets, quadruplets, etc. in the final level scheme, and states with integral S (e.g., for the C atom) will normally give rise to singlets, triplets, etc. Exceptions arise where two sublevels are incompletely filled, as in our $n_1 = 2$ case, where effectively only singlets and triplets occur, due to the decoupling of the spin of the $2s$ electron. Also $L = 0$ states will always be singlets. Hence, in the same spectrum we may observe singlets, doublets, and triplets in the fine structure.*

Instead of taking the LS coupling scheme as intermediate steps in the classification of states, one may use the jj coupling scheme. Since in atomic physics the spin–orbit force is much weaker than the effective spin–spin interaction of particles in the same orbit (arising from the Coulomb repulsion of the electrons and the wavefunction antisymmetrization), the LS approximation sequence is more realistic, in the sense that here the configuration mixing is rather small. In nuclear physics the jj coupling scheme is a better approximation procedure because of the exceptionally strong spin–orbit forces.

If we use the jj scheme for our B atom (see Table II), we classify the individual particle states according to their angular momentum j, which characterizes the representations of $SU(2)$. The spin–orbit force will favor smaller j values. The levels with $n_1 = 1, 2, 3$ related to representations of $G_1 = [O(3)]^3 \otimes [SU(2)]^3$ hence will split up according to their content with respect to the subgroup $G_2' = [SU(2)]^3$. States with several particles in the same level j, however, also transform into each other under the $(2j + 1)$-dimensional unitary group $U(2j + 1)$. Due to the antisymmetry of the total wavefunction, they even transform irreducibly under these transformations. Hence, we may equally well characterize the relevant subgroup by

$$G_2'' = \prod_j U(2j + 1)$$

where again completely filled or empty shells should be omitted.

As a next step we consider the behavior of the levels under the subgroup

$$G_3' = \prod_j Sp(2j + 1)$$

with $Sp(2j + 1)$ the symplectic group in $(2j + 1)$ dimensions. It is characterized by the seniority quantum number v_j, which refers to the number of nonsaturated electrons in the j shell. (It differs from the seniority v_l in the LS scheme, because in the present case $J = 0$ pairs are omitted. This is

---

* Actually our hierarchy of level splittings is not finished here, if the atomic nucleus also carries spin. Then we observe in addition a hyperfine splitting, which, however, is only appreciable for the S states.

TABLE II

n	n_l	j	$v_j(v)$	J (lowest order)
$(1)^2(2)^2$ [56]	$(2p)^3$ [20]	$(2p_{3/2})^3$ [4]	$(2p_{3/2})^3, v_{3/2} = 1\ (v=1)$ [4]	$(\frac{3}{2}\frac{3}{2})_{3/2}$ [4]
		$(2p_{1/2})(2p_{3/2})^2$ [12]	$(2p_{1/2})(2p_{3/2})^2, v_{3/2} = 0\ (v=1)$ [2]	$(\frac{1}{2}\frac{3}{2})_{1/2}$ [2]
			$(2p_{1/2})(2p_{3/2})^2, v_{3/2} = 2\ (v=3)$ [10]	$(\frac{1}{2}\frac{3}{2})_{5/2}$ [6]
		$(2p_{1/2})^2(2p_{3/2})$ [4]	$(2p_{1/2})^2(2p_{3/2}),\ (v=1)$ [4]	$(\frac{1}{2}\frac{3}{2})_{3/2}$ [4]
	$(2s)(2p)^2$ [30]	$(2s)(2p_{3/2})$ [12]	$(2s)(2p_{3/2})^2, v_{3/2} = 0\ (v=1)$ [2]	$(\frac{1}{2},\frac{3}{2}\frac{3}{2})_{1/2}$ [2]
		$(2s)(2p_{1/2})(2p_{3/2})$ [16]	$(2s)(2p_{3/2})^2, v_{3/2} = 2\ (v=3)$ [10]	$(\frac{1}{2},\frac{3}{2}\frac{3}{2})_{5/2}, (\frac{1}{2},\frac{3}{2}\frac{3}{2})_{3/2}$ [6+4]
			$(2s)(2p_{1/2})(2p_{3/2}),\ (v=3)$ [16]	$(\frac{1}{2},\frac{1}{2}\frac{3}{2})_{5/2}, (\frac{1}{2},\frac{1}{2}\frac{3}{2})_{3/2}$ [6+4]
		$(2s)(2p_{1/2})^2$ [2]	$(2s)(2p_{1/2})^2,\ (v=1)$ [2]	$(\frac{1}{2},\frac{1}{2}\frac{3}{2})_{3/2}, (\frac{1}{2},\frac{1}{2}\frac{3}{2})_{1/2}$ [4+2]
	$(2s)^2(2p)$ [6]	$(2s)^2(2p_{3/2})$ [4]	$(2s)^2(2p_{3/2}),\ (v=1)$ [4]	$(\frac{1}{2},\frac{1}{2}\frac{1}{2})_{1/2}$ [2]
		$(2s)^2(2p_{1/2})$ [2]	$(2s)^2(2p_{1/2}),\ (v=1)$ [2]	$(\frac{1}{2}\frac{1}{2},\frac{3}{2})_{3/2}$ [4]
				$(\frac{1}{2}\frac{1}{2},\frac{1}{2})_{1/2}$ [2]
$O(4)^3$ $\otimes [SU(2)]^3$	$[O(3)]^3 \otimes [SU(2)]^3$	$\prod_j U(2j+1)$	$\prod_j Sp(2j+1)$	$SU(2)$

obvious from the group-theoretical point of view.) Due to the electron–electron interaction, levels with $J = 0$ pairs are pushed up.

The long-range part of the electron–electron interaction will then in lowest approximation split these levels according to their total angular momentum J, and in higher approximations also mix different configurations in such a way as to produce the same final levels as obtained by the LS scheme. The levels again will transform irreducibly under the fundamental group $G_5 = SU(2)$.

The fine structure of the levels before configuration mixing is determined by the total angular momentum J_j of the subshells. In our case we obtain singlets and doublets. Due to the configuration mixing, singlets and doublets may combine to triplets, or two singlets to doublets, as can be easily seen by comparing the tables.

If we were directly confronted with the energy spectrum of the B atom without the knowledge of its structure and the spectrum of the H atom, we would probably be quite puzzled with its interpretation. Perhaps we would start to understand this spectrum by studying, at first, the stationary states of a single "particle" in an appropriate potential. This will not go wrong. The lowest levels of the B atom are, indeed, essentially levels of a single particle with spin $\frac{1}{2}$, because the two filled ($1s$ and $2s$) shells will (practically) not participate. The next levels we could interpret in the same way, but to explain the fine structure we would have to postulate a spin 0 or 1 for the "particle." The occurrence of these abnormal multiplets here in our example are, of course, related to the omission of the very weak spin–spin interaction between electrons of different shells. As a consequence the atom can be divided into an active part consisting of the two $2p$ electrons and a core-atom which contains an inactive $2s$ electron. Due to this $2s$ electron the core-atom is a degenerate doublet. The appearance of the abnormal multiplets is directly related to this degeneracy.

After we have succeeded in constructing such a phenomenological one-particle description of the levels, we would emphasize their striking physical similarity and observe that all these levels may be conceived as different components of a 56-plet, which belongs to a very high symmetry group G_0. In addition to the exact quantum numbers J, M we would introduce new quantum numbers L, S_l, v_l, n_l, n to characterize these states. The symmetries connected with these quantum numbers would, however, be badly violated. Correspondingly, there would be no exact conservation laws for these quantities if we study B scattering processes. Whenever, for example, an electron flips its spin in this process (and this will happen, depending on the relative strength of the spin–orbit force), angular momentum will be shifted from the intrinsic to the orbital motion, and hence S and L conservation will be violated. They will change by

$\Delta S = \pm 1$, $\Delta L = \pm 1$, and, with much smaller probability, by two units, if we encounter a double spin flip.

4. Symmetries Broken by the Ground State

In the last section we discussed an example in atomic physics where high approximate symmetries occur as a consequence of the particular structure of the system. Many other examples can be cited of such symmetries which are not reflected in the underlying dynamics. Structural symmetries of the discussed type are actually the consequence of an approximate dynamical independence of certain subsystems within the total system. This independence allows us to approximately apply the symmetry operations of the dynamics on these subsystems alone. The violation of this symmetry is then brought about by the residual interaction of this subsystem with the rest system.

These subsystems may have a simple physical interpretation, such as the single electrons in the B atom. Some of its levels, in particular the lowest ones in our example, can be directly related to these one-particle excitations. Some subsystems, however, may also be of a rather abstract nature, for example, the "intrinsic spin system" characterized by the total intrinsic spin S, which is effectively decoupled from the "orbital system." It may be convenient to speak quite generally about the energy spectrum of "quasi-particles" with spin S and orbital angular momentum L, where only in special cases these quasi-particles will be nearly identical with physical particles.

In the above interpretation broken symmetries occur as a consequence of residual interactions of approximately independent subsystems (quasi-particles). The exact symmetries can always be detected by measurements on the whole system, which is the only natural system for a local system. They are reflected in the exact energy degeneracies of the levels and in the exact conservation laws in reactions. If our dynamical system, however, gets very large, e.g., represents a macroscopic object—a crystal, a liquid, a ferromagnet, etc.—the situation may be quite different. In this case it appears quite natural to make observations simply on a local system (in particular, if we, with our measuring equipment, are confined to a certain part of the large system), and suppress the reaction of the large body or replace it by certain average forces.

We may, for example, imagine a crystal lattice of Be atoms (with filled $1s$ and $2s$ shells), and add a single electron at some place in a $2p$ level to form a Be$^-$ ion, the spectrum of which should be similar to our B spectrum. We further assume that the overlap of the wavefunctions between neighboring Be atoms are big enough to allow this $2p$ electron to move through the lattice. (We will not be concerned at this point about the real physical situation.) The electron will then appear like a spin $\frac{1}{2}$ quasi-particle with energies related to a certain band of linear momenta. The lowest eigenstate of the crystal as a whole, of course, will be an exact eigenstate of energy and momentum (namely $\mathbf{P} = 0$) due to the translational invariance of the interaction, but not the eigenstates of the quasi-particles, for which the translational invariance is "broken" by the crystal lattice, which is only translational invariant under certain displacements in certain directions. The spin of the quasi-particles, however, will not feel the influence of the lattice, because it is constructed from spinless Be atoms in their ground state.

We may very well imagine also a situation where the lattice points in the crystal still have spin properties (e.g., if, in all the Be atoms of our lattice, one of the $2s$ electrons is excited to a $2p$ state such that the Be$^-$-ion state would now be similar to the $n_1 = 2$ level in the B atom). If the spin directions of these lattice points are not completely averaged out by zero-point fluctuations, a preferred ordering of the spins, as in a ferro- or antiferromagnet, may be established by the interactions of the lattice electrons, which leads to a polarization of the crystal in a certain direction. The levels of the additional electron will now split up on account of the residual interaction with this polarization of the rest system. Although the dynamics are rotational invariant, and hence the energy of the total system is independent of its orientation, the energy of the quasi-particles depends on their spin orientation, since the rotation symmetry is "broken" by the nonzero magnetization of the crystal.

Our two examples of "broken" symmetries from a dynamical point of view are actually of the same nature as our former structural symmetries. Their apparent difference from the latter arises solely through the physical interpretation: In one case we take the total system, in the other case the subsystem, as the natural physical object. If we wish to extract the fundamental symmetries of the dynamics, we should, however, only consider isolated objects, i.e., objects which are exactly independent from their "surroundings," i.e., the total system. The latter consideration may, however, be very inconvenient or even physically meaningless. For example, if we deal with a very large—let us say, even an infinitely large—ferromagnet as a ground state for our system, then the classification of states in terms of the total spin would be completely uninteresting, because the spin of all systems would be infinite and hence ill-defined. Consequently spin conservation would also be meaningless. To observe angular momentum conservation in this case, one has to find some way to show that in a local reaction a change of angular momentum in a certain volume element is always accompanied by a transport of angular momentum through the surface of the volume element. It would, however, be quite difficult to observe such a leakage directly, if it is not carried by a particle.

There is, however, an important property of such an unsymmetrical ground state. According to the Goldstone theorem,[6] and in the absence of long-range interactions, there will occur low energy excitations of the ground state, or "zerons," which, in the relativistic case, correspond to zero mass particles. To demonstrate this,[7] let us consider the case of a finite crystal. The exact ground state of the crystal $|G\rangle$ would be an exact eigenstate of the total linear momentum $\mathbf{P} = 0$, and with a certain energy E_G. Due to the uncertainty principle, however, the center of mass of the system would be completely uncertain. In fact, this ground state corresponds physically to a coherent superposition of an infinite number of crystals with all possible positions of their center of mass. Because of this position smearing, the local structure of the crystal will be invisible in the simplest matrix elements, e.g. the mean density $\langle G| \rho(x) |G\rangle$ of the lattice atoms (it will be a constant), and would only show up in higher point functions, e.g., in the density correlations $\langle G| \rho(x)\rho(x') |G\rangle$ between two locations x and x'. This formal ground state, however, is not very physical. It does not correspond to a crystal of our macroscopic world, which is characterized by the expectation value of its center of mass being localized around some definite value, e.g., the origin. For this physical ground state $|0\rangle$ the mean density $\langle 0| \rho(x) |0\rangle$ will show the periodic behavior of the lattice (it violates translational invariance). Such a ground state, however, can only be brought about by a superposition of states with different momentum $\mathbf{P}$, which may be chosen such that at least the expectation value of the momentum operator $\langle 0| \mathbf{P} |0\rangle$ vanishes. This state then will have an energy

$$E_0 = E_G + \frac{1}{2M} \langle 0| \mathbf{P}^2 |0\rangle$$

(with M the total mass of the crystal) which is slightly higher than the formal ground state energy. Only in the limit of an infinite crystal, where also $M \to \infty$, will these states be energetically degenerate with the formal ground state. In this limit we may even exactly localize the center of mass, which should be interpreted as a localization of the crystal lattice. To obtain this ground state formally, one way is to calculate the exact ground state in the presence of a weak lattice periodic potential, which favors the particular lattice location with respect to all the others, and then go to the limit of vanishing potential.

Due to the translational invariance of the dynamics, it is obvious that this ground state is infinitely degenerate, because there is an infinite number of crystals with different locations of the center of mass. The transition between these different degenerate states, i.e., a certain rigid translation of the crystal, however, is only a possible operation for a finite crystal. For an infinite crystal an infinite number of atoms have to be moved. As a consequence, matrix elements of a finite product of operators between the original and the translated state vanish. (Mathematically speaking the two states are in different Hilbert spaces which belong to different inequivalent representations of the operator algebra.) Therefore the degeneracy is only formally present, because, after a certain ground state is chosen, we never can reach any of the other possible ones. An operation in which only a finite number of lattice points are translated is, however, a possible operation. If we make this translation sufficiently smooth over a long range—let us say, periodically with a wavelength λ— this will not cost much energy, and, in fact, with $\lambda \to \infty$ should simply go over in the rigid translation, for which no energy is necessary. Hence, due to the (formal) degeneracy of the ground state we obtain (physical) long-wavelength excitation modes, the energy of which vanishes for $\lambda \to \infty$; they correspond to the phonons (actually, a very particular type) in the crystal. They are, so to say, the localized "step operators" which connect degenerate states. The appearance of such "zerons" can be shown to be the general consequence of any degenerate, or better asymmetrical, ground state if, in particular, long-range forces are excluded. This is known as the Goldstone theorem.[6]

Perhaps it should be pointed out that the ground state of the crystal is not merely fixed by dynamics but also depends on its preparation. It is not only possible to locate the crystal in different fixed ways, but we may also use a coherent or incoherent superposition of many such states, in particular our formal ground state $|G\rangle$. These states, however, can only be constructed in a reducible Hilbert space, which is an infinite direct sum of Hilbert spaces constructed from ground states with different fixed locations. However, there is nothing gained theoretically by using such complicated representations.

The ferromagnet is another lucid example for an asymmetrical ground state.[6-8] The physical ground state is characterized by a macroscopic polarization, e.g., in the z direction. If N is the number of spin $\frac{1}{2}$ atoms in the lattice, this state would have spin $J = N/2$ with magnetization $M = J = N/2$, and hence shall be denoted by $|JJ\rangle$. Ground states oriented in any other direction will belong to the same energy, due to the rotational invariance of the dynamics. They are superpositions of states $|JM\rangle$ with $-J \leq M \leq J$. The ferromagnet is therefore degenerate with respect to spin orientation. In particular, the state $|JJ - 1\rangle$ will have the same energy, which can be generated from $|JJ\rangle$ by the "step operator" or "spin-flip operator"

$$J_- = \tfrac{1}{2} \sum_n \sigma_-^{(n)}$$

Similarly, the step operator

$$J_+ = \tfrac{1}{2} \sum_n \sigma_+^{(n)}$$

will restore the original state.

For an infinite ferromagnet, the total spin and the magnetization will both be infinite. An application of a finite number of step operators will change the magnetization only by a finite amount, and hence, in the usual description, these ground states are identified. A ground state which differs from the original one by a finite, although arbitrarily small, angle in orientation can only be obtained by application of an infinite number of spin-flip operators, which again leads out of the Hilbert space. The localized spin-flip operators however, are well defined and produce observable effects. They are the Bloch spin waves, the magnons, and represent the Goldstone zerons in this case.

If we add an extra electron of spin $\tfrac{1}{2}$ to this system, its energy level will be split, due to the effective interaction of the electron with the lattice polarization vector which favors parallel orientation. Hence the level with total spin $J + \tfrac{1}{2}$ will be somewhat lower than the level with $J - \tfrac{1}{2}$. If J becomes very large we may disregard the infinite spin of the ground state and label these levels simply as an electron level with spin $j = \tfrac{1}{2}$ and $m = \pm\tfrac{1}{2}$. We may say that the energy degeneracy of the quasi-particle levels is removed as a consequence of the violation of rotational symmetry by the ground state.

Actually the situation is more complicated, because the original ferromagnet also has excited states in which the alignment of spins is incomplete. These levels belong to states with lower total spin, i.e., $J - 1, J - 2$, etc., and are higher in energy by an amount which depends on the spin interaction of the lattice electrons. The additional electron will also lead to split levels, e.g., in case of the $J - 1$ level, a lower $J - \tfrac{1}{2}$ level and a higher $J - \tfrac{3}{2}$ level. We therefore observe that the total spin is now insufficient to completely characterize our levels. There are two different $J - \tfrac{1}{2}$ levels which can only be distinguished by interior spin quantum numbers. In one case essentially the spin of the extra electron is flipped, in the other case the spin of a lattice electron. With increasing strength of the electron–lattice spin interaction, in comparison with the lattice–lattice interaction, these two levels will get mixed (see Fig. 1), and —as Biritz,[7] for instance, has explicitly shown for a one-dimensional model—the two doublets will eventually arrange into a singlet and a triplet, i.e., into an abnormal multiplet, similar to the ones we obtained for the $n_1 = 2$ levels of the B atom, and for a similar reason. These anomalous states, so to say, "branch off" a spin $\frac{1}{2}$ of the ground state; the electrons moving through the lattice "ride on a spin wave of spin $\frac{1}{2}$." The situation is complicated further by the appearance of the magnons which lead to a continuum of states. Some of the states, like the triplet in the Biritz model, become unstable under magnon emission, and all will be "dressed" by them. We will not go further into this.

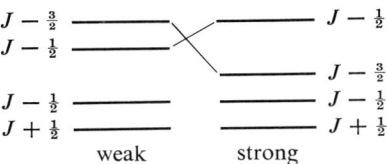

electron–lattice spin–spin interaction

Fig. 1. Electron level splitting.

We have demonstrated in this section that an exact symmetry of the dynamics may be masked if the ground state of a large, or even infinite, system is not invariant under the symmetry group. In the absence of long-range interactions, this asymmetry of the ground state leads (in the infinite case) to gapless excitations (Goldstone zerons) of the system.

5. Interpretation of the Broken Symmetries in Elementary Particle Physics

Our discussion of the symmetries exhibited by the energy spectrum of atoms and of systems involving many atoms has revealed some striking similarities with the elementary particle spectrum, and one may therefore ask whether a similar mechanism may be applicable for its interpretation. In Section 1 we have already argued that, despite many important differences between relativistic and nonrelativistic local quantum mechanics, the actual situation may be, after all, quite similar. A direct comparison of atomic and elementary particle physics, however, seems to be difficult, because the atom possesses a dominant force center which also plays an important role in the approximation scheme, whereas nothing of this kind is apparent in elementary particle dynamics.

Nuclear physics characterizes an intermediate situation between atomic and elementary particle physics, in the sense that there exist strong interactions among many, but a finite number of, particles without a dominating force center. The remarkable success of the nuclear shell model has shown that even in such systems an independent particle (or better, quasi-particle) description is meaningful to some extent. This surprising fact is probably essentially caused by the Fermi statistics of the interacting particles. If the interaction between fermions does not become too singular, they try to stay apart, due to Pauli's exclusion principle, and the interaction of a single fermion for small energies is effectively limited by the presence of the others.[10] However, to a larger degree than in the atom, the quasi-particles are different from the physical particles, the nucleons, because a large proportion of their pair interaction is subtracted and incorporated into the average shell model potential, which in general will be momentum dependent. Hence, by the choice of an appropriate average potential and the corresponding ground-state configuration, the extremely nonlinear nuclear dynamics can be approximately linearized in the sense that the remaining part of the pair interaction can be treated by some perturbation procedure. In this many-quasi-particle description the aspect of approximate higher symmetries, as they occurred in the atoms in the many-particle description, is again of importance. The introduction of seniority, reduced isotopic spin, and other quantum numbers was quite successful in nuclear physics.[5]

We therefore may conjecture that in elementary particle physics a similar description will be possible. There the vacuum must play the same role as the nuclear ground state in nuclear physics, the core-atom in atomic physics, and the lattice in solid-state physics. Elementary particles then would correspond to quasi-particles or collective excitations of this vacuum. Because of the translational invariance of the vacuum, an infinitely extended nucleus (nuclear matter) or an infinite solid could actually serve as the closest analog for the elementary particle situation. Hence, we see that the broken symmetries in elementary particle physics may either be nonfundamental, structural symmetries, or fundamental symmetries broken by the ground state, the vacuum. In a relativistic theory, due to the additional causality requirement, the Goldstone theorem predicts in the latter case as "zerons" certain mass-zero particles, at least if no long-range interactions are present (which are always related to gauge fields).

One of the most important approximate symmetries in elementary particle physics is the isospin symmetry, which characterizes strong interactions and, in the phenomenological description, is broken essentially by the electromagnetic interactions (and, to much less an extent, by weak interactions). On the other hand, the connection of this violation with mass-zero particles, the photon, or a long-range interaction, the Coulomb interaction, seems to strongly suggest an interpretation which postulates an asymmetry of the vacuum with respect to the isospin group, perhaps in analogy to a ferromagnet.[1] The investigation of this conjecture has met quite a number of difficulties, because it becomes quite clear from the Goldstone theorem that the photon, because of its spin, cannot be directly connected with the "zerons." The zerons have only spin properties if the Lorentz group is broken, which certainly is not what one wants. In addition, photons also seem to have the wrong isospin properties, because the zerons in an isoferromagnet are localized isospin-flip operators, and hence transform like τ_+ or τ_-, if τ designates the isospin operators. Objects of this type are positively or negatively charged and hence differ from the neutral photon, which has a $(1 + \tau_3)$ property.

However, a closer study reveals that the situation is probably more complicated. One finds that the isoferromagnet is not the correct analog model for elementary particle physics, because it not only violates the isospin group but also CPT, which should be strictly valid on account of the exact particle–antiparticle symmetry in nature. We can see that immediately, if we imagine a π isotriplet in an isoferromagnetic vacuum. In first approximation it will split up linearly into three separate states, and hence will produce different masses for the antiparticles π^+, π^-. A better analog for the vacuum would be a kind of double ferromagnet, constructed from a particle ferromagnet polarized in one direction and a coinciding antiparticle ferromagnet oriented in the opposite direction. Such a system would transform into itself under CPT. A system of this type can be obtained by assuming isospin–isospin forces between the lattice particles, which favor energetically an alignment (triplet structure) between equal particles but an antialignment (singlet structure) between unequal particles. This situation would be rather similar to an antiferromagnet. The exact ground state of such a system is very complicated, because it is a compromise of two conflicting tendencies: the ferromagnetic ordering between equal particles and the formation of local singlets of unequal particles (i.e., a one-to-one mixture of an antialigned particle–antiparticle pair and its inverse). The symmetry is broken by the preferred orientation of the particle and antiparticle subsystems. The total isomagnetization of the ground state will be zero ($J_3 = M = 0$). However, it will be a certain mixture of states with different isospin. Unlike the ferromagnet which carries definite spin and hence transforms according to an irreducible representation of the rotation group, our particle–antiparticle "antiferromagnet" belongs to a reducible representation of the isospin group. It is rather plausible that the Goldstone zerons in this case should be neutral (flipping of particle–antiparticle pairs) and that they are "flippers" between different isospin states, i.e., different irreducible representations. The operator $(1 + \tau_3)$, indeed, would be neutral and also mix states which differ by an isospin one. It is a projection operator which senses wrongly oriented particles in the lattice. These questions are being investigated by Biritz and Yamazaki.

With respect to the problem of spin of the photon, a solution may be sought in the following direction. It seems rather conclusive that the spin of the Goldstone zerons should be zero if only isospin is broken, but there appears, at present, no convincing reason why the zerons should not be connectable with the scalar part of gauge fields, which for a vector field in the Coulomb gauge, for example, do produce a physical effect, namely the long-range Coulomb interaction. One has the strong impression that it should always be possible to transfer the "zeron property" of spinless particles to long-range forces of gauge fields of arbitrary spin but equal internal quantum numbers. The corresponding transverse modes, like the photon, would then follow only as a second step by locality and Lorentz invariance. The mathematical tool to obtain such a generalization is either to work in a theory with indefinite metric (e.g., Gupta-Bleuler representation), or to give up the requirement of manifest Lorentz invariance. If one restricts oneself to a theory with positive definite metric and manifest Lorentz invariance, the theory can only produce as "zerons" the only possible "longitudinal" object of mass zero, which transforms irreducibly under the Poincaré group, namely the spin-zero particle. Investigations by Higgs[11] and Guralnik, Hagen, and Kibble,[12] who explicitly introduced gauge fields of mass zero coupled to the current which is connected with the broken symmetry, led to the interesting conclusion that, indeed, the original spin-zero Goldstone zeron can be avoided, but unfortunately, however, that by eliminating these the mass-zero gauge particles are transformed into massive particles and the original long-range forces disappear completely from the scene. The mass-zero character of the zerons has slipped into mass-zero gauge particles, which decouple completely from physics because of the gauge invariance of all observables. If this conclusion were general and final, it would indeed be fatal for our conjecture. However one observes that in these models with gauge fields, one fails to produce a coupled mass-zero particle because the broken symmetry transformations lost their meaning in the subspace of the Hilbert space which contains all the physical states. Therefore the symmetry is only formally and not physically broken, and one should not expect this to produce anything of physical relevance. The situation seems, however, to be different in cases where a non-Abelian group is only partially broken, as is the case with the isospin group. There is, therefore, still great hope that the photon, or rather the Coulomb interaction, will appear in connection with an isospin asymmetrical vacuum of a dynamical theory which contains the isospin group as an exact symmetry. A lot still has to be done in this respect.

The reason the vacuum should have such an asymmetrical structure may have a dynamical origin, as in a finite ferro- or antiferromagnet, but the particular property may also be one out of several possible dynamically consistent boundary conditions which characterize the global structure of the actual world. The isospin asymmetry, for example, may reflect the dominance of neutrons over protons in the universe (one probably then also has to assume a similar density of antimatter with dominating antineutrons), and hence may be only explained by cosmology. The principle of Mach, if it is at all valid, would similarly express such a symmetry violation (here probably scale invariance) from "outside."

If we assume that the isospin asymmetry of the vacuum can be established, it should then, in principle, be possible to observe normal and abnormal quasi- particle multiplets. Perhaps one may also express this by stating that the "dressed" particles will not necessarily have the same intrinsic properties as the "bare" particles, if the vacuum has properties. A spinor field, e.g., with spin $\frac{1}{2}$ and isospin $\frac{1}{2}$, as it is introduced in Heisenberg's nonlinear spinor theory, hence cannot only create particles with integer spin and isospin (even products of ψ) or half-integer spin and isospin (odd products of ψ) from the vacuum, but also abnormal states with integer spin, half-integer isospin (even products of ψ) or vice versa (odd products of ψ) by "branching off" an isospin $\frac{1}{2}$ from the ground state, in the sense in which it was discussed above. The abnormal modes could then be identified with the strange particles.[13,14] Let us visualize this in a model.

We take the very idealized model of the double isoferromagnet of lattice particles L of isospin $\frac{1}{2}$ (charge 0, -1) and antiparticles of isospin $\frac{1}{2}$ (charge $+1, 0$) which is realized by strong aligning forces between equal particles and very weak antialigning forces between unequal particles. The isospin of all L particles shall be up (L^0) and of all $\bar{L}$ particles down ($\bar{L}^0$), so that the lattice is neutral. We bring into this lattice a nucleon which, as a particle, shall interact weakly with only the L lattice. As a consequence the nucleon level splits up favoring energetically the up orientation (p) in comparison with the down orientation (n). We will assume, now, that there exists another mode with somewhat higher energy, where the nucleon interaction with the lattice is relatively stronger than the lattice interactions. In this case we obtain a strange isospin triplet (Σ^+, Σ^0, Σ^-) and a singlet (Λ) as indicated earlier (Fig. 1). With a more realistic ground state, where also the antialigning particle–antiparticle forces are fully considered, all strange particle modes will involve important participation of the lattice. If a normal mode transforms into an abnormal mode, the disturbance in the lattice must be compensated. Since isospin is locally conserved, this can only be guaranteed by a pair creation of strange particles (with opposite strangeness). In our picture this would mean that any local change in the lattice participation will be accompanied by a similar change in the antilattice, in order to leave the vacuum the same. For a simple formulation of this somewhat complicated situation the concept of a "spurion" and "antispurion"—i.e., objects of isospin $\frac{1}{2}$, hypercharge ± 1, but vanishing energy, momentum, and spin—was introduced. Strange particles are "bound states" containing such spurions. In strong interactions the spurion number is conserved.

With strange particles being interpreted as the abnormal multiplets of an isospin asymmetrical vacuum, there may be no need to fundamentally introduce higher symmetry groups such as $SU(3)$. The absence of zero-mass particles or observable long-range forces in connection with the strong violation of $SU(3)$ appears to us a strong indication (which is questioned by many people on the basis of the investigations of Higgs and others) that this symmetry should be interpreted as only a structural symmetry. In fact, it was shown that the four generators of an exact internal symmetry $U(2)$ (hypercharge and isospin) together with the four spurion operators (S^0, S^-, $\bar{S}^0$, $\bar{S}^+$) may imitate, in some sense, the eight generators of $SU(3)$.[14] However, this has still to be investigated much more thoroughly before any conclusion can be drawn.

We wish to close with only a few remarks concerning the weak and PC-violating interactions. A detailed study of these with respect to our conjecture of the vacuum has not been undertaken as yet.

Weak interactions also violate isospin symmetry, but also the third component of isospin, strangeness, and parity. If our conjecture is correct, then the isospin violation in weak interactions (the appearance of only charged currents) should somehow be related to the electromagnetic interaction. We will not attempt here to make proposals for such a relation. Certainly, a solution of this problem will necessitate a better understanding of the leptons and their symmetry properties.[15]

The violation of strangeness and the third component of isospin, however, has the appearance of an unobservable emission of a spurion. In fact, the concept of a spurion was originally introduced by Wentzel in this context. To understand these processes, therefore, one must find some interpretation for such an emission process in our vacuum picture. Intuitively, it means that hypercharge-isospin properties are transferred to the vacuum as a whole (because the spurion is completely spread out over all space). This is difficult to understand because of the causality requirement. Perhaps the Mössbauer effect in a crystal can serve here as an analogy, because in this case the recoil momentum in a local nuclear γ emission appears also to be instantaneously transferred to the crystal, and not merely with sound velocity, as one would expect. Weisskopf[16] has shown that, strangely enough, there is no observable violation of causality due to the large zero-point fluctuations which a local measurement of this propagation encounters. It should also be remarked that there are two zero-charge spurions in our description, S^0 and $\bar{S}^0$, or the C-symmetric and anti-symmetric combinations

$$S_+ = \frac{1}{\sqrt{2}}(S^0 + \bar{S}^0) \quad \text{and} \quad S_- = \frac{1}{\sqrt{2}}(S^0 - \bar{S}^0)$$

Because they are assumed to be unaffected by space properties, they can also be considered even or odd under PC. In weak interactions only S_+ should be emitted, to conserve PC in the observable remaining system. However, there is no obvious reason why we should not have processes in which S_- is emitted. These processes would violate PC. Perhaps this could render an explanation for a weak PC-violating interaction. In our description they then should only occur in strangeness-changing (spurion-emitting) processes. However, we have now entered the field of wild speculation and hence should stop here.

To summarize, we wish to emphasize that the goal of this paper was mainly to remind us of the natural complexity of spectra of many-particle systems and point out their apparent similarity to the mass spectrum of elementary particles. If this apparent similarity can be interpreted as a fundamental analogy, we may, in principle, have a key to an understanding of the complexity of elementary particle physics, in particular regarding their hierarchy of interactions. We do not claim that our investigations do prove very much in this respect, but they may be considered as an encouraging starting point in the exploitation of this exciting possibility. The comparison with atomic physics, however, may also indicate that the computational effort to obtain reasonable accuracy in numerical predictions can be quite formidable.

References

1. W. Heisenberg, *Rev. Mod. Phys.*, **29**, 269 (1957); *Proc. Ann. Intern. Conf. High Energy Phys. CERN*, 1958, 119. H. P. Dürr, W. Heisenberg, H. Mitter, S. Schlieder, and K. Yamazaki, *Z. Naturforsch.*, **14a**, 441 (1959). W. Heisenberg, *An Introduction to the Unified Theory of Elementary Particles*, Wiley, London–New York, 1967; German edition, S. Hirzel Verlag, Stuttgart, 1967.
2. E. Noether, *Nach. Kgl. Ges. Wiss. Göttingen* 235 (1918). E. L. Hill, *Rev. Mod. Phys.*, **23**, 253 (1951).
3. G. Chikovani et al., *Phys. Letters*, **22**, 233 (1964).
4. W. Pauli, *Z. Physik*, **36**, 336 (1926). V. Fock, *ibid.*, **98**, 145 (1935).
5. E. P. Wigner, *Group Theory and Atomic Spectra*, Academic Press, New York, 1959; U. Fano and G. Racah, *Irreducible Tensor Sets*, Academic Press, New York, 1959; A. de-Shalit and I. Talmi, *Nuclear Shell Theory*, Academic Press, New York, 1963.

6. J. Goldstone, *Nuovo Cimento*, **19**, 154 (1961). J. Goldstone, A. Salam, and S. Weinberg, *Phys. Rev.*, **127**, 965 (1962). S. A. Bludman and A. Klein, *ibid.*, **131**, 2364 (1963).
7. H. Wagner, *Z. Physik*, **195**, 273 (1966).
8. W. Heisenberg, in *Proceedings of Seminar on Unified Theories of Elementary Particles*, Max-Planck-Institut für Physik und Astrophysik, München, July 1965, p. 409.
9. H. Biritz, *Nouvo Cimento*, **47**, 581 (1966).
10. L. C. Gomes, J. D. Walecka, V. F. Weisskopf, *Ann. Phys. (N.Y.)*, **3**, 241 (1958).
11. P. W. Higgs, *Phys. Letters*, **12**, 132 (1964); *Phys. Rev.*, **145**, 1156 (1966).
12. G. S. Guralnik, C. R. Hagen, and T. W. B. Kibble, *Phys. Rev. Letters*, **13**, 585 (1964); T. W. B. Kibble, *Proc. Intern. Conf. Elementary Particles*, 1965, p. 19.
13. H. P. Dürr and W. Heisenberg, *Z. Naturforsch.*, **16a**, 726 (1961); *Nuovo Cimento* **37**, 1446 (1965); *ibid.*, **37**, 1487 (1965). H. P. Dürr and J. Géhéniau, *ibid.*, **28**, 132 (1963).
14. D. Kastler, D. Robinson, and A. Swieca, *Commun. Math. Phys.*, **2**, 108 (1966); H. Ezawa and J. A. Swieca, DESY (Deutsches Elektronen Synchroton) preprint 66/38 (1966).
15. H. P. Dürr, W. Heisenberg, H. Yamamoto, and K. Yamazaki, *Nuovo Cimento*, **38**, 1220 (1965).
16. V. F. Weisskopf, *Lectures in Theoretical Physics*, Vol. III, University of Colorado, Boulder, 1960 (Interscience, New York, 1961), p. 79.

On the Theory of Near-Adiabatic Transitions*

KENNETH M. WATSON†

*Institute of Geophysics and Planetary Physics,
University of California, La Jolla, California*

I. Introduction

Edward Teller has had a long interest in, and has made many contributions to, the understanding of molecular and atomic physics. It is thus a considerable pleasure to have this modest contribution to an old problem included in the volume honoring his 60th birthday.

The problem addressed here is that of atomic and molecular reactions in the *near-adiabatic* limit. In the strictly adiabatic limit the relative velocity of the colliding particles is sufficiently slow compared with electronic orbital velocities that electronic transitions do not occur (except in very special circumstances involving degeneracies). In the *near-adiabatic* limit, the particle velocities are such that electronic transitions may occur during a collision, but with small probability. We can typically characterize the near-adiabatic limit by the condition

$$v \ll e^2/\hbar \tag{1.1}$$

where v is the relative velocity of the colliding atoms or molecules. [In explicit cases, more detailed conditions than (1.1) may be needed.]

Most discussions of atomic and molecular collisions use some form of semiclassical description of the heavy particle motion. For the validity of our analysis we require

$$v \gg (m/M)(e^2/\hbar) \tag{1.2}$$

where m is the electron mass and M is the reduced mass of the heavy particles. The inequalities (1.1) and (1.2) bound the energies to which our discussion can be applied.

We shall be concerned with transitions between pairs of discrete levels only of the colliding atoms or molecules. In the near-adiabatic limit

---

* This work was supported in part by a grant from the Office of Scientific Research, U.S. Air Force.

† On leave from the Department of Physics, University of California, Berkeley.

such transitions occur much more readily between pairs of levels which lie relatively close in energy—and, in particular, at those points on the trajectory at which the separation of levels is least. The classic treatment of this problem is that of Landau,[1] Zener,[2] and Stueckelberg.[3] In this discussion it is supposed that two atoms (they may be molecules, but for convenience we shall refer to them as *atoms*) A and B collide. The internal energy, $w_0(r)$, of these atoms as a function of internuclear separation r is illustrated in Fig. 1.* At large distances the atoms are in the state of

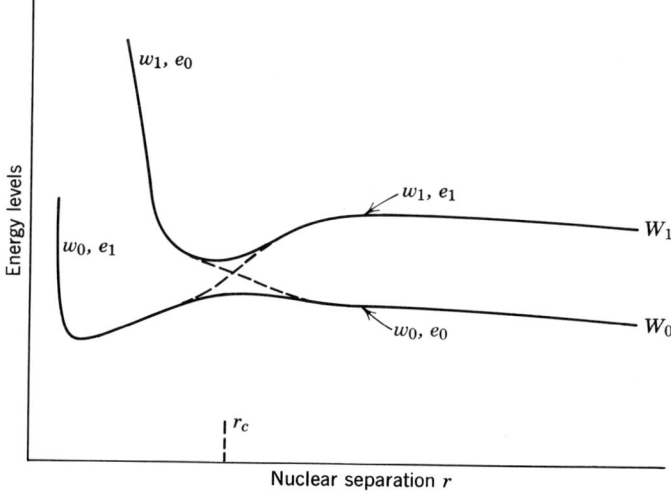

Fig. 1. Typical energy level diagram illustrating Landau-Zener model.

energy W_0, and a second level of energy W_1 lies above W_0. At some intermediate separation r_0 these levels cross, as indicated by the dotted lines in Fig. 1. The respective energies of these crossing lines are denoted as $e_0(r)$ and $e_1(r)$. It is supposed that the interaction between the atoms gives rise to a perturbation interaction $V_{01}(r)$, which modifies the energy level diagram to give the levels indicated by the solid lines in Fig. 1 and labeled as $w_0(r)$ and $w_1(r)$.

The calculation supposes that r is given as a function of t, the time, and the *time-dependent* Schrödinger equation is integrated through the "pseudo-crossing point" r_c. The probability that a transition from w_0 to

---

* For the case of molecules r must in general be supplemented by other parameters describing orientation, etc.

w_1 occurs on going through r_c is given by the celebrated Landau-Zener formula.*

$$P = \exp\left[-\frac{2\pi}{v}\frac{|V_{01}|^2}{\left|\frac{de_0}{dr} - \frac{de_1}{dr}\right|}\right] \qquad (1.3)$$

Here v is again the relative velocity and all quantities are evaluated at $r = r_c$.

When the colliding particles emerge, they again pass through the radial point $r = r_c$. The final probability that they are left in the state of energy w_1 is

$$\mathscr{P} = 2P(1 - P) \qquad (1.4)$$

Here the factor of 2 accounts for the two passages through r_c, and $(1 - P)$ corrects for the fact that two transitions would lead back to the initial state of energy w_0.

There has developed a very extensive literature addressed to rederiving, modifying, or using the above model.† Surprisingly, the published derivations appear almost universally to use the time-dependent Schrödinger equation, integrating along an orbit $r(t)$.‡ The purpose of the present discussion is to study the Landau-Zener and some related reactions using modern stationary state scattering theory.§ The advantages of such a formulation include great flexibility in treating a broad class of reactions and the possibility of a quantitative analysis, with real molecules, such as has not been done with the time-dependent approach. Indeed, we shall see that the Landau-Zener formula (1.3) may be derived in the first Born approximation.

II. Stationary State Formulation

We suppose that two atoms,‖ A and B, collide, their internuclear separation being called **r**. As a first example, let us suppose that only

---

* A more careful description of the conditions under which Eq. (1.3) is derived is given in Section VI.

† This literature is reviewed by Bates, reference 4, p. 608.

‡ A nice survey of time-dependent perturbation methods is given by Coulson and Zalewski.[5]

§ For a general description of modern techniques in scattering theory, see, for example, reference 6.

‖ The corresponding description for colliding molecules requires only an appropriate change in notation.

excitation, without rearrangement, occurs. The reaction is then of the form

$$A + B \rightarrow A^* + B \tag{2.3}$$

etc. The Hamiltonian for the internal states of A and B, when they are separated by a large distance r, is written as h. The eigenstates and eigenvectors of h are written as g_n and W_n, respectively†:

$$hg_n = W_n g_n \qquad n = 0, 1, \ldots \tag{2.2}$$

The interaction between A and B is written as V, which permits us to define the *adiabatic* Hamiltonian as

$$h_a \equiv h + V \tag{2.3}$$

This is assumed to have the respective eigenstates and eigenvalues φ_α and w_α:

$$h_a \varphi_\alpha = w_\alpha \varphi_\alpha, \qquad \alpha = 0, 1, \ldots \tag{2.4}$$

The quantities h and $g_0, g_1, \ldots$ depend on the electron coordinates $\xi_1 \cdots \xi_N$. The quantities $V, h_a, \varphi_0, \ldots$ depend on these and also *parametrically* on r. The energies $w_\alpha(\mathbf{r})$ are functions of $\mathbf{r}$.

Since V vanishes as $r \rightarrow \infty$, we may choose the quantum labels α and n to be identical. Then

$$\lim_{r \to \infty} \varphi_n = g_n$$

$$\lim_{r \to \infty} w_n = W_n \tag{2.5}$$

When electronic rearrangement occurs, so the reaction products C and D are not the same as A and B, reaction (2.1) is rewritten in the form‡

$$A + B \rightarrow C + D \tag{2.6}$$

In this case we again write h for the Hamiltonian of the two separated atoms A and B, assuming this to have the form

$$h = T + V_1 \tag{2.7}$$

with eigen-equations

$$hg_n = W_n g_n \tag{2.8}$$

as in (2.2). The interaction between A and B is written as V_2, so the adiabatic Hamiltonian is

$$h_a = H + V_2 \tag{2.9}$$

† We shall be concerned here with transition between discrete states g_n only.
‡ A thorough discussion of such reactions is given in Chapters 4 and 5 of reference 6.

As in (2.4) this has the eigen-equations

$$h_\alpha \varphi_\alpha = w_\alpha \varphi_\alpha \qquad (2.10)$$

Now, the Hamiltonian for the internal states of atoms C and D, when they are separated to large distances, is written as h'. We assume this to have the form

$$h' = T + V_2 \qquad (2.11)$$

and to have eigenfunctions g'_m and eigenvalues W'_m:

$$h'g'_m = W'_m g'_m \qquad (2.12)$$

Let us suppose that we are studying transitions among discrete states $g_{n1}, g_{n2}, \ldots$ and $g'_{m1}, g'_{m2}, \ldots$. Then we may assume that the levels α have been chosen to coincide with $n_1, n_2, \ldots$ and $m_1, m_2, \ldots$ so that as $r \to \infty$,

$$\varphi_{n1}, \varphi_{n2}, \ldots \to g_{n1}, g_{n2}, \ldots$$

$$\varphi_{m1}, \ldots \to g'_{m1}, \ldots$$

$$w_{n1}(\mathbf{r}), \ldots \to W_{n1}, \ldots$$

$$w_{m1}(\mathbf{r}), \ldots \to W'_{m1}, \ldots \qquad (2.13)$$

Finally, the total Hamiltonian H of the interacting atoms is

$$H = h_a + K \qquad (2.14)$$

where K is the kinetic energy operator for relative motion of the two atoms,

$$K = -\frac{1}{2M} \nabla^2 \qquad (2.15)$$

Here M is their reduced mass* and ∇ is the gradient operator acting on $\mathbf{r}$.

The scattering is supposed to proceed from some initial state χ_a, to some final state χ_b, both corresponding to large separations r of the atoms. The initial state χ_a is taken to be

$$\chi_a = \varphi_0 \lambda_\mathbf{p} \qquad (2.16a)$$

where

$$\lambda_\mathbf{p} = (2\pi)^{-3/2} e^{i\mathbf{p}\cdot\mathbf{r}} \qquad (2.16b)$$

describes a plane incident wave of relative momentum $\mathbf{p}$ and φ_0 describes an internal state g_0 with energy W_0 of the separated atoms. The final state

$$\chi_b = \varphi_1 \lambda_\mathbf{k} \qquad (2.17)$$

* We are here neglecting any change in M associated with transfer of electrons between the atoms.

describes a final relative momentum **k** and internal state corresponding to one of the conditions (2.5) or (2.13). The energy of the system is

$$E = \varepsilon_p + W_0 \tag{2.18a}$$

where

$$\varepsilon_p = p^2/2M \tag{2.18b}$$

The Lippmann-Schwinger scattering equation for the wavefunction ψ_a^+ is then

$$(E + i\eta - K - h_a)\psi_a^+ = i\eta\chi_a. \tag{2.19}$$

Here η is the *small* positive parameter of scattering theory which is set equal to zero at an *appropriate* step* in solving (2.19).

To simplify Eq. (2.19), we introduce the expansion

$$\psi_a^+ = \sum_\alpha \varphi_\alpha \Psi_{a\alpha}^+(\mathbf{r}) \tag{2.20}$$

and then form the scalar product with respect to a given φ_α. The resulting equation has the form

$$(E + i\eta - K - w_\alpha)\Psi_{a\alpha}^+ - \sum_\beta \Delta_{\alpha\beta}\Psi_{a\beta}^+ = i\eta\delta_{\alpha 0}\lambda_p \tag{2.21}$$

Here

$$\Delta_{\alpha\beta}(\mathbf{r}, \nabla) \equiv (\varphi_\alpha, [K, \varphi_\beta]) \tag{2.22}$$

where **r** is just a parameter when the scalar product is formed with respect to the electron coordinates

$$\xi_1 \cdots \xi_N$$

To further simplify the scattering equation, we write

$$w_\beta(\mathbf{r}) = W_\beta + \mathscr{V}_\beta(\mathbf{r}) \tag{2.23a}$$

where

$$W_\beta = \lim_{r \to \infty} w_\beta(\mathbf{r}) \tag{2.23b}$$

corresponds to a W_n or W'_m for the appropriate β. The scattering potential $\mathscr{V}_\beta(\mathbf{r})$ is assumed to vanish faster than $\mathcal{O}(r^{-1})$ as $r \to \infty$. This lets us rewrite (2.21) as the integral equation

$$\Psi_{a\alpha}^+ = \delta_{\alpha 0}\lambda_p + \frac{1}{E + i\eta - K - W_\alpha} \sum_\beta [\mathscr{V}_\alpha \delta_{\alpha\beta} + \Delta_{\alpha\beta}]\Psi_{a\beta}^+ \tag{2.24}$$

The scattering matrix T_{ba} is

$$T_{ba} = \sum_\beta (\lambda_\mathbf{k}, [\mathscr{V}_1 \delta_{1\beta} + \Delta_{1\beta}]\Psi_{a\beta}^+) \tag{2.25}$$

where $a = (0, \mathbf{p})$ and $b = (1, \mathbf{k})$, as mentioned earlier.

* See, for example, Chapters 3, 4, and 5 of reference 6.

An alternate form of Eq. (2.21) may be obtained using the *coherent states* $\Psi_{c\alpha p}^{\cdot +}$:

$$[E + i\eta - w_\alpha - K]\Psi_{c\alpha q}^{\cdot +} = i\eta \lambda_q \qquad (2.26a)$$

or

$$\Psi_{c\alpha q}^{\cdot +} = \lambda_q + \frac{1}{E + i\eta - W_\alpha - K} \mathscr{V}_\alpha \Psi_{c\alpha q}^{\cdot +} \qquad (2.26b)$$

where $\varepsilon_q + W_\alpha = E$. Using Eq. (2.26a) with $\mathbf{q} = \mathbf{p}$ to eliminate $i\eta\lambda_p$ from Eq. (2.21), we obtain the integral equation

$$\Psi_{a\alpha}^{+} = \delta_{a0}\Psi_{c\alpha p}^{\cdot +} + \frac{1}{E + i\eta - K - w_\alpha} \sum_\beta \Delta_{\alpha\beta} \Psi_{a\beta}^{+} \qquad (2.27)$$

The scattering matrix (2.25) can now be expressed as*

$$T_{ba} = (\Psi_{c l \mathbf{k}}^{\cdot -}, \mathscr{V}_0 \lambda_\mathbf{p})\delta_{10} + \sum_\beta (\Psi_{c l \mathbf{k}}^{\cdot -}, \Delta_{1\beta}\Psi_{a\beta}^{+}) \qquad (2.28)$$

For scattering with a change of state or rearrangement, for which $\delta_{10} = 0$, this reduces to

$$T_{ba} = \sum_\beta (\Psi_{c l \mathbf{k}}^{\cdot -}, \Delta_{1\beta}\Psi_{a\beta}^{+}) \qquad (2.29)$$

In the *Born approximation*, we may keep only the first term on the right in Eq. (2.27) to write this as

$$T_{ba} \cong (\Psi_{c l \mathbf{k}}^{\cdot -}, \Delta_{10}\Psi_{c 0 \mathbf{p}}^{\cdot +}) \qquad (2.30)$$

The differential cross section for scattering from state $a = (0, \mathbf{p})$ to $b = (1, \mathbf{k})$, with $\mathbf{k}$ lying in the solid angle interval $d\Omega$, is†

$$\frac{d\sigma}{d\Omega} = (2\pi)^4 M^2 \left(\frac{k}{p}\right) |T_{ba}|^2 \qquad (2.31)$$

When there are spin (or internal angular momentum) degeneracies, suitable sums and averages over there may be performed in Eq. (2.31).

The "potential" $\Delta_{\alpha\beta}$ of Eq. (2.22) represents the interaction responsible for scattering with change of state in Eq. (2.24) and (2.27). This may be written as

$$\Delta_{\alpha\beta} = -\frac{1}{2M}[(\varphi_\alpha, \nabla^2 \varphi_\beta) + 2(\varphi_\alpha, \nabla\varphi_\beta) \cdot \nabla] \qquad (2.32a)$$

Here ∇ is again the gradient operator acting on $\mathbf{r}$ and in $\nabla^2 \varphi_\beta$ and $\nabla\varphi_\beta$ the ∇ does not act on any other quantity than φ_β. For most applications

---

* This is an example of the famous "two-potential" problem. See, for example, Eq. (5.103) on p. 203 of reference 6.

† See, for example, Eq. (3.137), p. 90 of reference 6.

anticipated, the first terms in (2.32a) can be neglected compared with the second, so

$$\Delta_{\alpha\beta}(\mathbf{r}, \nabla) \simeq -(1/M)(\varphi_\alpha, \nabla \varphi_\beta) \cdot \nabla \qquad (2.32b)$$

The magnitude of this quantity may be estimated as follows. We take, for $r \approx a$ (the Bohr radius)

$$|(\varphi_\alpha, \nabla \varphi_\beta)| \simeq \mathcal{O}(a^{-1})$$

and

$$|\nabla/M| \simeq \mathcal{O}(v)$$

where v is the relative velocity of the colliding particles. Then in units of the Rydberg (Ry), we have

$$\frac{|\Delta_{\alpha\beta}|}{\mathrm{Ry}} = \mathcal{O}\left(\frac{v}{c}\frac{\hbar c}{e^2}\right) \qquad (2.33)$$

When $r \approx a$, we estimate

$$\mathscr{V}_\alpha = \mathcal{O}(\mathrm{Ry}) \qquad (2.34)$$

Thus, $\Delta_{\alpha\beta}$ may be considered to be a small perturbation in Eq. (2.24) when the relative velocity v is *small* compared with electronic orbital velocities ($\simeq e^2/\hbar$); that is,

$$v \ll e^2/\hbar \qquad (2.35)$$

which is the already assumed condition (1.1).

We note that for applications it may be convenient to include the diagonal elements $\Delta_{\alpha\alpha}$ of Δ in the definition of the potentials $\mathscr{V}_\alpha$. This then has the advantage that the $\Delta_{\alpha\beta}$ are associated with changes of internal state only, while the $\mathscr{V}_\alpha$ describe the *elastic* scattering. Since for cases of interest to us here, $|\Delta_{\alpha\alpha}|$ may be considered as small compared to $|\mathscr{V}_\alpha|$, simply dropping the terms $\Delta_{\alpha\alpha}$ from Eqs. (2.24) and (2.27) is probably justified.

III. The Eikonal Approximation

The eikonal approximation provides simple convenient expressions for the coherent wavefunctions $\Psi^{\prime\pm}_{c\alpha q}$ [Eqs. (2.26)]. Indeed, in this approximation the $\Psi^{\prime\pm}_{c\alpha q}$ have many of the pleasant properties of plane waves. As we shall see, the eikonal approximation can be used when condition (1.2) is satisfied.

We describe briefly here Weinberg's[7] presentation of the eikonal approximation,* which is applicable to a wide class of scatterings (including nonlocal and spin–orbit interactions). If we set $\eta = 0$ and drop labels

---

* See also, references 8 and 9, where further references to this approximation may be found.

on the wavefunction Ψ'_c, Eq. (2.26) is of the form

$$D(\mathbf{r}, -i\nabla)\Psi'_c = 0 \tag{3.1}$$

When we can drop the $\Delta_{\alpha\alpha}$ and assume no spin–orbit scattering, D has the form

$$D(\mathbf{r}, -i\nabla) = \nabla^2 + p^2 - 2M\mathscr{V}(r) \tag{3.2}$$

where $\mathscr{V}$ is a just function of r.

In this present study we shall assume no spin–orbit interactions, although these present no great complication with Weinberg's method. Then, the first eikonal approximation assumes Ψ'_c to have the form

$$\Psi'_c = (2\pi)^{-3/2} e^{iS(\mathbf{r})} \tag{3.3a}$$

where the *eikonal* S is expressed as a path integral:

$$S(\mathbf{r}) = \int^{\mathbf{r}} \varkappa(\mathbf{x}) \cdot d\mathbf{x} \tag{3.3b}$$

with

$$\varkappa(\mathbf{x}) = \nabla S(\mathbf{x}) \tag{3.3c}$$

The first approximation to (3.1) is obtained on setting

$$D(\mathbf{r}, -i\nabla)\Psi'_c \cong D(\mathbf{r}, \varkappa(\mathbf{r}))\Psi'_c = 0 \tag{3.4}$$

or*

$$D(\mathbf{r}, \varkappa(\mathbf{r})) = 0 \tag{3.5}$$

To define the vector $\varkappa$ one introduces a parameter τ and considers $\mathbf{r}$ and $\varkappa$ to be functions of τ:

$$\mathbf{r} = \mathbf{r}(\tau)$$
$$\varkappa = \varkappa(\tau)$$

This functional dependence is defined when the trajectory equations

$$\frac{d\mathbf{r}}{d\tau} = \frac{\partial D}{\partial \varkappa}$$
$$\frac{d\varkappa}{d\tau} = -\frac{\partial D}{\partial \mathbf{r}} \tag{3.6}$$

are combined with (3.5) and the condition that the trajectory be one of a set each of which passes through a prescribed surface parallel to the normal at that point. The integral (3.3b) is finally defined by specifying that the path of integration be one of the trajectories so defined.

* With spin interactions, (3.5) is replaced by det $D = 0$ in Weinberg's theory.

To construct the wavefunction $\Psi_{cq}{}^+(\mathbf{r})$, corresponding to an initial momentum $\mathbf{p}$, we refer to Fig. 2. The proper boundary condition on this is that in region I, corresponding to the incident, unscattered wave,

$$\Psi'_{cp}{}^+(\mathbf{r}) = (2\pi)^{-3/2} e^{i\mathbf{p}\mathbf{r}} \tag{3.7}$$

This determines the lower limit of integration in Eq. (3.3b). For a given impact parameter $\boldsymbol{\rho}$, the eikonal $S_p{}^+(\mathbf{r})$ is then defined along a trajectory as determined by Eqs. (3.6). The "plus" solution has the form

$$\Psi'^+_{c\alpha p}(\mathbf{r}) = (2\pi)^{-3/2} \exp\left[iS_{p\alpha}{}^+(\mathbf{r})\right] \tag{3.8}$$

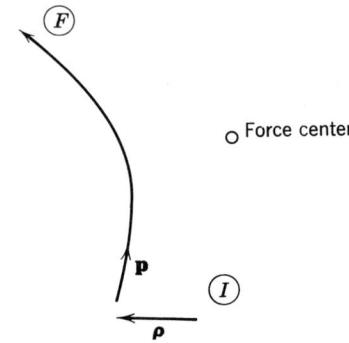

Fig. 2. Illustration of the wavefunction $\Psi_{cp}{}^+$ for an initial momentum $\mathbf{p}$.

The "minus" solution may be easily obtained from the relation (the time reversal operator T is just the operation of complex conjugation when there are no spin interactions).

$$\Psi'^-_{c\beta k}(\mathbf{r}) = T[\Psi^+_{c\beta(-\mathbf{k})}(\mathbf{r})] = (\Psi^+_{c\beta(-\mathbf{k})}(\mathbf{r}))^* \tag{3.9}$$

The wavefunction $\psi^+_{c\beta(-\mathbf{k})}$ is represented in Fig. 3. In the prescattering region I it has the asymptotic form

$$\Psi'^+_{c\beta(-\mathbf{k})}(\mathbf{r}) = (2\pi)^{-3/2} \exp\left[i(-\mathbf{k}) \cdot \mathbf{r}\right] \tag{3.10}$$

The eikonal for this wavefunction is

$$S^+_{\beta(-\mathbf{k})}(\mathbf{r}) = \int^{\mathbf{r}} \boldsymbol{\varkappa}^-(\mathbf{x}) \cdot d\mathbf{x} \tag{3.11}$$

where $\boldsymbol{\varkappa}^-(\mathbf{x})$ is the "momentum" at a point $\mathbf{x}$ on the trajectory of Fig. 3.

The time reversed trajectory has $\boldsymbol{\varkappa}(\mathbf{x}) = -\boldsymbol{\varkappa}^-(\mathbf{x})$ at every point on the trajectory, as is illustrated in Fig. 4. On introducing this into Eq. (3.9), we obtain

$$\Psi'^-_{c\beta k}(\mathbf{r}) = (2\pi)^{-3/2} \exp\left[-iS^+_{\beta(-\mathbf{k})}\right] = (2\pi)^{-3/2} \exp\left[iS^-_{\beta k}\right] \tag{3.12a}$$

where
$$S_{\beta k}^-(\mathbf{r}) = \int^r \varkappa(\mathbf{x}) \cdot d\mathbf{x} \qquad (3.12b)$$

The asymptotic form of this in region F is
$$\Psi^*_{c\beta k}(\mathbf{r}) = (2\pi)^{-3/2} e^{i\mathbf{k}\cdot\mathbf{r}} \qquad (3.13)$$

This corresponds to the waves which have *emerged* from the scattering.

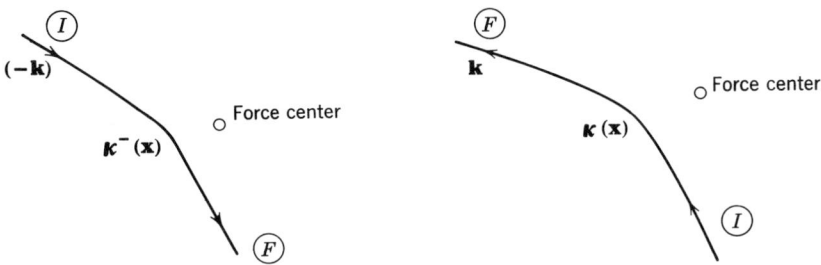

Fig. 3. Illustration of the wavefunction $\Psi^*_{c_p(-\mathbf{k})}$.

Fig. 4. Illustration of the wavefunction $\Psi^*_{c\mathbf{k}}{}^-$.

The second approximation to the wavefunction (3.3a) follows from writing
$$\Psi^*_{cp}(\mathbf{r}) = (2\pi)^{-3/2} N(\mathbf{r}) e^{iSp(\mathbf{r})} \qquad (3.14)$$

where $N(\mathbf{r})$ is a slowly varying function of $\mathbf{r}$. This function can usually be estimated fairly easily. Indeed, it often may be set equal to unity, or somewhat better, written as
$$N(\mathbf{r}) \approx (p/\varkappa(r))^{1/2} \qquad (3.15)$$

The condition for the validity of the eikonal approximation (which qualitatively speaking is the condition that $N(\mathbf{r})$ not be of very great importance in Eq. (3.14)) is that
$$\varkappa(r) \gg \left| \frac{d}{dr} \ln \mathscr{V}(r) \right| \qquad (3.16)$$

If we take
$$\left| \frac{d}{dr} \ln \mathscr{V} \right|^{-1} = \mathcal{O}(a)$$

where a is the Bohr radius, condition (3.16) becomes
$$v \gg \frac{m}{M} \times \frac{e^2}{h} \qquad (3.17)$$

which is just (1.2). As remarked in Section I, conditions (1.1) and (1.2) place upper and lower bounds on the energies to which our discussion is applicable.

IV. Scattering in the Born Approximation

We have seen from Eq. (2.33) that when our condition (1.1) is satisfied the interaction $\Delta_{\alpha\beta}$ in Eq. (2.27) is weak. This suggests that the Born approximation (2.30) for T_{ba} will be adequate.* Use of the eikonal wave functions in the form (3.14) then lets us write

$$T_{ba} = (2\pi)^{-3} \int N_b(\mathbf{r}) N_a(\mathbf{r}) \Delta_{10}(\mathbf{r}) \exp\left[i(S_a^+ - S_b^-)\right] d^3r \quad (4.1)$$

Here we have written a and b for the index pairs $(0, \mathbf{p})$ and $(1, \mathbf{k})$, respectively. We have

$$S_a^+ = \int^\mathbf{r} \boldsymbol{\varkappa}_a(\mathbf{x}) \cdot d\mathbf{x}$$

$$S_b^- = \int^\mathbf{r} \boldsymbol{\varkappa}_b(\mathbf{x}) \cdot d\mathbf{x} \quad (4.2)$$

as defined in Section III, and according to Eq. (2.32b),

$$\Delta_{10}(\mathbf{r}) \cong -i \frac{\boldsymbol{\varkappa}_a(\mathbf{r})}{M} \cdot (\varphi_1, \nabla \varphi_0) \quad (4.3)$$

We can simplify the form of Eq. (4.1) by defining

$$\varphi(\mathbf{r}) \equiv S_a^+(\mathbf{r}) - S_b^-(\mathbf{r})$$
$$\Delta(\mathbf{r}) \equiv N_b(\mathbf{r}) N_a(\mathbf{r}) \Delta_{10}(\mathbf{r}) \quad (4.4)$$

Then

$$T_{ba} = (2\pi)^{-3} \int d^3r\, \Delta(\mathbf{r}) e^{i\varphi(\mathbf{r})} \quad (4.5)$$

The point of this writing is that *rapidly* varying quantities are contained in $\varphi(\mathbf{r})$ [varying our distances like p^{-1}] and *slowly* varying quantities are contained in $\Delta(\mathbf{r})$ [varying our distances like a, the Bohr radius].

We shall make no attempt here to discuss Eq. (4.5) in a general fashion. We shall restrict ourselves to the case that there are one or more surfaces on which the moments $\boldsymbol{\varkappa}_a$ and $\boldsymbol{\varkappa}_b$ become nearly equal in magnitude. Associated with such surfaces there are expected to be stationary

---

* We shall see that in the near-adiabatic limit the matrix elements of Δ are even smaller than given by our estimate (2.33).

phase points for $\varphi(\mathbf{r})$. We shall assume that the principal contribution to T_{ba} comes from the neighborhoods of these points. If we let $\mathbf{r}_1, \mathbf{r}_2, \ldots, \mathbf{r}_L$ be convenient reference points (at or near the stationary phase points), we have

$$T_{ba} = \sum_{l=1}^{L} (2\pi)^{-3} \int_l d^3r\, \Delta(\mathbf{r}) e^{i\varphi(\mathbf{r})} \qquad (4.6)$$

where $\int_l d^3r$ represents an integral over a neighborhood of the point $\mathbf{r}_l$.

Let us consider the integral (4.6) near a particular point $\mathbf{r}_l$:

$$I_l \equiv \int_l d^3r\, \Delta(\mathbf{r}) e^{i\varphi(\mathbf{r})} \qquad (4.7)$$

The trajectories for the initial and final states are illustrated in Fig. 5. They intersect at the point $\mathbf{r}$, their tangents forming an angle ω. That

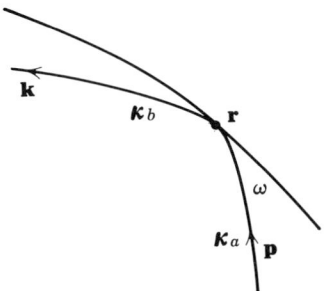

Fig. 5. Illustration of trajectories for Eq. (4.6).

surface $S_a{}^+(\mathbf{r}) = $ constant which passes through $\mathbf{r}_l$ will be chosen as a reference surface. We let s measure distance from this surface along a given trajectory. Locally rectangular coordinates λ_1, λ_2 are now introduced with $\mathbf{r}_l + \hat{\mathbf{n}}s$ (here $\hat{\mathbf{n}}$ is the local normal to the surface) as origin and lying in the surface $S(\mathbf{r}_l + \hat{\mathbf{n}}s) = $ constant.

Now, for small enough $\boldsymbol{\lambda}$ we have

$$\varphi(\mathbf{r}_l + \hat{\mathbf{n}}s + \boldsymbol{\lambda}) \simeq \varphi(\mathbf{r}_l + \hat{\mathbf{n}}s) + (\boldsymbol{\varkappa}_a - \boldsymbol{\varkappa}_b) \cdot \boldsymbol{\lambda} \equiv \varphi_l(s) - \boldsymbol{\varkappa}_b \cdot \boldsymbol{\lambda} \qquad (4.8)$$

where

$$\varphi_l(s) \approx \varphi(\mathbf{r}_l + \hat{\mathbf{n}}s) \qquad (4.9)$$

and we have used the condition that $\hat{\mathbf{n}} \cdot \boldsymbol{\varkappa}_a = 0$. Because of our conclusion that $\Delta(\mathbf{r})$ varies slowly over distances like $\varkappa_b{}^{-1}$, we have

$$\int d^2\lambda\, \Delta(\mathbf{r}) e^{i\varphi(\mathbf{r})} \simeq \Delta(\mathbf{r}_l) \left[\int d^2\lambda\, e^{-i\varkappa_b \cdot \boldsymbol{\lambda}} \right] e^{i\varphi_l(s)} \simeq 0$$

unless $\omega = 0$. Thus, we find as one condition on the $\mathbf{r}_l$ that at each of these points $\omega = 0$. This means that there is no discontinuity in the direction of the orbit—hardly surprising for a semiclassical theory of orbits! It is of course the conclusion that Δ varies slowly over a de Broglie wavelength which leads to this.

To evaluate the $\boldsymbol{\lambda}$ integral we must keep terms of order λ^2 in φ. These terms can be expressed as functions of the principal radii of curvature of the surfaces of constant S_a^+ and S_b^-. To consider an example, for the special case of cylindrical symmetry about the direction of $\mathbf{p}$ (which would obtain with spherical symmetry of the interactions), with λ_1 in the radial direction and λ_2 in the azimuthal direction, we have

$$\varphi(\mathbf{r}_l + \hat{\mathbf{n}}s + \boldsymbol{\lambda}) = \varphi_l(s) - \frac{\varkappa_b}{2}\frac{\lambda^2}{R_l} + \mathcal{O}(\lambda^3)$$

Here R_l depends on the principal radii of curvature of the two surfaces at $\mathbf{r}_l + \hat{\mathbf{n}}s$. Then

$$\int d^2\lambda e^{i\varphi(\mathbf{r}_l + \hat{\mathbf{n}}s + \boldsymbol{\lambda})} \simeq e^{i\varphi_l(s)} e^{-i(\pi/2)} \left[\frac{\pi R_l}{\varkappa_b}\right]$$

$$\equiv e^{i\varphi_l(s)} 2\pi A_l \qquad (4.10)$$

Here ρ_l is the distance of $\mathbf{r}_l + \hat{\mathbf{n}}s$ from the symmetry axis and A_l has the dimensions of an area.

The result (4.10) lets us rewrite Eq. (4.6) in the form*

$$T_{ba} \equiv \sum_l (2\pi)^{-2} A_l J_l \qquad (4.11)$$

where

$$J_l = \int_l ds \Delta_l(s) e^{i\varphi_l(s)} \qquad (4.12)$$

and we have written $\Delta(\mathbf{r})$ as $\Delta_l(s)$ in the vicinity of the point $\mathbf{r}_l$.

Since we have argued that Δ varies slowly with s compared with φ, it might be concluded that one could remove $\Delta_l(s)$ from the integral of (4.12). The reason we do not is that for the particular model described in Section VI, $\Delta_l(s)$ has a singularity at the point of stationary phase for $\varphi(s)$.†

We note that it may be impossible to meet the conditions $\omega = 0$ at each point $\mathbf{r}_l$ at which $\varphi(s)$ is stationary. Stationary phase points for

---

* Strictly speaking, A_l depends on s. Because this is slowly varying on the de Broglie wavelength scale, we have removed it from the integral over s.

† A more elaborate calculation than that reported here indicates that the quantity A_l is not singular for this example. Here our casual evaluation of oscillating integrals can also be justified by doing the s-integral before the λ-integral.

which ω cannot be chosen equal to zero should be omitted from the sum over l in (4.11). The conditions $\omega = 0$ will of course change with scattering angle (the angle between **k** and **p**).

V. Case with Saddle Point

We now illustrate the result of the last section with a simple example. To make the visualization easy, we assume the energies w_0 and w_1 to be spherically symmetric. The eikonal momenta $\varkappa_a(r)$ and $\varkappa_b(r)$ as determined from Eq. (3.5) are given by

$$\varepsilon_{\varkappa_a} + w_0(r) = \varepsilon_{\varkappa_a} + W_0 + \mathscr{V}_0(r) = E$$
$$\varepsilon_{\varkappa_b} + w_1(r) = \varepsilon_{\varkappa_b} + W_1 + \mathscr{V}_1(r) = E \qquad (5.1)$$

where [as in Eq. (2.18b)]

$$\epsilon_\varkappa = \varkappa^2/2M \qquad (5.2)$$

As in Fig. 1, we shall suppose that for all r

$$w_1(r) > w_0(r)$$

This implies that

$$\varkappa_a(r) > \varkappa_b(r) \qquad (5.3)$$

Also, as illustrated in Fig. 1, we suppose that $[\varkappa_a(r) - \varkappa_b(r)]$ has a minimum at $r = r_0$ (close to r_c), such that $[\varkappa_a(r_0) - \varkappa_b(r_0)] \ll \varkappa_a(r_0)$. We shall suppose that the trajectory passes through r_0 twice, satisfying the condition $\omega = 0$ both times. This means that we have $l = 1, 2$ in (4.11).

Let us select one of these points and drop the l index. With proper choice of origin, we write

$$\varkappa_a(r) - \varkappa_b(r) = d + qs^2 + \cdots \qquad (5.4)$$

as a Taylor expansion along the trajectory. From Eq. (5.3) it follows that

$$d, q > 0 \qquad (5.5)$$

By integration, we obtain from (5.4)

$$\varphi(s) = \varphi(0) + ds + \frac{q}{3} s^2 + \cdots \qquad (5.6)$$

where $\varphi(0)$ is a constant. Finally, we assume that the function $\Delta_l(s)$ varies slowly enough to be removed from the integrand in Eq. (4.12), which now becomes

$$J = \Delta \int ds e^{i\varphi(s)} \qquad (5.7)$$

with Δ written for $\Delta_l(0)$.

In the spirit of keeping our illustration simple we shall assume that terms of the order s^4 and higher may be neglected in Eq. (5.6).* Then, we see that $\varphi(s)$ has a saddle point at

$$s = e^{i\pi/2}(d/q)^{1/2} \tag{5.8}$$

through which the contour can be displaced to give

$$\int ds e^{i\varphi(s)} \simeq e^{i\varphi(0)} \frac{2\sqrt{\pi}}{(qd)^{1/4}} e^{-g} \tag{5.9}$$

where

$$g \equiv (d^3/q)^{1/2} \tag{5.10}$$

The condition for validity of the saddle-point integration is that

$$g \gg 1 \tag{5.11}$$

This implies that the integral (5.9) is small, or that we are well down into the regime of near-adiabatic energies.

Using Eqs. (5.7) and (5.9) and putting l labels on q, d, and g, we obtain from (4.11)

$$T_{ba} = \sum_{l=1}^{2} \frac{A_l \Delta_l}{2\pi^{3/2}(q_l d_l)^{1/4}} e^{-g_l} e^{i\varphi_l} \tag{5.12}$$

where φ_l is the appropriate value of $\varphi(0)$ in (5.6) for $l = 1, 2$. For this simple example with spherical symmetry the quantities A_l, Δ_l, q_l, d_l, and g_l would be the same for both l values, except for possible asymmetry introduced by the curvature of the orbits.

For a qualitative description of the cross section we keep only one of the l terms in (5.12) and write

$$\varkappa_a - \varkappa_b = \frac{\varkappa_a^2 - \varkappa_b^2}{\varkappa_a + \varkappa_b}$$

$$\simeq \frac{1}{v}(\varepsilon_{\kappa_a} - \varepsilon_{\kappa_b})$$

$$= \frac{1}{v}[w_1 - w_0] \equiv \frac{\Delta w}{v} \tag{5.13}$$

where v is the relative velocity at r_0. Then

$$d \simeq \Delta w/v$$

$$q \simeq \Delta w/2vb^2 \tag{5.14}$$

* This need not be the case, of course. The numerical value of our final cross section is sensitive to such details.

NEAR-ADIABATIC TRANSITIONS

where b is a measure of the "width" of the minimum in $\varkappa_a - \varkappa_b$. This gives

$$g \cong \frac{2\sqrt{2}}{3}\left(\frac{\Delta w}{\hbar}\frac{b}{v}\right) \tag{5.15}$$

From Eqs. (2.33) and (4.10) we have the estimates

$$\Delta_l \cong v/a$$
$$A_l \cong a^{3/2}/\varkappa^{1/2} \tag{5.16}$$

This lets us simplify Eq. (5.12) to the form

$$T_{ba} \simeq \frac{a}{\varkappa}\left(\frac{v}{a}\right)\frac{b}{g^{1/2}} e^{-g}$$

The cross section (2.31) is then

$$\frac{d\sigma}{d\Omega} \simeq \left(\frac{k}{p}\right)\frac{1}{g} b^2 e^{-2g} \tag{5.17}$$

This overly simplified description has been in agreement, of course, with the qualitative features of the classic descriptions of adiabatic and near-adiabatic scattering. Expression (5.15) is the familiar quantity which substituted into (5.11) gives the textbook criterion for adiabatic scattering. In particular, if we say that $\Delta w \approx e^2/b$, condition (5.11) gives the inequality (1.1).

Our discussion does exhibit, however, a delicate dependence of near-adiabatic scattering on details—such as curvature of orbits, the detailed form of potential curves, etc.

VI. The Landau-Zener Model

As a second illustration of our method we discuss the Landau-Zener model. Although this model suffers from being somewhat artificial (for application to real atoms or molecules), it does permit an explicit calculation to be made.

For simplicity we again assume spherical symmetry and suppose that only two internal states need be considered. When the atoms are separated and noninteracting, these are g_0 and g_1, as determined from Eq. (2.2):

$$hg_n = W_n g_n \tag{6.1}$$

The corresponding adiabatic states of the interacting atoms are determined from Eq. (2.4):

$$(h + V)\varphi_\alpha = w_\alpha(r)\varphi_\alpha \tag{6.2}$$

These pairs of states can be related by the unitary transformation

$$\varphi_\alpha = \sum_{n=1}^{2} U_{n\alpha}(r) g_n \qquad (6.3)$$

Using Eqs. (6.1) and (6.3) we can write (6.2) in the form

$$(W_n \delta_{nn'} + V_{nn'}) U_{n'\alpha} = w_\alpha U_{n\alpha} \qquad (6.4)$$

where

$$V_{nn'} \equiv (g_n V g_{n'}) \qquad (6.5)$$

Corresponding to the example illustrated in Fig. 1, we shall suppose that

$$W_1 > W_0$$
$$w_1(r) > w_0(r) \qquad (6.6)$$

Finally, using Eq. (6.3) we can write the expression (4.3) for Δ_{10} as

$$\Delta_{10}(r) = -iv \left[U_{01}^* \frac{\partial U_{00}}{\partial s} + U_{11}^* \frac{\partial U_{10}}{\partial s} \right] \qquad (6.7)$$

where $v \equiv \varkappa_a(r)/M$.

Equation (6.4) provides expressions for the U's and w's. To write these, let us introduce the definitions

$$e_0 \equiv W_0 + V_{00}, \qquad e_1 \equiv W_1 + V_{11}$$
$$\gamma \equiv 2V_{10} = 2V_{01}$$
$$D \equiv (e_1 - e_0)^2 + \gamma^2 \qquad (6.8)$$

Then, we find

$$w_{1,0} = \tfrac{1}{2}[e_1 + e_0 \pm \sqrt{D}]$$

$$U_{00} = \frac{D_0}{(D_0^2 + \gamma^2/4)^{1/2}}, \qquad U_{10} = \frac{\gamma/2}{(D_0^2 + \gamma^2/4)^{1/2}}$$

$$U_{01} = \frac{D_1}{(D_1^2 + \gamma^2/4)^{1/2}}, \qquad U_{11} = \frac{\gamma/2}{(D_1^2 + \gamma^2/4)^{1/2}} \qquad (6.9)$$

where

$$D_{1,0} = w_{1,0} - e_1 = \tfrac{1}{2}[e_0 - e_1 \pm D^{1/2}] \qquad (6.10)$$

With these relations we can put Eq. (6.7) into the form

$$\Delta_{10} = \frac{iv\gamma}{2D^{1/2}} \frac{\partial \ln D_0}{\partial s} \qquad (6.11)$$

The Landau-Zener model now assumes that

$$y \equiv e_0 - e_1 = \beta s, \qquad \beta > 0 \qquad (6.12)$$

as illustrated in Fig. 1. This "crossing of unperturbed levels" e_0 and e_1 assures us that $(w_1 - w_0)$ is a minimum at $s = 0$.

It should be observed that this model with crossing "unperturbed levels" is not a unique way to get the difference $(w_1 - w_0)$ to exhibit a minimum. Also, the introduction of e_0 and e_1 as energies has more mathematical than physical significance. The energies w_1 and w_0 do have physical significance, since for slow collisions they can be observed.

As before, we assume that near $s = 0$

$$[\varkappa_a - \varkappa_b] \ll \varkappa_a$$

so

$$\varkappa_a - \varkappa_b \simeq \frac{1}{v}[w_1 - w_0] = \frac{1}{v}D^{1/2} \qquad (6.13)$$

Thus, with Eq. (6.12), we obtain (to within an additive constant)

$$\Phi(y) \equiv \varphi(s) = \frac{1}{v}\int^s (\beta^2 s'^2 + \gamma^2)^{1/2}\, ds'$$

$$= \frac{1}{2v\beta}\{y(y^2 + \gamma^2)^{1/2} + \gamma^2 \ln [y + (y^2 + \gamma^2)^{1/2}]\} \qquad (6.14)$$

Equation (6.12) lets us evaluate (6.11) as

$$\Delta(s) \simeq \Delta_{10} = -iv\beta\gamma/2D(y) \qquad (6.15)$$

with

$$D(y) = y^2 + \gamma^2 \qquad (6.16)$$

[Here we have ignored the eikonal correction factors N_a and N_b in Eq. (4.4).] At either of the "crossing points" the integral (4.12) is

$$J = \frac{iv\gamma}{2}\int \frac{dy}{D(y)} e^{i\Phi(y)} \qquad (6.17)$$

The above integral is easily evaluated in the near-adiabatic limit. The contour is displaced into the upper half y plane and a branch cut established along the positive imaginary axis in the interval

$$i\gamma < y < i\infty$$

It is convenient to transform to a new variable z,

$$y = z + A, \qquad A = \gamma e^{i(\pi/2)} \qquad (6.18)$$

In the cut half y plane

$$-\frac{3\pi}{2} < \arg z < \frac{\pi}{2} \qquad (6.19)$$

We then have
$$D(y) = z^2 + 2Az \quad (6.20)$$
and near $z = 0$
$$\Phi(y) = \Phi(A) + \frac{(2A)^{1/2}}{v\beta}\left[\frac{2}{3}z^{3/2} + \frac{z^{5/2}}{50A} + \cdots\right] \quad (6.21)$$
with
$$\Phi(A) = \frac{\gamma^2}{2\beta v}\left[i\frac{\pi}{2} + \ln \gamma\right] \quad (6.22)$$

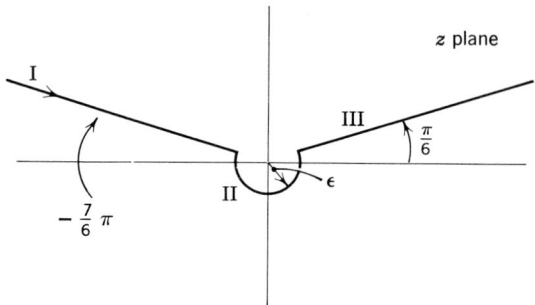

Fig. 6. Contour of integration for Eq. (6.17).

The appropriate contour of integration in the z plane for evaluating Eq. (6.17) is shown in Fig. 6. It consists of three parts:

$$I: \arg z = -\tfrac{7}{6}\pi, \quad \varepsilon < |z| < \infty$$

$$II: -\frac{7}{6}\pi < \arg z < \frac{\pi}{6}, \quad |z| = \varepsilon$$

$$III: \arg z \frac{\pi}{6}, \quad \varepsilon < |z| < \infty \quad (6.23)$$

Here ε is considered to be an arbitrarily small positive quantity.

The quantity J is now easily evaluated in the form (valid for $G \gg 1$)
$$J = -i\frac{\pi}{3} v e^{-G} e^{i\varphi(0)} \times \{1 + \mathcal{O}(G^{-2/3})\} \quad (6.24)$$
where
$$G = \frac{\pi}{4\beta v}\gamma^2 \quad (6.25)$$
and $\varphi(0)$ is a phase factor arising from the unevaluated constant of integration in Eq. (6.14).

We note that

$$P = e^{-2G}$$

is just the Landau-Zener probability as given in Eq. (1.3). We also note that the quantity (6.24) is primarily determined by the value of $\varphi(s)$ at the stationary phase point [where $\varphi(s) = \Phi(A)$]. This is in agreement with the result of Section V, where the stationary phase point was assumed to be a saddle point. In the example of this section the stationary phase point was a branch point and the Taylor expansion not possible. Although the results of this and the last section are similar, the explicit expressions would lead to significant numerical differences in the reaction cross section.

For a qualitative description of the cross section, we again ignore the second crossing point and use Eqs. (4.11) and (5.16) to obtain

$$\frac{d\sigma}{d\Omega} = \left(\frac{k}{p}\right) a^2 e^{-2G} \qquad (6.26)$$

This is similar to Eq. (5.17).

VII. Conclusions

We have discussed the use of stationary state scattering theory for studying atomic and molecular reactions in the near-adiabatic limit. The application of modern scattering theory to these phenomena provides the possibility for a flexibility and precision which is difficult to achieve with the older methods.

The Born approximation to Eq. (2.27) has been used here. The argument given was that the matrix elements of $\Delta_{\alpha\beta}$ are small in the near-adiabatic limit. Because the Green's function $(E + i\eta - \varkappa - w_\alpha)^{-1}$ and the matrix elements of $\Delta_{\alpha\beta}$ are relatively simple in the near-adiabatic limit, corrections to the Born approximation can be examined in some detail. For example, the second Born approximation contains (among other things*) the correction factor $(1 - P)$ of Eq. (1.4). The second Born corrections are, of course, small in the near-adiabatic limit.

References

1. L. D. Landau, *Phys. Z. Sowjetunion*, **2**, 46 (1932).
2. C. Zener, *Proc. Roy. Soc. (London) Ser. A*, **137**, 696 (1932).
3. E. C. G. Stueckelberg, *Helv. Phys. Acta*, **5**, 369 (1932).

* This includes some phrase corrections. See, for example, reference 10.

4. D. R. Bates, Ed., *Atomic and Molecular Processes*, Academic Press, New York, 1962.
5. C. A. Coulson, and K. Zalewski, *Proc. Roy. Soc.* (*London*), *Ser. A*, **268,** 437 (1962).
6. M. L. Goldberger and K. M. Watson, *Collision Theory*, Wiley, New York, 1963.
7. S. Weinberg, *Phys. Rev.*, **126,** 1899 (1962).
8. C. Critchfield, *Am. J. Phys.*, **32,** 542 (1964).
9. D. R. Bates and A. R. Holt, *Proc. Roy. Soc.* (*London*) *Ser. A*, **292,** 168 (1966).
10. D. R. Bates, H. C. Johnston and I. Stewart, *Proc. Phys. Soc.* (*London*), **84,** 517 (1964).

Some Solutions of the Classical Isotopic Gauge Field Equations

T. T. WU*

Harvard University, Cambridge, Massachusetts

and

C. N. YANG

Institute for Theoretical Physics, State University of New York,
Stony Brook, New York

I. Introduction

It was pointed out[1] a number of years ago that an *isotopic gauge* can be defined in analogy with the usual electromagnetic gauge, and that the concept of local isotopic gauge invariance leads to a gauge field $\mathbf{b}_\mu$. The equations describing $\mathbf{b}_\mu$ in interaction with any source of isotopic spin are essentially uniquely determined, much like the equations describing the electromagnetic field A_μ in interaction with the electric charge. In the absence of any external sources of isotopic spin, the $\mathbf{b}_\mu$ field interacts with itself, since the $\mathbf{b}_\mu$ field possesses an isotopic spin and hence is self-generating. In this latter characteristic, the $\mathbf{b}_\mu$ field is different from the electromagnetic field, which is described by linear equations in the absence of other fields. (The nonlinear equations describing the self-generating $\mathbf{b}_\mu$ field is in some respects[2] similar to the equations of general relativity.)

We seek in this paper to find a solution of the (unquantized) $\mathbf{b}_\mu$ field in the absence of other interacting fields. Our aim is then similar to that of Born and Infeld,[3] except that they started with equations which were written down on a more or less *ad hoc* basis.

II. A Special Type of Solution

The equations for the $\mathbf{b}_\mu$ field are[1]

$$\mathbf{f}_{\mu\nu} = \mathbf{b}_{\mu,\nu} - \mathbf{b}_{\nu,\mu} - \mathbf{b}_\mu \times \mathbf{b}_\nu \tag{1}$$

$$\mathbf{f}_{\mu\nu,\nu} + \mathbf{b}_\nu \times \mathbf{f}_{\mu\nu} = 0 \tag{2}$$

$$\mathbf{b}_{\mu,\mu} = 0 \tag{3}$$

* National Science Foundation Senior Postdoctorate Fellow.

We have chosen units for $\mathbf{b}_\mu$ so that the coupling constant ε is equal to $\tfrac{1}{2}$. We further adopt the convention that $x_4 = ict$, and that $_{,\mu}$ means differentiation with respect to x_μ. Also subscripts μ or ν run 1 to 4, while others run 1 to 3.

Solutions in which the $\mathbf{b}_\mu$ field lies in one isotopic direction are easily found, since for them the nonlinear terms vanish, and the $\mathbf{b}_\mu$ field equations reduce to that for the free electromagnetic field. But such solutions are of no interest to us here.

To find some special solutions of Eqs. (1)–(3) we look for a static case, so that

$$b_4 = 0, \qquad b_{i,4} = 0 \tag{4}$$

We write the components of $\mathbf{b}_\mu$ as $b_{\mu\alpha}$, where $\alpha = 1, 2, 3$ designates the isotopic spin index.

We shall seek for a solution of the following form

$$b_{11} = b_{22} = b_{33} = 0, \qquad b_{12} = -b_{21} = x_3 f(r)/r, \text{ etc.}$$

i.e.,

$$b_{i\alpha} = \varepsilon_{i\alpha\tau} x_\tau f(r)/r \tag{5}$$

where r is the length of (x_1, x_2, x_3). Equation (3) is then automatically satisfied and (1) and (2) reduce to

$$f'' + \frac{2}{r} f' - (1 + rf)\left(\frac{2f}{r^2} + \frac{f^2}{r}\right) = 0 \tag{6}$$

where the prime means d/dr.

Writing

$$\Phi(r) = 1 + rf(r) \tag{7}$$

one has

$$r^2 \Phi'' - \Phi(\Phi^2 - 1) = 0 \tag{8}$$

To study this equation we observe that putting

$$r = e^\xi \tag{9}$$

one has

$$\frac{d^2\Phi}{d\xi^2} - \frac{d\Phi}{d\xi} = \Phi(\Phi^2 - 1) \tag{10}$$

or

$$\frac{d\Phi}{d\xi} = \psi \tag{11a}$$

$$\frac{d\psi}{d\xi} = \psi + \Phi(\Phi^2 - 1) \tag{11b}$$

In the Φ-ψ plane (phase plane), $(d\Phi/d\xi, d\psi/d\xi)$ defines a vector field given by the right-hand side of (11a, b). The curves that are tangent to this vector field are the solutions we desire. The vector vanishes at exactly three points: $(\Phi, \psi) = (0, 0), (1, 0)$, and $(-1, 0)$.

It is not difficult to study the vector field. One finds that there are only five solutions of Eqs. (11a, b) which are finite for all $0 < r < \infty$:

(a) $\Phi = 0$ (12a)

(b) $\Phi = +1$ (12b)

(c) $\Phi = -1$ (12c)

(d) $\Phi = 1 - \dfrac{c}{r} + O\left(\dfrac{1}{r^2}\right)$ as $r \to \infty$, $(c > 0)$; $\Phi \to 0$ as $r \to 0$; (12d)

(e) The same as (d) except Φ changes sign. (12e)

To discuss the meaning of these solutions, we split the 6-vector $\mathbf{f}_{\mu\nu}$ into "electric" and "magnetic" components:

$$\mathbf{E}_j = i\mathbf{f}_{j4}, \qquad -\varepsilon_{ijk}\mathbf{H}_k = \mathbf{f}_{ij}$$

For the special type of solution satisfying (5),

$$\begin{aligned}\mathbf{E}_j &= 0, \\ -H_{j\alpha} &= -\delta_{j\alpha}\Phi'/r + x_j x_\alpha r^{-3}[\Phi' - r^{-1}\Phi^2 + r^{-1}] \\ &= -\delta_{j\alpha}\psi/r^2 + x_j x_\alpha r^{-4}[\psi - \Phi^2 + 1]\end{aligned} \quad (13)$$

It is clear that for (12b) and (12c), $H_j = 0$ and the solutions are merely complicated ways of describing the vacuum $\mathbf{f}_{\mu\nu} = 0$. The nontrivial solutions are thus the three tabulated below.

1. *Solution (12a)*

$$\Phi = 0, \qquad f = -r^{-1}$$
$$E_{j\alpha} = 0, \qquad H_{j\alpha} = -x_j x_\alpha r^{-4} \quad (14)$$

2. *Solution (12d)*

This solution has the following asymptotic behavior:

$r \to \infty$,

$$\Phi = 1 - \dfrac{c}{r} + O\left(\dfrac{1}{r^2}\right), \qquad f = -\dfrac{c}{r^2} + O\left(\dfrac{1}{r^3}\right)$$

$$E_{j\alpha} = 0, \qquad H_{j\alpha} = \dfrac{-c}{r^3}\left[-\delta_{j\alpha} + \dfrac{3x_j x_\alpha}{r^2}\right] + O\left(\dfrac{1}{r^4}\right) \quad (15)$$

$r \to 0$,

$\Phi \to 0$, $\psi \to 0$, as oscillatory functions of r with minima and maxima $= O(r^{1/2})$.

$$E_{j\alpha} = 0, \qquad H_{j\alpha} = -x_j x_\alpha r^{-4} + O(r^{-3/2}) \quad (16)$$

This solution is actually a one-parameter family of solutions with the parameter c of the dimension of a length. Numerical results are given in Table I.

TABLE I. Φ as a Function of r for Solution (12d),[a] with $c = 1$

r	Φ
∞	1
9.880×10	9.898×10^{-1}
1.095×10	9.136×10^{-1}
3.297	7.510×10^{-1}
1.098	4.584×10^{-1}
6.141×10^{-2}	-9.229×10^{-2} = min.
1.617×10^{-3}	1.498×10^{-2} = max.
4.296×10^{-5}	-2.441×10^{-3} = min.
1.142×10^{-6}	3.980×10^{-4} = max.
3.036×10^{-8}	-6.489×10^{-5} = min.
8.070×10^{-10}	1.058×10^{-5} = max.
2.15×10^{-11}	-1.725×10^{-6} = min.

[a] This table is obtained by numerical integration. As $r \to 0$, Φ oscillates with damped amplitude. The first four minima and first three maxima are tabulated.

3. Solution (12e)

This solution has the following asymptotic behavior:

$r \to \infty$,

$$\Phi = -1 + \frac{c}{r} + O\left(\frac{1}{r^2}\right), \quad f = \frac{-2}{r} + \frac{c}{r^2} + O\left(\frac{1}{r^3}\right)$$

$$E_{j\alpha} = 0, \quad H_{j\alpha} = \frac{-c}{r^3}\left[\delta_{j\alpha} + \frac{x_j x_\alpha}{r^2}\right] + O\left(\frac{1}{r^4}\right) \tag{17}$$

$r \to 0$,

$\Phi, \psi \to 0$ as oscillatory functions of r with maxima and minima = $O(r^{1/2})$.

$$E_{j\alpha} = 0, \quad H_{j\alpha} = -x_j x_\alpha r^{-4} + O(r^{-3/2}) \tag{18}$$

This solution is again a one-parameter family of solutions with the parameter c of the dimension of a length.

Notice that the three types of solutions (12a), (12d), and (12e) share the same dominant asymptotic form as $r \to 0$.

III. Energy

The Hamiltonian of the **b** field was given in reference 1. For the present cases,

$$H = \frac{1}{4}\int H_{j\alpha}H_{j\alpha}\,d^3x = \pi\int_0^\infty [2\Phi'^2 + (\Phi^2 - 1)^2 r^{-2}]\,dr \qquad (19)$$

This integral is divergent at $r \simeq 0$.

If one replaces $\Phi \to \Phi + \delta\Phi$ in (19) for variations $\delta\Phi$ which are zero in an interval $r = 0 \to r_0$, (19) gives a stationary value at the solutions Φ discussed in Sections II-1, II-2, and II-3, above [cf. Eq. (8)]. The second variation gives

$$\delta^2 H = \pi \int_{r_0}^\infty [2(\delta\Phi')^2 + (\delta\Phi)^2(6\Phi^2 - 2)r^{-2}]\,dr \qquad (20)$$

For sufficiently small r_0, $6\Phi^2 - 2 \sim -2$ near $r = r_0$ and (20) is not positive definite.

IV. Source

Are the solutions above really sourceless? The answer is clearly yes except at $r = 0$. For $r = 0$, however, the solutions are singular and this question will have to be examined in greater detail.

Another way of asking the same question is whether the solutions exhibited above do satisfy the field equation (2). To discuss this we define

$$-\mathbf{J}_\mu = \mathbf{f}_{\mu\nu,\nu} + \mathbf{b}_\nu \times \mathbf{f}_{\mu\nu} \qquad (21)$$

Equation (5) leads to

$$J_{4\alpha} = 0, \qquad J_{i\alpha} = \varepsilon_{i\alpha\tau}x_\tau r^{-1}J \qquad (22)$$

where

$$J = -f'' - \frac{2}{r}f' + (1 + rf)\left(\frac{2f}{r^2} + \frac{f^2}{r}\right) \qquad (23)$$

At $r \neq 0$, this is clearly zero, by Eq. (6). At $r \simeq 0$, the dominant term in f is $-r^{-1}$ for all three types of solutions (12a), (12d), and (12e). Thus

$$J = -4\pi\delta^3(x) \qquad (24)$$

i.e.,

$$J_{4\alpha} = 0, \qquad J_{i\alpha} = -4\pi\varepsilon_{i\alpha\tau}x_\tau r^{-1}\delta^3(x) \qquad (25)$$

The source function (25) is, in the sense of Dirac's definition of δ functions or in the sense of the theory of distributions, equal to zero. We thus conclude that the solutions indeed represent classical sourceless gauge fields.

V. Total Isotopic Spin

The total isotopic spin was given in reference 1:

$$\mathbf{T} = \int \mathbf{b}_\nu \times \mathbf{f}_{4\nu}\, d^3x$$

which is equal to zero for the present solutions.

Generalizations of the above solutions are in progress.

Acknowledgment

One of the authors (T. T. Wu) would like to take this opportunity to thank Professor A. Pais and the Rockefeller University for the hospitality extended him during his visit when this work was carried out.

References

1. C. N. Yang and R. L. Mills, *Phys. Rev.*, **96,** 191 (1954).
2. R. Utiyama, *Phys. Rev.*, **101,** 1597 (1956).
3. M. Born and L. Infeld, *Proc. Roy. Soc. (London), Ser. A*, **143,** 410; **144,** 425; **147,** 522 (1934); **150,** 141 (1935).

Eigenvalues of Casimir Operators*

CHARLES L. CRITCHFIELD

University of California, Los Alamos Scientific Laboratory,
Los Alamos, New Mexico

The original Casimir operator is the invariant bilinear form composed of generators of a semi-simple Lie group.[1] The idea has been generalized to include all invariant operators of homogeneous degree that commute with the generators of the group.[2] In the following we shall be concerned primarily with operators of degree 4 and lower for the general linear group. The results apply, of course, to the unitary groups, $U(N)$, since their generators are the same as for $GL(N)$.

The Casimir operator encountered most frequently in mathematical physics is the quadratic form of generators of $O(3)$. The latter are essentially components of orbital angular momentum of a single particle. Let

$$\mathbf{l} = \mathbf{r} \times \nabla \quad (1)$$

The invariant operator is

$$\mathbf{l}^2 = r^2 \nabla^2 - (r, \nabla)^2 - (r, \nabla) \quad (2)$$

Operating upon a harmonic polynomial of degree n, $H_i(n)$, $\mathbf{l}^2$ has the well-known eigenvalues

$$\mathbf{l}^2 H_i(n) = -n(n+1) H_i(n) \quad (3)$$

when $\nabla^2 H_i(n) = 0$. The $2n+1$ independent polynomials $H_i(n)$ constitute a basis for an irreducible representation of the orthogonal group in three dimensions.

Irreducible bases of O_3 may also be constructed from a given number of particles all of which are in p-waves. The generators are then most conveniently expressed in the form given by Racah.[3] Let $x^\varkappa, y^\varkappa, z^\varkappa$ be the coordinates of the $\varkappa$th particle and define

$$x_1^\varkappa = (1/\sqrt{2})(x^\varkappa + iy^\varkappa); \quad x_0^\varkappa = z^\varkappa; \quad x_{-1}^\varkappa = (1/\sqrt{2})(x^\varkappa - iy^\varkappa) \quad (4)$$

Then the generators have the form

$$X_k^i = -X_i^k = \sum_\varkappa \left(x_i^\varkappa \frac{\partial}{\partial x_{-k}^\varkappa} - x_k^\varkappa \frac{\partial}{\partial x_{-i}^\varkappa} \right) \quad i, k = 0, \pm 1 \quad (5)$$

---

* Work performed under the auspices of the U.S. Atomic Energy Commission.

The diagonal generator is $X_{-1}{}^1$. A substate of P particles may be labeled by giving the number ν_1 of particles in x_1, ν_0 the number in x_0 and ν_{-1} in x_{-1}, viz., $[\nu_1, \nu_0, \nu_{-1}]$

$$\nu_1 + \nu_0 + \nu_{-1} = P$$

The eigenvalue of $X_{-1}{}^1$ is then $\nu_1 - \nu_{-1}$:

$$X_{-1}{}^1[\nu_1, \nu_0, \nu_{-1}] = (\nu_1 - \nu_{-1})[\nu_1, \nu_0, \nu_{-1}] \tag{6}$$

The operator $X_0{}^1$ changes each $x_0{}^\varkappa$ that occurs into an $x_1{}^\varkappa$ and takes the sum over $\varkappa$. It also changes each $x_{-1}{}^\varkappa$ into $x_0{}^\varkappa$ (and changes the sign) so that it is a form of the differential step-up operator. Applying $X_0{}^1$ repeatedly produces a polynomial upon which $X_0{}^1$ gives only zero and the corresponding nonvanishing set of ν_i is the highest weight. As is well known, each irreducible representation belongs to a single highest weight and the latter is equivalent to a set of quantum numbers among which is the total angular momentum.

We now wish to apply a similar method to the general linear group of point transformations in N dimensions. We label the coordinates x_N, $x_{N-1}, \ldots, x_1$ and the corresponding highest weight, $\{\lambda_N, \lambda_{N-1}, \ldots \lambda_1\}$. A polynomial of degree P

$$\lambda_N + \lambda_{N-1} + \cdots + \lambda_1 = P \tag{7}$$

is considered as describing a system of P particles so that the diagonal generators have the form

$$E_k{}^k = \sum_{\varkappa=1}^{P} x_k{}^\varkappa \frac{\partial}{\partial x_k{}^\varkappa} \quad \text{(no sum on } k\text{)} \quad k = 1, 2, \ldots N \tag{8}$$

The simplest, nontrivial Casimir operator for the general linear group is then

$$C_1^{(N)} = \sum_{k=1}^{N} E_k{}^k \tag{9}$$

One readily shows that $C_1^{(N)}$ commutes with the complete set of generators

$$E_l{}^k = \sum_{\varkappa=1}^{P} x_k{}^\varkappa \frac{\partial}{\partial x_l{}^\varkappa} \tag{10}$$

For a given highest weight

$$\lambda_N \geq \lambda_{N-1} \geq \cdots \geq \lambda_1 \tag{11}$$

its eigenvalue is obviously the sum P, Eq. (7).

The Casimir operator of degree 2 may be written

$$C_2^{(N)} = E_k{}^k E_k{}^k + E_l{}^k E_k{}^l \quad k \neq l \tag{12}$$

Summation over repeated indices is assumed.

Let the contribution to Eq. (12) which comes from the first term when acting upon a given highest weight be $f(E_k{}^k E_k{}^k)$. Then, each $E_k{}^k$ produces a factor λ_k and

$$f(E_k{}^k E_k{}^k) = \sum_{1}^{N} \lambda_k^2 \tag{13}$$

The second term in Eq. (12) contains operators

$$E_l{}^k E_k{}^l = \sum_{\varkappa} x_k{}^{\varkappa} \frac{\partial}{\partial x_l{}^{\varkappa}} \sum_{\nu} x_l{}^{\nu} \frac{\partial}{\partial x_k{}^{\nu}}$$

According to the meaning of the highest weight, the λ_k particles with coordinate x_k are in a symmetric state relative to the exchange of labels, $\varkappa$. If $k > l$ then λ_l with x_k are in an antisymmetric state with those with x_l and the operator $x_l(\partial/\partial x_k)$ gives zero for each of those. The permissible steps thus allow $\lambda_k - \lambda_l$ transitions followed by their inverses giving

$$f(E_l{}^k E_k{}^l) = \sum_{k=1}^{N} \sum_{l=1}^{k-1} (\lambda_k - \lambda_l) \tag{14}$$

The eigenvalue of $C_2^{(N)}$, which is denoted $[C_2^{(N)}]$, is

$$[C_2^{(N)}] = \sum_{k=1}^{N} \lambda_k^2 + \sum_{k=1}^{N} \sum_{l=1}^{k-1} (\lambda_k - \lambda_l) \tag{15}$$

We can now make use of the theorem

$$\sum_{k=1}^{N} \sum_{l=1}^{k-1} h(l) = \sum_{k=1}^{N} (N - k) h(k) \tag{16}$$

and find

$$[C_2^{(N)}] = \sum_{k=1}^{N} [\lambda_k^2 - (N - 2k + 1)\lambda_k] \tag{17}$$

If, in place of the highest weight we define the set of "hooks," η_k,

$$\eta_k \equiv \lambda_k + k - 1 \tag{18}$$

Eq. (17) becomes a form of symmetric function

$$[C_2^{(N)}] = \sum_{k=1}^{N} [\eta_k^2 - (N - 1)\eta_k] + \binom{N}{3} \tag{19}$$

where $\binom{N}{i}$ stands for the usual binomial coefficient. In this notation we see that

$$[C_1^{(N)}] = \sum_{k=1}^{N} \eta_k - \binom{N}{2} \tag{20}$$

The operator of degree 3 may be written

$$C_3^{(N)} = E_k{}^k E_k{}^k E_k{}^k + E_k{}^k E_l{}^k E_k{}^l + E_l{}^k E_k{}^l E_k{}^k + E_l{}^k E_l{}^l E_k{}^l + 2E_l{}^k E_m{}^l E_k{}^m$$
$$k > l > m \quad (21)$$

It is assumed that one operates upon a given highest weight and the various contributions are denoted $f(E_k{}^k E_k{}^k E_k{}^k)$, $2f(E_k{}^k E_l{}^k E_k{}^l)$, etc. The second and third terms have the same value, so the eigenvalue is

$$[C_3^{(N)}] = f(E_k{}^k E_k{}^k E_k{}^k) + 2f(E_k{}^k E_l{}^k E_k{}^l) + 2f(E_l{}^k E_m{}^l E_k{}^m) + f(E_l{}^k E_l{}^l E_k{}^l) \quad (22)$$

with

$$f(E_k{}^k E_k{}^k E_k{}^k) = \sum_{k=1}^{N} \lambda_k{}^3 \quad (23a)$$

$$f(E_k{}^k E_k{}^k E_l{}^l) = \sum_{k=1}^{N} \sum_{l=1}^{k-1} \lambda_k (\lambda_k - \lambda_l) = \sum_{k=1}^{N} (k-1)\lambda_k{}^2 - \sum_{1}^{N} \sum_{1}^{k-1} \lambda_k \lambda_l \quad (23b)$$

$$f(E_l{}^k E_m{}^l E_k{}^m) = \sum_{1}^{N} \sum_{1}^{k-1} \sum_{1}^{l-1} (\lambda_k - \lambda_l)$$

$$= \sum_{1}^{N} \sum_{1}^{k-1} (l-1)(\lambda_k - \lambda_l)$$

$$= \sum_{1}^{N} [\tfrac{1}{2}(k-1)(k-2) - (N-k)(k-1)]\lambda_k \quad (23c)$$

$$f(E_l{}^k E_l{}^l E_k{}^l) = \sum_{1}^{N} \sum_{1}^{k-1} (\lambda_k - \lambda_l)(\lambda_l + 1)$$

$$= \sum_{1}^{N} [(k-1)\lambda_k - (N-k)(\lambda_k{}^2 + \lambda_k)] + \sum_{1}^{N} \sum_{1}^{k-1} \lambda_k \lambda_l \quad (23d)$$

In the last term of Eq. (23) the counting operator $E_l{}^l$ finds one more occupation number than that in the highest weight. Collecting terms we find

$$[C_3^{(N)}] = \sum_{1}^{N} \{\lambda_k{}^3 - (N - 3k + 2)\lambda_k{}^2$$
$$- [(2k-1)N - 3k^2 + 3k - 1]\lambda_k\} - \sum_{1}^{N} \sum_{1}^{k-1} \lambda_k \lambda_l \quad (24)$$

In order to convert Eq. (24) into an expression in terms of the hooks we need

$$\sum_{1}^{N} \sum_{1}^{k-1} \lambda_k \lambda_l = \sum_{1}^{N} \sum_{1}^{k-1} [\eta_k \eta_l - (k-1)\eta_l - (l-1)\eta_k + (k-1)(l-1)]$$

$$= \sum_{1}^{N} \sum_{1}^{k-1} \eta_k \eta_l - \sum_{1}^{N} [\tfrac{1}{2}N(N-1) - k + 1]\eta_k$$
$$+ \tfrac{1}{24}N(N-1)(N-2)(3N-1) \quad (25)$$

We then find

$$[C_3^{(N)}] = \sum_1^N [\eta_k^3 - (N-1)\eta_k^2 + \tfrac{1}{2}(N-1)(N-2)\eta_k]$$
$$- \sum_1^N \sum_1^{k-1} \eta_k \eta_l - \binom{N}{4} \quad (26)$$

In Eq. (25) a new form appears, namely,

$$\sum_1^N \sum_1^{k-1} \eta_k \eta_l$$

and it is expressible, of course, in terms of the sums of powers of the hooks.

$$\sum_1^N \sum_1^{k-1} \eta_k \eta_l = \frac{1}{2}\left\{\left(\sum_1^N \eta_k\right)^2 - \sum_1^N \eta_k^2\right\} \quad (27)$$

For the operator of degree 4 we have as many as four indices but these may be reduced to single and double sums. A generalization of Eq. (16) is required, viz.,

$$\sum_1^N \sum_1^{k-1} g(k) h(l) = \sum_1^N [G(N) - G(k)] h(k) \quad (28)$$

where

$$G(k) = \sum_{i=1}^k g(i)$$

Details of the calculation will be omitted. In terms of the highest weight we get

$$[C_4^{(N)}] = \sum_1^N \{\lambda_k^4 - (N-4k+3)\lambda_k^3 - [N(3k-2) - 6k^2 + 8k - 3]\lambda_k^2$$
$$- [(N-k)(3k^2 - 3k + 1) - (k-1)^3]\lambda_k\}$$
$$+ \sum_1^N \sum_1^{k-1} \{(N - 2k - 2l + 2)\lambda_k \lambda_l - \lambda_k^2 \lambda_l - \lambda_k \lambda_l^2\} \quad (29)$$

Finally, in terms of hooks

$$[C_4^{(N)}] = \sum_1^N \left\{\eta_k^4 - (N-1)\eta_k^3 + \binom{N-1}{2}\eta_k^2 - \binom{N-1}{3}\eta_k\right\} + \binom{N}{5}$$
$$+ \sum_1^N \sum_1^{k-1} [(N-2)\eta_k \eta_l - \eta_k^2 \eta_l - \eta_k \eta_l^2] \quad (30)$$

Again, the binary symmetric functions may be expressed in terms of the sums of powers.

Equations (19), (20), (26), and (30) reveal a simple general equation which summarizes them all.

$$[C_n^{(N)}] = \sum_{p,\varphi} (-1)^{n+p} \binom{N-\varphi}{n+1-p-\varphi} \sigma_p(\varphi) \qquad n \le 4 \qquad (31)$$

where φ is unity for sums of powers and p is the degree of the function, i.e.,

$$\sigma_p(1) = \sum_{k=1}^{N} \eta_k^p \qquad (32)$$

Otherwise,

$$\sigma_0(0) = 1$$

$$\sigma_2(2) = \sum_{1}^{N} \sum_{1}^{-1} \eta_k \eta_l$$

$$\sigma_3(2) = \sum_{1}^{N} \sum_{1}^{k-1} (\eta_k^2 \eta_l + \eta_k \eta_l^2)$$

Presumably, with the proper choice of definitions for the $\sigma_p(\varphi)$ the eigenvalues of $C_n^{(N)}$ can be put in the form of Eq. (31) for higher n. However, the calculation becomes rapidly more unwieldy and the results have been established only through $n = 4$.

The general method may be applied, however, to the case of a completely symmetrical state and any value of n. The highest weight contains all zeros except one which we denote by Λ, i.e., $\{\Lambda, 0, \ldots, 0\}$. Evaluating the functions $f(E_k^k E_k^k \cdots E_k^k)$, $f(E_k^k E_k^k \cdots E_l^k E_k^l)$, etc., we get the series

$$\begin{aligned}[][C_n^{(N)}(\Lambda)] &= \Lambda^n + (n-1)(N-1)\Lambda^{n-1} \\ &\quad + \tfrac{1}{2}(n-1)(n-2)(N-1)^2 \Lambda^{n-2} + \cdots \\ &= \Lambda(\Lambda + N - 1)^{n-1} \end{aligned} \qquad (33)$$

which is the formula obtained originally by Louck.[4] The corresponding hooks are

$$\eta_N = \Lambda + N - 1 \qquad \eta_k = k - 1, \qquad k < N \qquad (34)$$

and Eq. (33) becomes

$$[C_n^{(N)}(\Lambda)] = \eta_N^n - (N-1)\eta_N^{n-1} \qquad (35)$$

which when taken in conjunction with Eqs. (31) and (34) implies a number of relations among the symmetric functions of integers, at least for $n \le 4$.

Attempts to establish the general validity of the form of Eq. (31), i.e., for $n > 4$, have not been fruitful although particular examples for $n = 5$ and 6 indicate that the form holds with reasonable definitions of the more complex symmetric functions of the hooks.

Eigenvalues of the Casimir operators for $U(2)$ are of importance to many-electron term values in atoms. To a very good approximation the atomic states are characterized by a definite total spin S which in the notation above is $S = \frac{1}{2}(\lambda_2 - \lambda_1)$, i.e., for a highest weight $\{\lambda_2, \lambda_1\}$. Owing to the Pauli principle the λ_2 electrons with spin up are in antisymmetrical space states and hence the $\frac{1}{2}\lambda_2(\lambda_2 - 1)$ Coulomb interactions among them produce a lower energy than if there were no "exchange" force. Similarly the $\frac{1}{2}\lambda_1(\lambda_1 - 1)$ interactions among electrons with spin down tend to be lower than average. On the other hand, there are λ_1 interactions in which an electron with spin up is in a symmetric space state with an electron with spin down, and these will have higher than average Coulomb energy by approximately the same absolute value. All other pairs will have an average value so that the effect of symmetry upon the term value is proportional to

$$\Delta E \sim -\tfrac{1}{2}\lambda_2(\lambda_2 - 1) - \tfrac{1}{2}\lambda_1(\lambda_1 - 1) + \lambda_1$$
$$= -\tfrac{1}{2}[\lambda_2^2 + \lambda_1^2 - \lambda_2 - 3\lambda_1]$$
$$= -\tfrac{1}{2}[\lambda_2^2 + \lambda_1^2 + \lambda_2 - \lambda_1] + (\lambda_2 + \lambda_1) \tag{36}$$
$$= -\tfrac{1}{2}C_2^{(2)} + C_1^{(2)} \tag{37}$$

as we can verify from Eqs. (17) and (20). In terms of $S = \frac{1}{2}(\lambda_2 - \lambda_1)$ and $n = \lambda_2 + \lambda_1$ (here n stands for the total number of electrons) we find

$$\Delta E \sim -S(S + 1) - \tfrac{1}{4}n(n - 4)$$

which, when divided by the total number of electron pairs, $\frac{1}{2}n(n - 1)$, is just Dirac's formula for X_{12} which gives the effect of the exchange energy.[5]

The treatment of electrons in atoms was extended to nuclei by Wigner.[6] Here a nucleon can be not only in either of two spin states but also in either of two charge states. The appropriate unitary transformations are then those of $U(4)$. Assuming each nucleon to interact under the same conditions with every other nucleon, the exchange effects will again be given by the symmetry character of the space wavefunction. Since nucleons obey the Pauli principle this number is just the negative of that for the "spin" wavefunction where "spin" comprises all four states.

Wigner introduces a set of quantum numbers based on the highest weight $\{\lambda_4, \lambda_3, \lambda_2, \lambda_1\}$ which are defined as follows

$$\begin{aligned} A &= \lambda_4 + \lambda_3 + \lambda_2 + \lambda_1 \\ S &= \tfrac{1}{2}(\lambda_4 + \lambda_3 - \lambda_1 - \lambda_1) \\ T &= \tfrac{1}{2}(\lambda_4 - \lambda_3 + \lambda_2 - \lambda_1) \\ Y &= \tfrac{1}{2}(\lambda_4 - \lambda_3 - \lambda_2 + \lambda_1) \end{aligned} \tag{38}$$

In terms of the hooks

$$A = \eta_4 + \eta_3 + \eta_2 + \eta_1 - 6$$
$$S = \tfrac{1}{2}(\eta_4 + \eta_3 - \eta_2 - \eta_1 - 4)$$
$$T = \tfrac{1}{2}(\eta_4 - \eta_3 + \eta_2 - \eta_1 - 2) \quad (39)$$
$$Y = \tfrac{1}{2}(\eta_4 - \eta_3 - \eta_2 + \eta_1)$$

The fact that the eigenvalues are symmetric functions of hooks then means that they are also symmetric functions of $A + 6$, $S + 2$, $T + 1$, and Y. The value of $C_1^{(4)}$ is just A, of course, and

$$\sum_1^4 \eta_k = A + 6$$
$$\sum_1^4 \eta_k^2 = \tfrac{1}{4}(A + 6)^2 + (S + 2)^2 + (T + 1)^2 + Y^2 \quad (40)$$
$$C_2^{(4)} = \tfrac{1}{4}A^2 + S(S + 4) + T(T + 2) + Y^2$$

which can be combined with $C_1^{(4)}$ to obtain the symmetry character which is $-\tfrac{1}{2}C_2^{(4)} + 2C_1^{(4)}$. Since comparison of term values is mainly among states with the same number of nucleons, the part of interest is that independent of A.

It seems quite likely that the assumption of an equal opportunity for interaction between all pairs of nucleons is not as good as it is for pairs of electrons because of the long-range force between the latter. On the other hand, there is prominent evidence among nuclei of symmetry effects in the pairing energy. It may be expected that, as the complexities of shell formation, Coulomb energies, and spin-dependent forces become better understood, the relative importance of the symmetry character can be ascertained. It is conceivable that three- and four-body interactions are functions of the eigenvalues of the Casimir operators of degree 3 and 4, respectively.

For the operator of degree 3 we require

$$\sum_1^4 \eta_k^3 = \tfrac{1}{16}(A + 6)^3 + \tfrac{3}{4}(A + 6)[(S + 2)^2 + (T + 1)^2 + Y^2]$$
$$+ 3(S + 2)(T + 1)Y$$
$$\sum_1^4 \sum_1^{k-1} \eta_k \eta_l = \tfrac{3}{8}(A + 6)^2 - \tfrac{1}{2}[(S + 2)^2 + (T + 1)^2 + Y^2]$$

giving us the result:

$$C_3^{(4)} = \tfrac{1}{16}A^3 + (\tfrac{3}{4}A + 2)[S(S + 2) + T(T + 1) + Y^2]$$
$$+ 3(S + 2)(T + 1)Y \quad (41)$$

The expression for $C_4^{(4)}$ is readily calculated by similar methods. Other applications of the formulas, Eqs. (31) and (32), may be found in the particle groups $U(3)$ and $U(4)$ which lead to the same eigenvalues as the corresponding general linear group, since the generators are identical. The corresponding values for $SL(N)$ and $SU(N)$ are obtained by setting $C_1^{(N)} = 0$ whenever it occurs.

References

1. H. Casimir, *Koninkl. Akad. Wetenschap. Te Amsterdam, Proc.*, **34,** 844 (1931).
2. L. M. Gelfand, *Mat. Sb.*, **26,** 103 (1950).
3. G. Racah, *Lecture Notes on Group Theory and Spectroscopy*, Institute for Advanced Study, Princeton, N.J., 1951.
4. J. D. Louck, *J. Math. Phys.* **6,** 1786 (1965); cf. Eq. (27).
5. P. A. M. Dirac, *Quantum Mechanics*, Oxford University Press, London, 1958, p. 223.
6. E. P. Wigner, *Phys. Rev.*, **51,** 106 (1937).

Strange Matter

JOHN ARCHIBALD WHEELER

*Palmer Physical Laboratory,
Princeton, New Jersey*

"It is strange about everything, it is strange about pictures, a picture may seem extraordinarily strange to you and after some time not only does it not seem strange but it is impossible to find what there was in it that was strange."
GERTRUDE STEIN, Picasso

How many basic laws does it take to encompass all of physics? Is the number greater than we imagine? Or less? No new law has come to light since electromagnetism (1864), special and general relativity (1905 and 1915), and the quantum principle (1900–1925); regularities, yes; symmetries, yes; but no evidence for a new law in more than four decades of search. Is the fantastic thought conceivable that we, still looking, already have in our hands all of the basic laws that there are? And if by a miracle of nature and of man's enterprise we should be in this happy position, how can we capitalize on it?

J. J. Berzelius asked a similar question in 1819: Does not "all chemical combination depend solely on two opposing forces, positive and negative electricity"; or, in a word, do not chemical forces originate exclusively from electrostatic interactions? The idea was attractive. It motivated many chemical researches. Nevertheless, the results of several decades of work led the world of chemistry overwhelmingly to reject the electrostatic hypothesis. How can two oxygen atoms attract when two like charges repel? And how can a single law of force possibly account for interactions so completely different in strength and specificity as ionic forces, valence forces, homopolar forces, and van der Waals forces? Plainly chemistry is chemistry, and electrostatics is electrostatics! To reverse this negative judgment took decades more. It was not enough to recognize the electron as the active dynamic entity in the atom and in the molecule. One had also to bring to light the deeper content of Planck's

1900 quantum principle: stationary states, characteristic energies, and probability amplitudes. Today no one doubts that Maxwell theory and the wave equation between them account for everything from the tetrahedral bonds of carbon to the superconductivity of tin, from the crystal structure of the antimonides to the transition temperature of DNA; for all the familiar physical properties of solids, liquids, and gases; in a work, for all the features of molecular "chemistry," in the extended sense of the word "chemistry."

Throughout this marvelous mixture of atomic physics, solid state science, and molecular magic, in the search for understanding one no longer has to ask, "What is the basic law?" The challenge is more sophisticated: find the way from law to the explanation of regularity! Whether the regularity first comes to light in observation or in the output of labyrinthine calculation is of little moment. Why are so many molecules bent that might otherwise be linear? The Jahn-Teller effect! Why do so many crystals have a lesser symmetry than one might have anticipated? The Jahn-Teller effect! What lies behind superconductivity? The Bardeen-Cooper-Schrieffer coupling between pairs of electrons! Why is the photochemistry of polyatomic molecules qualitatively different from that of diatomic molecules? The Teller crossing of potential energy surfaces in polyatomic molecules!

Strange matter! Why is it so strange? The answers come to us, not as equations, but as concepts. The heart of each idea lends itself to statement in three clear sentences. Each concept is unique. Each has its own indispensable part in laying the machinery of nature open to our understanding. Who fully responds to the richness and strangeness of matter who does not have the twenty-one most precious concepts threaded on a string to tell over reflectively one by one as he ponders on a new strangeness?

The discovery of the neutron in 1932 made it natural to believe that nuclear physics was about to repeat the history of atomic and molecular physics. The dynamic entities having been identified—as the electron had been identified three and a half decades earlier—it only remained to find the law of force and apply the principles of quantum mechanics in order to have the whole subject reduced to reason. Our hopes of those times could be summarized in two cries: "Give me nucleon–nucleon scattering cross sections and I will give you the law of force," and "Give me a good machine and I will give you the cross sections." Now, three and a half decades further down the road, we still do not know the "law of force." We even doubt that that term has meaning for objects that give off mesons and transmute into hyperons under high velocity impact. Thus deprived of a "proper foundation," nuclear physics nevertheless grew and flourished, until today in many respects data on energy levels and transformation cross sections are not only more extensive for nuclei than for atoms and molecules, but also better understood. Exact knowledge of the law of force turned out not to be essential. More important proved to be concepts like "compound nucleus," "level width," "reaction channel," "individual particle states," and "pairing energy"; and other organizing ideas like "nuclear deformation," "fissility parameter," "rotational band," "coupling parameter," "strength function," and "imaginary potential"; and a score of additional concepts, some of which are still in course of active elucidation. The character of the spectrum of energy levels is found in many nuclei to be changed drastically by minor changes in the value of one or another coupling parameter. To have calculated the value of such a parameter with the requisite precision from first principles would have been a difficult matter even had one known any "law of force" between nucleons.

If without a precise knowledge of the force between nucleons nuclear chemistry has become an advanced science, it can hardly be regarded as amazing that "elementary particle chemistry" is fast becoming a science of comparable sophistication, even though we know neither the underlying dynamic entity nor its law of motion. Marvelous symmetries tie together most of the elementary particles. The so-called "bootstrap" approach relates the properties of one particle to those of other particles without ever pretending to, or even needing to, supply a rationale for the existence of particles in the first place. And if nuclear chemistry as it develops shows less and less sign of ever uncovering anything like a simple basic law of interaction between nucleons, even less does "elementary particle chemistry" promise to uncover in and by itself the common ingredient of elementary particles and its dynamic law.

It is no surprise to find it difficult to work down from the sophisticated to the simple. Who, trying to understand the complexity of the work-hardening of metals at the scale of centimeters, would have conceived of an explanation in terms of dislocations at a scale of 10^{-4} cm? And who, seeing a dislocation at a scale of 10^{-4} cm would have proffered as explanation atoms at the scale of 10^{-8} cm? Historically the path of explanation ran the other way. It was not

$$1 \text{ cm} \rightarrow 10^{-4} \text{ cm} \rightarrow 10^{-8} \text{ cm}$$

but

$$10^{-8} \text{ cm} \rightarrow 10^{-4} \text{ cm} \rightarrow 1 \text{ cm}$$

Who would use the complex properties of DNA as a way to unravel the structure of the atom? And who does not recall that all one knew about atoms at the scale of 10^{-8} cm did not break the secret of their structure?

For that it took Rutherford and Marsden and experiments at the scale of 10^{-13} cm. How baffling to work down from the composite to the elementary! And how much more easily progress has gone working up from the elementary to the composite.

Starting from the present "bottom," we have been able to see why layer after layer of structure is forced on matter. Successive steps in the saturation of the primary forces leave ever smaller residual forces, and these residuals organize matter into successively higher levels of structure. Put together a few dozen nucleons. They stay together by reason of the nucleonic attractions. By comparison with these short range attractions the coulomb forces are weak residuals. Weak though it is on this scale of comparison, the electric force from the nucleus dominates the atom, holds a family of electrons in orbit, and supports this new level of structure. Yet it does not cancel the electric forces at every place and every time to have the electrons balanced in number against the nuclear protons. Residual electric forces remain and give rise to chemistry. Chemical forces in their turn are saturated as the atoms unite into molecules. At this stage the residual van der Waals forces can at last make themselves felt, acting between the molecules, and helping to organize still higher levels of structure.

As the size of the structure goes up, gravitation, at first negligible, becomes appreciable and then dominant. Molecules to crystals, crystals to rocks, rocks to planets—then stars, globular clusters, galaxies, and clusters of galaxies. Static structures are replaced by dynamic systems. The greatest of these systems, the universe itself, expanding now, predicted to recontract, sets a term to the life of all the others. Dynamics, and a dynamics powered by gravitation, overwhelms in the end all static structure.

This universe of remarkable structures and beautiful physical effects one can divide in imagination into forty precincts, each a decade wide, reaching together all the way from 10^{-13} to 10^{28} cm. Each precinct has its own specialities, from the primordial cosmic fireball radiation at 10^{28} cm to dislocations spaced every 10^{-4} cm, and from the collapse of a white dwarf star at 10^9 cm to the fission of a uranium nucleus at 10^{-12} cm. In all this wealth of phenomena, not one single effect has been discovered since 1925 which has led to a new law of physics, and no single finding has been obtained which is generally recognized to be incompatible with existing law.

Four decades is a long time to wait for the discovery of a new law, and four decades more might be too long a wait before reconsidering the "plan" of physics—the implicit doctrine that there are endless overarching laws of physics still "around the corner"—or six, or one. What is the alternative? If by almost unbelievable good fortune we already have in our hands all of the basic laws that there are, then what words could be more appropriate than those of the engine inventor, John Kris: "Start her up and see why she don't run!"

What engine do we own, what "plan" of physics do we have to serve as our guide, if relativity—or Einstein's "geometrodynamics"—plus electromagnetism plus the quantum principle do constitute the complete background of all we see? No serious suggestion along this line has ever been put forward other than the "space theory of matter" of Clifford and Einstein. It envisages particles, not as foreign objects immersed in geometry, but as structures manufactured out of geometry, no other building material being available. Their vision, translated into today's terms, is more specific, not least because the addition of the quantum principle has given us a whole new concept of what Einstein's geometrodynamics is and says[1]:

(*1*) The geometry of space is not static, but dynamic.

(*2*) "Spacetime" is a *classical* deterministic dynamical history of space.

(*3*) Space in its dynamic evolution according to the laws of quantum mechanics bursts out of bounds of spacetime into superspace (Fig. 1).

(*4*) Superspace is a topological manifold any one "point" of which symbolizes an entire curved closed spacelike three-dimensional manifold ("3-geometry" or "configuration of space" or "geometrodynamical coordinate").

(*5*) A subset of these 3-geometries, of these "points" in superspace, constitutes the "YES" 3-geometries of a classical history of space, H. Thus, these "YES" 3-geometries stack up to make a 4-geometry, as automobile fenders of gradually varying shapes stack up to make a pile. However, the pile constitutes in effect a one-parameter family of fenders, whereas classical spacetime constitutes much more than a one-parameter family of 3-geometries. Spacelike slices can be cut through spacetime with enormous freedom, and with one of these "slices," one of these 3-geometries, curving with respect to another, and crossing and interweaving with it, in the greatest variety of ways. This interweaving, not merely of two "YES" 3-geometries, but of all the multitudinous variety of "YES" 3-geometries, imparts to the resulting 4-dimensional structure an inviolable rigidity which is summarized in the brief and simple phrase "4-geometry."

(*6*) If "YES" is painted on some of the 3-geometries of superspace, and "NO" on the enormously more numerous other 3-geometries of superspace, according as the 3-geometry in question can or cannot be obtained by a suitable slice through a given spacetime, then in turn that 4-geometry, that classical history of space, is obtained by solving

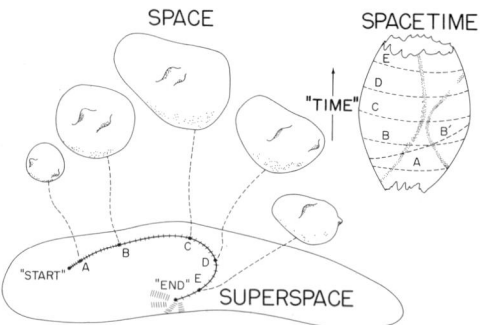

Fig. 1. Space, spacetime, and superspace depicted symbolically. Upper left: Five sample configurations (the three-dimensional geometries A, B, C, D, E) attained by space in the course of its expansion and recontraction. The two lumps symbolize the curvature of space in the vicinity of two stars, other stars being omitted to simplify the diagrams. The two stars move through space as the space itself expands and recontracts. Space, actually three-dimensional, is here drawn for convenience of representation as if two-dimensional ("surface of potato"). Upper right: Spacetime, the classical history of space. A cut through spacetime, such as A, gives a momentary configuration of space. [Spacetime is four-dimensional, but is shown as two-dimensional ("surface of egg") because of the limitations imposed by ink and paper. The same limitations force A, which is actually three-dimensional, to be represented in this particular diagram as if one-dimensional ("dashed curve running around egg"). The tracks of the two stars through spacetime show as two ridges ("world lines"; shaded in on "surface of egg").] Let the universe run through one conceivable dynamical history, the particular one indicated by this particular spacetime. This spacetime admits as spacelike slices, or as "YES" 3-geometries, all the 3-geometries A, B, C, D, E and many more besides, including for example a slice such as B'. Slice as one will through the given spacetime, one will be unable to get most conceivable 3-geometries. They are therefore appropriately called "NO" 3-geometries. The "YES" 3-geometries constitute a small subclass of the totality of all 3-geometries. [Given these "YES" 3-geometries, one has the wherewithal to reconstruct the entire spacetime, or 4-geometry. This spacetime is "rigid": it defines a definite time relationship between one event and another. One cannot define such time relationships nor even give any meaning to such terms as "before" and "after" when one deals with a random collection of 3-geometries. In contrast, the "YES" 3-geometries are distinguished by the fact that they *can* be reassembled into the particular 4-geometry or spacetime that happens to be illustrated.] Below: Superspace, the arena within which space undergoes its dynamic evolution (actually infinite dimensional; depicted here as two-dimensional). Each "point" in superspace (such as A) stands for an entire three-dimensional geometry (also denoted by A; "potato" at upper left). The "YES" 3-geometries constitute a submanifold of superspace. [This submanifold of superspace is, like superspace itself, infinite dimensional, because there is an infinity of ways in which one "YES" 3-geometry can differ from another. The infinite number of the "YES" 3-geometries, however, is of a lower order of infinity than the infinite number of all 3-geometries. Therefore the collection of "YES" 3-geometries is indicated symbolically in the diagram by a domain of dimensionality only half of that which the draftsman has been able to allocate to superspace itself—this domain being the one-dimensional smooth curve cutting across

Einstein's ten field equations (subject to the appropriate dynamical initial value conditions that serve to distinguish one complete classical history from all other complete classical histories).

(7) To approach closer to the spirit of quantum geometrodynamics while still for the moment remaining within the framework of classical general relativity, one can turn directly to superspace. Out of it, select the "YES" 3-geometries from the "NO" 3-geometries, without ever making any reference whatever to Einstein's ten field equations or such strictly classical concepts as "spacetime" and "4-geometry." Instead consider the Hamilton-Jacobi function

$$S(^{(3)}\mathscr{G}) \tag{1}$$

a function which takes on a real number value, this value varying smoothly as one passes from one "point" in superspace, or one 3-geometry, to another. This function is the analog of the Hamilton-Jacobi function of particle dynamics

$$S(x, t) = -Et + \int^{x} \{2m[E - V(x)]\}^{1/2} dx + \delta_E \tag{2}$$

in four respects. First, it gives the phase of the wave function in the semi-classical approximation when one later turns from classical theory to quantum theory; thus

$$\psi \sim \begin{pmatrix} \text{slowly varying} \\ \text{amplitude function} \end{pmatrix} \exp(iS/\hbar) \tag{3}$$

Second, it depends upon one or more adjustable parameters—one parameter in the problem with one degree of freedom; a continuous infinity of parameters in the problem of space with its continuous infinity of degrees of freedom. The function S is normally to be understood as depending upon these parameters even when, as above, they are not written down explicitly. Third, the condition of "constructive interference" picks out the "YES" points in the (x, t) plane

$$\frac{\partial S_E(x, t)}{\partial E} \begin{cases} = 0 \text{ for "YES" points } (x, t) \\ \neq 0 \text{ for "NO" points } (x, t) \end{cases} \tag{4}$$

---

two-dimensional sheet of paper!] In a classical dynamical history space remains sharply confined to such a submanifold, and the history has a sharp beginning and a sharp end. Not so in the real world of quantum physics. There the dynamics is symbolized by a wave with a finite spread. A succession of wave crests shows in the diagram. The wave character of dynamics (1) brings about fluctuations in the geometry of space at small distances, (2) prevents one from using such concepts as "spacetime," "time," "before," and "after" except in the approximation of classical physics, (3) completely alters the character of the final stages of gravitational collapse, and (4) brings about an inescapable coupling to other alternative histories of the universe (waves continuing past region of superspace marked "End").

or the "YES" 3-geometries in superspace from the "NO" ones

$$\frac{\partial S(^{(3)}\mathscr{G};\text{ complete set of adjustable parameters})}{\partial(\text{each adjustable parameter})} = 0 \text{ for "YES" }^{(3)}\mathscr{G}\text{'s} \quad (5)$$

The rationale for this classical condition lies in quantum mechanics. To provide a recognizable semblance to a classical description of motion, we construct a wave packet. For this purpose we do best to superpose waves with slightly different values of the relevant parameter or parameters; thus

$$\psi = \psi_E(x, t) + \psi_{E+\Delta E}(x, t) + \cdots \quad (6)$$

We identify the center of this wave packet with the point where the condition (4) of constructive interference is satisfied. Similarly the analogous condition (5) has really to do with the construction of a wave packet in superspace. The center of this wave packet as it evolves in superspace picks out the 3-geometries of the classical history H. Fourth, out of this condition of constructive interference we get directly the correct history—the "YES" points—without ever having to refer to any equations of motion. This favorable outcome is guaranteed because we insist—and it is our sole requirement—that S shall satisfy the Hamilton-Jacobi equation.

(8) The Hamilton-Jacobi equation has the familiar form

$$\frac{1}{2m}\left(\frac{\partial S}{\partial x}\right)^2 + V(x) = -\frac{\partial S}{\partial t} \quad (7)$$

in the problem of the particle. In geometrodynamics it reads

$$(\nabla S/\delta^{(3)}\mathscr{G})^2 + {}^{(3)}R = 0 \quad (8)$$

Here $^{(3)}R$ is the scalar curvature invariant of the 3-geometry at the point (x, y, z) of 3-space under investigation. The symbolic operator that precedes this term has a meaning, and in principle can be defined, free of all reference to coordinates. However, many investigators are accustomed to specify a 3-geometry more concretely, as for example by giving as a function of the three coordinates the six metric coefficients in the expression

$$ds^2 = g_{ij}\,dx^i\,dx^j \quad (^{(3)}\mathscr{G},\text{ not }^{(4)}\mathscr{G}!) \quad (9)$$

Then the Hamilton-Jacobi function (phase of a wave function in superspace!) becomes a functional of the six $g_{ij}(x,y,z)$; thus

$$S = S[g_{ij}(x)] \quad (10)$$

With the abbreviation $g = \det \|g_{ij}\|$, Eq. (8) now reads

$$\frac{1}{2g}(g_{ij}g_{jl} + g_{il}g_{jk} - g_{ij}g_{kl})\frac{\delta S}{\delta g_{ij}}\frac{\delta S}{\delta g_{kl}} + {}^{(3)}R = 0 \qquad (11)$$

This one equation, first given by Peres,[2] operates in superspace and captures the entire content of Einstein's ten vacuum field equations.

(*9*) Quantum geometrodynamics also operates in superspace. The wave equation, according to correspondence principle arguments, must be qualitatively of the form

$$-\nabla^2/(\delta \psi {}^{(3)}\mathscr{G})^2 + {}^{(3)}R\psi = 0 \qquad (12)$$

but problems about the precise form of the equation remain, connected with questions of factor ordering.[3] In addition one will recall that a knowledge of the Hamilton-Jacobi equation of relativistic particle dynamics did not and could not decide between the Klein-Gordon wave equation and the Dirac wave equation. Choices of analogous character lie open for analysis in quantum geometrodynamics. However, many qualitative and order-of-magnitude conclusions are independent of these questions of choice.

(*10*) In quantum geometrodynamics space does not have a unique geometry. Instead, there is a probability amplitude for it to have this, that, or the other 3-geometry.

(*11*) The 3-geometries that occur with significant probability amplitude are infinitely more numerous than one can accommodate in any one classical history or in any one 4-geometry. Thus there is no such thing as a unique spacetime in which every "event," past, present or future, has its determined place. And as spacetime is deprived of any well defined meaning, so is time. There is no "before," no "after," no "next." In principle there is nothing except the quantum mechanical evolution of the state function in superspace.

(*12*) For many purposes the spread of the quantum mechanical wave packet in superspace is sufficiently small so that it can be overlooked. Then the classical approximation makes sense, and it is reasonable to employ the concepts of spacetime and time.

(*13*) The characteristic spread of the wave packet about the classical history in superspace signifies a quantum mechanical fluctuation in the geometry of space at small distances. The scale of these fluctuations is governed by the Planck length

$$L^* = (\hbar G/c^3)^{1/2} = 1.6 \times 10^{-33} \text{ cm} \qquad (13)$$

It is remarkable that Planck stressed the importance of what is essentially this length as early as 1899,[4] before he had even discovered the quantum principle. It is the only length that one finds in all of nature, he emphasized, that is independent of the dimensions and rate of rotation of the particular planet upon which we happen to live, independent of all the complexities of solid state physics, independent of all details of the composition of atoms and molecules, and free of any reference to any property of any particle. If Planck gave us the length, quantum geometrodynamics tells us its message: fluctuations in geometry occur throughout all space. In a region of extension L the order of magnitude of the spread in the metric coefficients is

$$\delta g \sim L^*/L \tag{14}$$

the apparent local accelerations of gravity have fluctuations of the order

$$\delta \Gamma \sim L^*/L^2 \tag{15}$$

(with time measured in the units of length); and the true local measure of gravitation, the "tide producing acceleration" or the "Riemann curvature tensor" has an uncertainty of the order

$$\delta R \sim L^*/L^3 \tag{16}$$

Similar fluctuations have been known to occur in the electromagnetic field throughout all space ever since Einstein first taught us how to estimate their order of magnitude. Since World War II physics has achieved no triumph more striking than a powerful quantitative quantum electrodynamics to predict the effect of these field fluctuations upon the motion of an atomic electron and beautiful experiments that confirm these predictions to unprecedented accuracy. As contrasted to quantum electrodynamics where the fluctuations in potential and field

$$\delta A \sim (\hbar c)^{1/2}/L$$
$$\delta F \sim (\hbar c)^{1/2}/L^2 \tag{17}$$

are free of all reference to any characteristic length, quantum geometrodynamics shouts forth the Planck distance. Physics contains, it says, not forty precincts, but sixty. There are twenty more decades from 10^{-13} cm to 10^{-33} cm, each with its own characteristic phenomena, each with its novel physical effects, each with its own structure.

(14) Fluctuations become so dominant at distances of the order of the Planck length that the fine scale topology of space itself fluctuates (Fig. 2). Multiply connected geometries are overwhelmingly more numerous than singly connected geometries. Space is to be compared to nothing so much as a carpet of foam spread out upon the landscape. Viewed at everyday distances this carpet appears smooth. Upon closer inspection it is found to be filled with millions of small holes. Still closer

examination shows these bubbles continually bursting and new ones forming—an ever changing topology. If Clifford first, then Weyl, and later Einstein pondered the consequences of multiple connectivity for particle physics, it is remarkable that the space theory of matter in its present day quantum version leaves little escape from multiple connectivity.

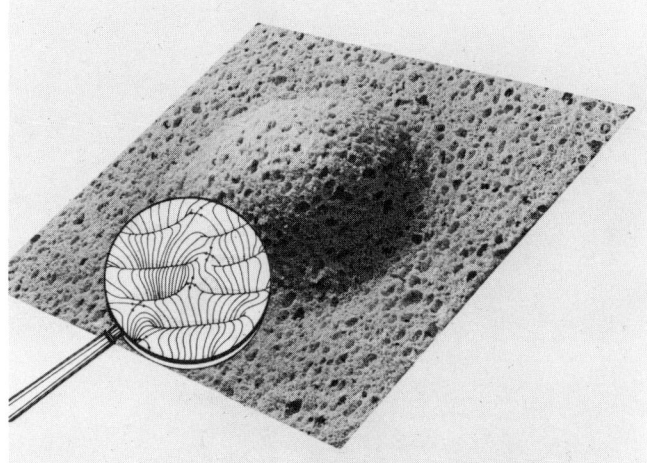

Fig. 2. Typical fluctuations in the geometry and topology of space at the Planck scale of distances in an idealized and symbolic representation.

And if electric lines of force trapped in the topology of multiply connected space manifest themselves as electric charge, conversely the existence of electric charge in nature is the strongest single piece of evidence we can cite today for the reality of these "wormholes" pervading all space.

(15) The effective density of any component of the energy associated with these fluctuations in geometry is of the fantastic order

$$\begin{pmatrix} \text{density of} \\ \text{mass-energy} \end{pmatrix} \sim \frac{\text{(Planck mass)}}{\text{(Planck length)}^3} \sim \frac{(\hbar c/G)^{1/2}}{(\hbar G/c^3)^{3/2}}$$

$$\sim \frac{10^{-5} \text{ g}}{(10^{-33} \text{ cm})^3} \sim 10^{95} \text{ g/cm}^3$$

By comparison the energy density of nuclear matter is completely negligible. A particle is as unimportant in the lively physics of the vacuum as a cloud is unimportant in the physics of the sky. See, yes. Watch its motion, yes. But count on it as the natural starting point for the description of the physics, no.

(*16*) A particle is not a local warp in the geometry of space. It is not a fluctuation in the geometry of space. Instead it is a change in the pattern of the fluctuations extending over a distance enormous in comparison with the characteristic scale of the fluctuations (Fig. 3). In brief, it is a quantum state of excitation of the geometry of space. It is a "geometrodynamical exciton."

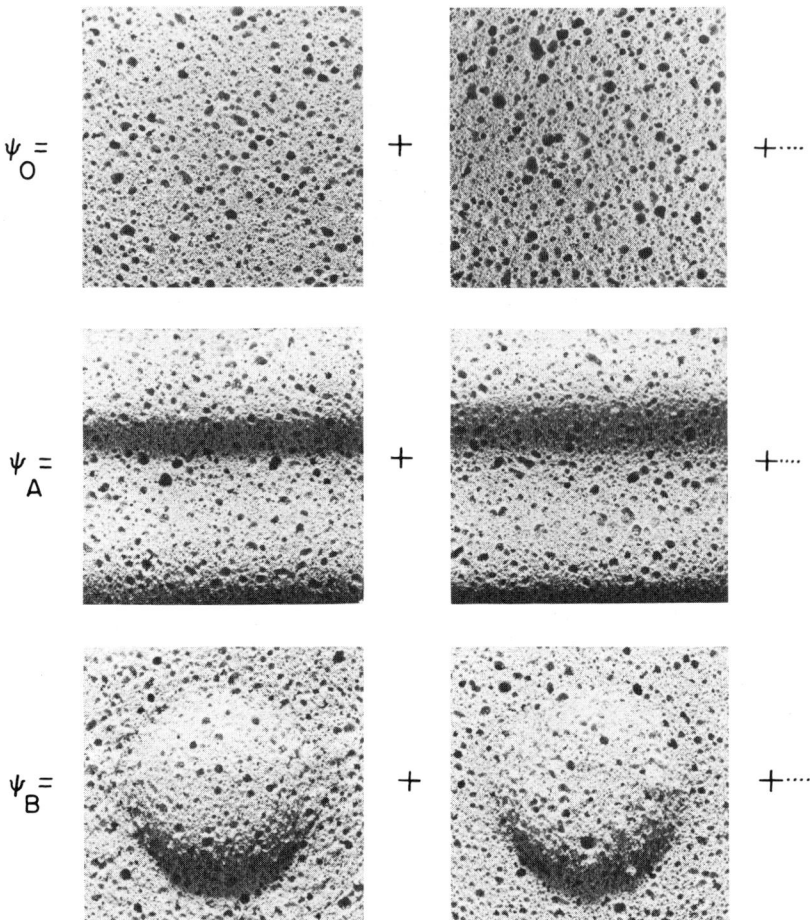

Fig. 3. Quantum state of the geometry of space, depicted schematically. ψ_O, space that looks flat at the dimensions of atoms and nuclei but has the inescapable quantum fluctuations in geometry at the Planck scale of distances. ψ_A, same, with a larger scale gravitational wave propagating through space. ψ_B, symbolic representation of a localized "quantum geometrodynamical exciton."

(*17*) Elementary particle couplings in all their variety, strong, intermediate and weak, and with all their specialities are geometrodynamical in origin, as chemical forces of the most diverse intensities and most marvelous directivities are electrical in origin. Chemical energies arise from percentagewise small alterations in the state of binding of electrons to nuclear charges. Elementary particle forces arise from percentagewise fantastically small alterations in the vacuum energy of the geometrodynamical fluctuations.

(*18*) The enormous scale of the elementary particles, $\sim 10^{-13}$ cm, as compared to the fundamental length of $\sim 10^{-33}$ cm, is not a quantity that can be calculated from purely dynamical arguments. It has to do with, and is governed by, the scale of the universe itself—for example, at the phase of maximum expansion—and may well have another value in a universe of another scale.

(*19*) The arena of quantum geometrodynamics is not "spacetime" or the "universe" but is far larger; it is superspace. Within superspace the wave function as it runs its course describes creation, expansion, recontraction and gravitational collapse for many alternative histories of the universe. No one of these "universes" can go through its dynamics in complete independence of the others. An inescapable coupling takes place between one such history and another in the final phases of gravitational collapse. The collapse itself drives all dimensions to lengths of the Planck order of magnitude. At this point in superspace quantum fluctuations in the geometry are no longer small scale effects superposed on a large scale classical history. Instead, they are everything. In this domain no choice is left as to whether one will use the concept of "time" (large scale effects) or give it up (small scale effects). Time has no meaning. Nor is there any way to say that one "history of the universe" follows after another.

(*20*) The gravitational collapse[5] of the core of a star in a supernova process provides a kind of "laboratory scale model" for studying the collapse of the universe itself. Depending upon the amount of matter present and the energy of the implosion, the infall of the core is predicted to halt with the formation of a neutron star, or go on to complete collapse. The resulting "black hole" emits no radiation but continues to exert its normal gravitational attraction. One of the most attractive methods of studying the collapse itself looks to gravitational radiation. For the possible pickup of such radiation J. Weber has built impressive detectors. The afterbody of the collapse appears best studied when it happens to be associated with a normal star in a double star system of the familiar kind. Matter streaming out of the normal star and falling onto the neutron star at nuclear velocity, or into the black hole at relativistic velocity, provides a probe and potential source of radiation full of interesting observational possibilities, as remarked especially by Schklovsky, Zel'dovich, and Novikov.

In conclusion, no consequence of the union of the quantum principle with Einstein's geometrodynamics is more revolutionary than superspace, and none seems more likely to have consequences for all of physics, from elementary particle physics to the dynamics of the universe. Quantum geometrodynamics would seem to be a time bomb at the heart of physics, exploding with new concepts as earlier the Schrödinger equation, much against the intention of Schrödinger himself, exploded into such new consequences as "probability amplitudes," "probability currents," "uncertainty principle," and "complementarity."

The most challenging of all the tasks of quantum geometrodynamics is evident: determine if the Clifford-Einstein "space theory of matter," updated, makes sense. And what a task! Starting from the bottom, from physics at the Planck scale of distances, described by a wave equation still to be sharpened in definition, trace out one by one in scrupulous detail twenty superimposed decades of physical consequences, physical effects, and physical structure. Perhaps tell over our twenty-one prayer beads as we work and ponder, and ponder and work! At the end, test for correspondence with the fantastic richness of elementary particle physics—with no checks available along the way at 10^{-28} cm or 10^{-23} cm or 10^{-18} cm: a nonstop flight to outdo all nonstop flights. Happily we are all colleagues together in the enterprise. Some of us work from 10^{-13} cm down to smaller distances, joined up in the most impressive combination of experiment and theory in human history. Others on a far more modest scale work up from 10^{-33} cm, attempting to see what of value quantum geometrodynamics has in store for us. Far away though it still appears, the goal of both is the same: the secret of strange matter.

Notes and References

1. Details, references to the extensive literature, and a fuller discussion of the basic ideas of the Clifford-Einstein space theory of matter in its present day form are to be found in J. A. Wheeler, "Superspace and the Nature of Quantum Geometrodynamics," in *Battelle Rencontres: 1967 Lectures in Mathematics and Physics*, C. DeWitt and J. A. Wheeler, Eds., Benjamin, New York, 1968; see also J. A. Wheeler, *Geometrodynamics*, Academic Press, New York, 1962, and "Geometrodynamics and the Issue of the Final State", in *Relativity, Groups and Topology*, C. DeWitt and B. DeWitt, Eds., Gordon and Breach, New York, 1964.
2. The step from the Einstein theory to the Hamilton-Jacobi equation was implicit in the work of many investigators, but was first clearly and explicitly made by

A. Peres, *Nuovo Cimento*, **26,** 53 (1962). For the reverse step—the derivation of the ten field equations from the Hamilton-Jacobi equation—see U. Gerlach, *Bull. Am. Phys. Soc.*, Paper DE 7, 340 (April, 1966).
3. See especially Bryce DeWitt, "Quantum Theory of Gravity," *Phys. Rev.*, **160,** 1113 (1967); **162,** 1195, 1239 (1967).
4. M. Planck, *Sitzber. Preussische Akad. Wiss. Berlin, Math.-Phys. Kl.*, 440 (**1899**).
5. For a survey, see, for example, B. K. Harrison, K. S. Thorne, M. Wakano, and J. A. Wheeler, *Gravitation Theory and Gravitational Collapse*, University of Chicago Press, Chicago, 1965.

Author Index

Numbers in parentheses are reference numbers and indicate that the author's work is referred to although his name is not mentioned in text. Italic numbers are the pages on which the complete references are listed.

A

Aamodt, R. E., 233(42), *265*
Adir, J., 130(14), *138*
Ahrens, T. J., 46, 55, 56, 57(4), 67(4), *71*
Alder, B. J., 79(8, 9), 94(29), *103, 104*
Alexeff, J., 246(67), *265*
Alfvén, H., 173, 178(20), *264*
Allis, W. P., 165(16), *264*
Alpher, R. A., 16, *22*
Al'tshuler, L. V., 44(3), 49(3), 67(3), *71*, 85(20), 94(31), *104*

B

Babykin, M. V., 246(66), *265*
Bachynski, M. P., 165(14), *264*
Baiborodov, Y. T., 250(71), *266*
Baker, W. O., 109, *117*
Balchan, A. I., 93, 94(28), *104*
Balebonov, V. M., 181(25), *264*
Barker, L. M., 101(35,36), *104*
Barr, W. L., 251(78), *266*
Bass, R. C., 46, 57(5), *71*
Bassett, W. A., 91(26), 99(26), *104*
Batchelor, G. K., 268(4), *271*
Bates, D. R., 329, 334(9), 347(10), *348*
Bauer, M., 285, 296(2), *300*
Belzer, J., 14(2), *22*
BenDaniel, D. J., 162(8), 165(16), 185(8), *264*
Benedick, W. B., 101(33), *104*
Berk, H., 237(49), *265*
Berlin, A. A., 111(6), *117*
Bernstein, I., 220(36), *265*
Besancon, J. E., 101, *104*
Beyster, J. R., 130(12), *138*
Bird, P. F., 51(13), *71*
Biritz, H., 318, *326*

Blacic, J. D., 29, *42*
Blatt, J. M., 291, *300*
Bloom, G. H., 102, *104*
Bludman, S. A., 316(6), 317(6), *326*
Bogolubov, N. N., 297, *300*
Book, D. L., 233(42), *265*
Born, M., 349, *354*
Bowman, C. D., 135(31,32), *139*
Brase, W. F., 61, *71*
Bridgman, P. W., 27, 28, 33, *42*
Brillouin, L., 75, *103*
Brunauer, S., 115, *117*
Brush, S. G., 74(1), 76(1), *103*
Burbidge, G., 21, *22*
Burbidge, M., 21, *22*
Burgess, T., 101(37), *104*
Butkovich, T. R., 50, 51, 54, *71*

C

Carr, M. L., 165(15), *264*
Carslaw, H. S., 29, *42*
Carter, N. L., 29(19), *42*
Casimir, H., 355, *363*
Chabai, A. J., 46, 57(5), *71*
Champetier, J. L., 101(34), *104*
Chandrasekhar, S., 81, *104*, 167, 222(39), *264, 265*
Chapman, S., 164(12), *264*
Cherry, J. T., 57, 66, *71*
Chikovani, G., 304, *325*
Christian, R. H., 94(29), *104*
Christofilos, N. C., 151(3), *264*
Chuck, W., 164(13), 165(13), *264*
Clark, S. P., Jr., 46(7), 66(7), 67(7), *71*
Clark, S. S., 130(14), *138*
Clendenen, R. L., 91(26), 99(26), *104*
Coffer, C. O., 130(11), *138*
Conte, S. D., 181(28), *264*

C

Coppi, B., 241(58), *241*
Coulson, C. A., 329, *348*
Cowling, T. G., 164(12), *264*
Critchfield, C., 334(8), *348*
Crowley, W. P., 271(9), *271*

D

Damm, C. C., 252(80), *266*
Debrunner, P., 91(26), 99(26), *104*
Dee, J. B., 130(11), *138*
DeWitt, B., 369(1), 373(3), *378, 379*
DeWitt, C., 369(1), *378*
De Witt, H. E., 83, *104*
Dickinson, W. C., 135(31,32), *139*
Dirac, P. A. M., 361, *363*
Dorsey, J. P., 130(16), *139*
Dory, R. A., 237(48), 251(74), *265, 266*
Dougdale, I. S., 49(9), *71*
Drickamer, H. G., 91(26), 93, 94(28), 99(26), *104*
Drummond, W. F., 244(62,64), *265*
Dürr, H. P., 302(1), 321(1), 323(13), 324(15), *325, 326*
Duff, R. E., 51(13), *71*
Dunlap, J. L., 251(74), *266*
Duvall, G. E., 55, *71*
Dyson, F., 119, 126, *138*

E

Edge, C. K., 91(26), 99(26), *104*
Ellis, R. E., 249(68), *265*
Elton, R. C., 249(70), *265*
Emmett, P. H., 115, *117*
Ewing, W. M., 66(26), *71*
Ezawa, H., 323(14), 324(14), *326*

F

Fainberg, Y. B., 251(76), *266*
Fano, V., 309(5), 320(5), *325*
Fermi, E., 171, *264*
Foote, J. H., 252(80), *266*
Ford, F. C., 249(68), *265*
Fowler, T. K., 234(45), 242(59), *265*
Fraunfelder, H., 91(26), 99(26), *104*
Frazer, A. H., 113(7), *113*
Fried, B. D., 181(28), 241(57), *264, 265*
Frieman, E., 220(36), *264*
Froehlich, R., 130(14), *138*
Fuller, P. J. A., 93, 94(27), 98, *104*

Furth, H. P., 221(38), 241(56,58), *265*
Futch, A. H., Jr., 252(80), *266*

G

Gamow, G., 11(1), 14(2,3), 16(6), *22*
Gardiner, A. L., 252(80), *266*
Garofalo, F., 29, *42*
Gavrino, P. P., 246(66), *265*
Gaylord, N. G., 109(3), *117*
Géhéniau, J., 323(13), *326*
Gelfand, L. M., 355(2), *363*
Gell-Mann, M., 241(57), *265*
General Dynamics Corp., 120(5), *138*
Gerlach, U., 373(2), *379*
Getty, W. O., 251(77), *266*
Gibson, W., 181(24), *264*
Goldberger, M. L., 329(6), 330(6), 332(3), 333(6), *348*
Goldstone, J., 316, 317, *326*
Gombas, P., 75(5), *103*
Gomes, L. C., 320(10), *326*
Graham, R. A., 85, *104*
Gray, P., 297, *300*
Griggs, D. T., 28, 29, *42*
Grine, D. R., 57(16), *71*
Grover, R., 76, *103*
Guest, G. E., 237(48), *265*
Guralnik, G. S., 322, *326*

H

Hagen, C. R., 322, *326*
Hamada, T., 134, *139*
Hamann, S. D., 94(30), *104*
Handin, J. H., 28, *42*, 57, *71*
Harris, E. G., 234, *265*
Harris, G. M., 82, 83, *104*
Harrison, B. K., 377(5), *379*
Haste, G. R., 251(74), *266*
Hastie, R. J., 158(7), *264*
Hawk, H. L., 46, 57(5), *71*
Heard, H. C., 29, *42*, 57, *71*
Heirtzler, J. R., 27(7), *42*
Heisenberg, W., 302, 317(8), 321(1), 324(15), *325, 326*
Heller, L., 133, 134, *139*
Herbst, R., 69(28), *71*
Herman, R. C., 16, *22*
Hess, H. H., 27(9), *42*
Higgins, G., 44(2), *70*
Higgs, P. W., 322, *326*

AUTHOR INDEX

Hill, E. L., 302(2), *325*
Hoffmann, F. de, 126(9), *138*
Hollenbach, R. E., 101(35), *104*
Holt, A. R., 334(9), *348*
Holzer, F., 44(1), 67(1), *70*
Horton, C. W., Jr., 233(43), *265*
Huddlestone, R. H., 201(32), *264*
Hurdlow, W. R., 66, *71*

I

Infeld, L., 349, *354*
Ingram, G. E., 85(21), *104*
Ioffe, M. S., 250, *266*

J

Jackson, J. D., 230(40), 241(57), *265*
Jaeger, J. C., 29, *42,* 63, *71*
Jamieson, J. C., 91(26), 99(26), *104*
Jardetzky, W. S., 66(26), *71*
Jensen, T. H., 257(83), *266*
Johnson, G., 44(2), *70*
Johnston, H. C., 347(10), *348*
Johnston, I. D., 134, *139*
Johnston, T. W., 165(14), *264*
Jones, W. D., 246(67), *265*
Jordan, J. R., 285(2), 296(2), *300*
Jordan, P. C., 285(2), 296(2), *300*
Jordan, W., 181(24), *264*

K

Kadomstev, B. B., 237(51), 244(65), *265*
Kastler, D., 323(14), 324(14), *326*
Kaufman, A. N., 222(39), *265*
Keeler, R. N., 79(8), 94, 102, *103, 104*
Keller, G., 14(2), *22*
Kennedy, G. C., 61, *71*
Kennel, C. F., 252(79), *266*
Kerst, D. W., 220(37), *264*
Kharchenko, I. F., 251(76), *266*
Kibble, T. W. B., 322, *326*
Killeen, J., 241(56), *265*
Kirzhnits, P. A., 76, *103*
Klein, A., 316(6), 317(8), *326*
Kofoed-Hansen, O., 257(83), *266*
Kolb, A. C., 249(70), *265*
Kompaneyets, A. S., 48, *71*
Koppel, J. U., 130(13), *138*
Kornilov, E. A., 251(76), *266*
Kovacic, P., 109(5), *117*

Kraichnan, R. H., 270(7), *271*
Krall, N. A., 186(26), 211(35), 237(52), 238(53), 239(54), 240(54), 241(26), *264, 265*
Kruskal, M., 220(36), *265*
Kubo, R., 296, *300*
Kuleshova, L. V., 94(31), *104*
Kulsrud, R., 220(36), *265*
Kulterman, R. W., 101(33), *104*
Kuo-Petravik, L. G., 251(75), *266*

L

Landau, L. D., 187, 209(27), 218(27), *264,* 328, *347*
Langmuir, I., 115, *117*
Larsen, D., 57, *71*
Larsen, E. S., 28(13), *42*
Lashinsky, H., 251(72), *266*
Lathrop, K. D., 130(17), *139*
Latter, R., 75(2), *103*
Lauer, E., 181(24), *264*
Lawson, A. W., 91(26), 99(26), *104*
Leclonche, Y., 101(34), *104*
Leith, C. E., 270(8), 271(10), *271*
Lenihan, S. R., 130(15), *138*
Leonard, S. L., 201(32), *264*
Lilley, E. M., 61, 62, *71*
Lilly, D. K., 270, *271*
Linton, M., 94(30), *104*
Little, E. M., 249(69), *265*
Long, F., 109(5), *117*
Longmire, C. L., 149(2), 150(2), 204(2,33), *264*
Louck, J. D., 360, *363*
Lupton, W. H., 249(70), *265*
Lutsenko, E. A., 251(76), *266*
Luttinger, J. M., 296, *300*

M

MacDonald, D., 49(9), *71,* 164(13), 165(13), *264*
McLean, E. A., 249(70), *265*
McQueen, R. G., 85(19), *104*
Magourik, J. N., 57, *71*
Maier-Leibnitz, H., 131, *139*
Malmberg, J. H., 195(29), 196, 251(29), *264*
Mark, H. F., 109(3), 111(6), *117*
Marks, R. E., 57(15), *71*

Matreyek, W., 109(4), *117*
Matthews, D. H., 27(4), *42*
Mayer, J. E., 285(2), 296(2), *300*
Mikhailovskii, A. B., 209(34), 218(34), 236(47), 237(50), *264, 265*
Mills, R. L., 349(1), 353(1), 354(1), *354*
Mitchell, A. C., 94, 102, *104*
Mitter, H., 302(1), 321(1), *325*
Montgomery, D. C., 163(11), 165(11), 253(11), *264*
Moses, K. G., 252(80), *266*
Muehlhause, C. O., 135, *139*
Murphy, E. G., 251(75), *266*

N

Nakajima, S., 296(6), *300*
Naliboff, Y. D., 130(13), *138*
Nedoseev, S. L., 246(66), *265*
Neidigh, R. V., 246(67), *265*
Neilsen, C. E., 251(73), *266*
Neilson, F. W., 101, *104*
Neumann, J. v., 268, *271*
Newton, R. G., 61, *71*
Nikolaev, R. M., 251(76), *266*
Noether, E., 302, *325*
Northrop, T. G., 178(22), 180, 181(22), *264*
Nottorf, R., 124(6), *138*

O

Okhawa, T., 220(37), *264*
O'Neil, T. M., 254(81), *266*
Onsager, L., 274, *300*
Osher, J. E., 252(80), *266*

P

Pape, N. R., 109(4), *117*
Pasquali, G. de, 91(26), 99(26), *104*
Paterson, M. S., 26(2), *41*
Pauli, W., 307, *325*
Pauthenet, R., 89, *104*
Pavlovskii, M. N., 94(31), *104*
Pearlstein, L. D., 237(49), 238(53), *265*
Pedenko, N. S., 251(76), *266*
Peed, W. F., 246(67), *265*
Penzias, A. A., 16(5), *22*
Peres, A., 373, *379*
Perkins, W. A., 251(78), *266*
Petravic, M., 251(75), *266*

Petrov, V. M., 250(71), *266*
Petschek, H. E., 252(79), *266*
Phillips, N. A., 268, *271*
Pines, D., 244(62), *265*
Pipkorn, D. N., 91(26), 99(26), *104*
Pitman, W. C., III, 27(7), *42*
Planck, M., 373, *379*
Plantevin, J. P., 101(34), *104*
Post, R. F., 152(4), 162(9), 169(18), 179(4), 185(9), 231(41), 234(46), 249(68), 252(80), 257(82), *264-266*
Postma, H., 251(74), *266*
Press, F., 66(26), *71*
Price, J. H., 93, 94(27), 98, *104*

Q

Quinn, W. F., 249(69), *265*

R

Racah, G., 309(5), 320(5), *325,* 355, *363*
Raeuchle, R. F., 124(6), *138*
Raleigh, C. B., 26, *41, 42*
Rapp, E., 57, *71*
Riazuddin, 133, *139*
Ribe, F. L., 249(69), *265*
Rice, S. A., 297, *300*
Richtmyer, R. D., 268, *271*
Roberts, J. A., 165(15), *264*
Robertson, A. J. B., 30, *42*
Robinson, D., 323(14), 324(14), *326*
Rogers, L. A., 49, 57(15,17), 67(17), *71*
Roll, P. G., 16(5), *22*
Rosenbluth, M. N., 164, 165, 186, 204(33), 211(35), 221(38), 231(41), 234(46), 237(49,52), 241(26,56,58), 243(61), 249(68), *264, 265*
Ross, M., 79, *103*
Rostoker, N., 186(26), 241(26), *264*
Rouse, C. A., 51(12), *71,* 74(1), 76(1), 82, *103, 104*
Royce, E. B., 85, *104*
Rudakov, L. I., 237(50), 246(66), *265*
Rudakov, L. J., 241(55), *265*
Rundle, R. E., 124(6), *138*
Ryle, M., 21, *22*

S

Sagdeev, R. Z., 241(55,58), 243(60), 244(63), *265*

AUTHOR INDEX

Sailor, V., 135(29,31), *139*
Salam, A., 316(6), 317(6), *326*
Sawyer, G. A., 249(69), *265*
Scalettar, R., 126, 128, *138*
Schlieder, S., 302(1), 321(1), *325*
Schmidt, G., 199(31), 201(31), *264*
Schmidt, M., 21, *22*
Schneider, R. E., 133, *139*
Schott, G. L., 51(13), *71*
Semasko, N. N., 181(25), *264*
Sessler, A. M., 251(73), *266*
de-Shalit, A., 309(5), 320(5), *325*
Shima, Y., 234(45), *265*
Shkarofsky, I. P., 165(14), *264*
Shoptaugh, J. R., Jr., 130(11), *138*
Shull, C. G., 124(6), *138*
Signell, P., 134, *139*
Sinclair, R. M., 251(75), *266*
Skoryupin, V. A., 246(66), *265*
Slaus, I., 133, *139*
Sleeper, H. P., Jr., 120(4), 130(4), *138*
Smagorinsky, J., 268, *271*
Smullin, L. O., 251(77), *266*
Solbolev, R. I., 250(71), *266*
Spitzer, L., 149(1), 152(1), 156(5), 161(1), 167, 173, *264*
Springer, D., 69(28), *71*
Springer, T., 132, *139*
Stahl, R. H., 120(4), 130(4), *138*
Standen, A., 106(2), *117*
Stephens, D. R., 61, 62, 66, *71*
Stewart, I., 347(10), *348*
Stirling, W. L., 246(67), *265*
Stix, T. H., 196(30), 199(30), 226(30), *264*
Stone, R. S., 120(4), 130(4), *138*
Stueckelberg, E. C. G., 328, *347*
Suits, G., 106(1), *117*
Swartz, M., 249(70), *265*
Sweetman, D. R., 251(75), *266*
Swieca, A., 323(14), 324(14), *326*
Sykes, L. R., 23, 25, *41*
Symon, K. R., 251(73), *266*

T

Takahashi, T., 91(26), 99(26), *104*
Talley, W., 44(2), *70*
Talmi, I., 309(5), 320(5), *325*
Taylor, J. B., 158(6,7), *264*

Teller, E., 11(1), *22,* 44(2), *70,* 75, *103,* 119, *138,* 180, *264*
Thaler, R. M., 133, *139*
Thiel, M. van, 79(8), *103*
Thompson, E., 251(75), *266*
Thompson, W. B., 177, *264*
Thorne, K. S., 377(5), *379*
Thunborg, S., Jr., 85(21), *104*
Tidman, D. A., 163(11), 165(11), 253(11), *264*
Timofeev, A. V., 236(47), 237(51), *265*
Todt, L. J., 130(14), *138*
Triplett, J. R., 130(13), *138*
Trulio, J., 82, *104*

U

Ulam, S., 17, *22*
Utiyama, R., 349(2), *354*

V

Vedel, J., 101(34), *104*
Vedenov, A. A., 244(63), *265*
Velikov, E. P., 244(63), *265*
Vine, F. J., 27(4,6,8), *42*

W

Wagner, H., 316(7), 317(7), *326*
Wakano, M., 377(5), *379*
Walecka, J. D., 320(10), *326*
Walsh, J. G., 62, 63(24), *71*
Wandel, C. F., 257(83), *266*
Watson, K. M., 222(39), *265,* 329(6), 330(6), 332(6), 333(6), *348*
Weertman, J., 29, *42*
Weinberg, A. M., 126(10), *138*
Weinberg, S., 316(6), 317(6), *326,* 334, *348*
Weisskopf, V. F., 320(10), 324, *326*
Weld, H. W., 241(57), *265*
Werth, G. C., 69(28), *71*
West, G. B., 120(3,4), 130(4,11), 137(33), *138, 139*
Wharton, C. B., 195(29), 196, 201(32), 251(29), *264*
Wheeler, J. A., 369(1), 377(5), *378, 379*
Whittemore, W. L., 120(3), 124(7,11), 135(30,31), *138, 139*
Wigner, E. P., 126(10), *138,* 309(5), 320(5), *325,* 361, *363*

Wiin-Nielsen, A., 270(6), *271*
Wilkinson, D. T., 16(5), *22*
Wilson, A. S., 124(6), *138*
Wilson, J. T., 27(5,6), *42*
Wilson, R., 134(27), *139*
Wilson, R. W., 16(5), *22*
Winslow, F. H., 109, *117*
Wollan, E. O., 124(6), *138*

Y

Yamamoto, H., 324(15), *326*
Yamazaki, K., 302(1), 321(1), 324(15), *325, 326*

Yang, C. N., 349(1), 353(1), 354(1), *354*
Yoder, N. R., 134, *139*
Yokota, M., 296(6), *300*
Young, M. P., 249(70), *265*
Young, R. A., 251(74), *266*
Yushmanov, E. E., 162(10), 185(10), *264*

Z

Zalewski, K., 329, *348*
Zavoiskii, E. K., 246(66), *265*
Zener, C., 328, *347*
Zwanzig, R., 296, *300*

Subject Index

A

ACPR (Annular Core Pulsed Reactor), 119
Action integral, 177, 180
Adiabatic invariant, 177
Adsorption, 115
Alfvén velocity, 201, 212, 223
Ambipolar field, 159, 169
Anisotropy, 89, 165, 225
ARGUS, 151, 181
Aromatized, 111
Astron, 151, 261
Atomic physics, 301, 305, 327
Atomic weapons, 2, 4, 15

B

Band structure, 77
Boltzmann equation, 163, 171, 210
Born approximation, 333, 338

C

Casimir operators, 355
Chemical physics, 3, 367
Collective motion, 8
Collisionless Boltzmann equation, 184
Collisional relaxation, 165
Compression, 45, 61, 170
Compton effect, 8
Conductivity, 94
Conservation laws, 8, 302, 305
Coulomb function, 134
Coulomb potential, 307, 322
Coulomb scattering, 133, 168
Crystals, 105, 124, 315
CTP theorem, 321
Controlled thermonuclear reactions, 5, 141
Cusps, 152, 259
Cyclotron wave, 199

D

Debye length, 143, 161, 194
Demagnetization, 87

Diamond, 106
Diffusion, 160, 243
Dispersion, 189, 193, 218, 225, 239
Dispersion equation, 207, 234
Distribution function, 163, 166
Drift, 123, 160, 171
Drift waves, 202, 204, 210
Dyson effect, 124, 125

E

Earthquakes, 23, 69
Effective range, 134
Eigenvalues, 295, 355, 362
Eikonal approximation, 334
Elasticity, 61, 87
Electric field, 158, 188, 351
Electromagnetic field, 146
Electromagnetic waves, 197
Enstrophy, 270
Entropy, 292
Equation of continuity, 149
Equation of motion, 174, 197, 189
Equation of state, 44, 48, 73, 77, 84, 149
Equilibrium, 74, 151, 157, 274
Ergodic theorem, 288
Expanding universe, 11

F

Ferromagnet, 317
Finite orbit effect, 208, 218
Fireball, 15
Firehose instability, 222
Flow laws, 29
Fluid model, 147
Flute instability, 208, 218
Fokker-Planck equation, 164
Force laws, 365
Fracture, 23, 61
Fusion, 254

G

Galaxy, 11
Geometrodynamics, 369
Graphite, 107, 120

Group theory, 307, 355
Group velocity, 189, 200
Gravitation, 11, 377
Guiding center, 183

H

Hamiltonian, 282, 331, 353
Hamilton-Jacobi function, 371
Hartree-Fock calculations, 80
Heat conduction, 30
Helmholtz free energy, 84, 275
High temperature plasma, 145, 249
Hooks, 357
Hubble's constant, 12, 18
Hydrodynamics, 7, 84, 267
Hydrogen bomb, 5
Hydromagnetics, 147
Hydromagnetic stability, 250
Hydromagnetic wave, 200, 204

I

Instability, 15, 27, 34, 40, 208, 212, 222, 227, 230, 237, 246, 251, 268
Ion acoustic wave, 196
Ion heating, 169
Isopin, 305, 320
Isotopic gauge, 349

J

Jahn-Teller effect, 366
jj coupling, 4, 311

L

Ladderized, 111
Landau damping, 188, 203, 212, 233
Landau-Zener model, 328, 343
Laser, 101
Lattices, 105, 108, 316
Levitron, 261
Liouville equation, 164, 186
Loss cone, 180, 230
Loss cone instability, 230
Low frequency universal instability, 237
LS coupling, 311

M

Magnetic axis, 156
Magnetic confinement, 149, 151, 162, 165, 179

Magnetic cooling, 3
Magnetic field, 151, 162, 188, 198, 351
Magnetic mirror 152, 157
Magnetic moment, 176, 179
Magnetism, 85
Magnetostatic equilibrium, 154, 158, 176
Maxwell's equations, 149, 155, 163, 223
Method of characteristics, 186
Mie-Grüneisen equation, 48, 97
Mirror field, 152, 183
Mirror instability, 222
Mirror machine, 157, 259
Molecular physics, 2
Monte Carlo method, 17, 79
Multipole, 152, 261

N

Near adiabatic limit, 327
Neutron flux, 119
Neutron-neutron interaction, 133
Nonequilibrium thermodynamics, 274, 282
Nonlinear effects, 186, 241, 268
Nuclear physics, 3, 132
Nucleon-nucleon interaction, 133
Numerical methods, 34, 51, 130, 267, 297

O

Octahedral stress, 57, 58
Onsager matrix, 278

P

Particle model, 145, 163
Particle physics, 132, 301, 305, 319
Particle size, 115
Partition function, 83
Phase space, 284
Phase velocity, 190, 198, 217
Phonons, 317
Pinch effect, 258, 260
Plasma oscillation, 189
Plasma physics, 83, 141
Plasma waves, 188
Plastic flow, 37
Plasticity, 28, 87
Plowshare, 6
Poisson's equation, 74, 186, 207
Poisson's ratio, 67
Polymers, 105

SUBJECT INDEX

Power of reactor, 127
Pressure, high, 4, 37, 44, 73, 92
Pressure tensor, 150
Protogalaxy, 13
Pulsed reactor, 120, 135
Pulse width, 128

R

Radiation, 14
Rankine-Hugoniot relationship, 45, 46, 47, 79
Rayleigh-Taylor instability, 217
Reactor transient, 123
Relativistic dynamics, 14, 303
Relaxation time, 168
Resonance, 303
Reversibility, 291
Reynolds number, 268
Rocks, 29, 44ff

S

Saddle point integration, 342
Saha equation, 51, 82
Scattering, 124, 161, 168, 329, 332, 338
Schrödinger equation, 73, 328
Selection rules, 3
Shear melting, 40
Shear strain, 27
Shell model, 4, 77
Shocks, 45, 85, 251
Simulation, 147, 267
Snapping, 27
Spin echo, 291
Spin–orbit coupling, 310
Stability requirements, 154, 220
Stars, 3, 13, 377
Stationary states, 313, 322
Statistical mechanics, 273
Statistical model, 74, 81
Stellarator, 152, 156, 162, 261
Stimulated Brillouin scattering, 101

Stress, 30, 43, 55
Symmetry, 301, 305, 306, 314, 319, 361
SU (3), 305

T

Temperature coefficient, 124
Theoretical physics, 7
Thermodynamics, 81, 273, 288
Theta pinch, 260
Thomas-Fermi equation, 48, 74
Tokomak, 261
Torus, 152, 154, 220, 258
Toroidal octupole, 221
Transport theory, 130, 286
Transition probability, 328
TRIGA reactor, 119
Turbulence, 8, 267

U

Understanding, 7
Uranium zirconium hydride, 119, 120

V

Velocity-dependent forces, 171
Viscosity, artificial, 268
Vlasov equation, 164, 171, 185, 203, 209, 253
Vorticity, 270

W

Wave-particle instabilities, 227
Whistler, 199

Y

Yield function, 57

Z

Zeron, 317, 321